THE FAST FOURIER TRANSFORM

E. ORAN BRIGHAM

E–Systems, Inc.

THE FAST FOURIER TRANSFORM

Prentice-Hall, Inc.

Englewood Cliffs, New Jersey

Library of Congress Cataloging in Publication Data

BRIGHAM, E. ORAN.
The fast Fourier transform.

Bibliography
1. Fourier transformations. I. Title.
QA403.B74 515'.723 73-659
ISBN 0-13-307496-X

15 14 13 12 11

Printed in the United States of America

PRENTICE-HALL INTERNATIONAL, INC., *London*
PRENTICE-HALL OF AUSTRALIA, PTY. LTD., *Sydney*
PRENTICE-HALL OF CANADA, LTD., *Toronto*
PRENTICE-HALL OF INDIA PRIVATE LIMITED, *New Delhi*
PRENTICE-HALL OF JAPAN, INC., *Tokyo*

To Vangee

CONTENTS

PREFACE

The Fourier transform has long been a principle analytical tool in such diverse fields as linear systems, optics, probability theory, quantum physics, antennas, and signal analysis. A similar statement is not true for the discrete Fourier transform. Even with the tremendous computing speeds available with modern computers, the discrete Fourier transform found relatively few applications because of the exorbitant amount of computation time required. However, with the development of the fast Fourier transform (an algorithm that efficiently computes the discrete Fourier transform), many facets of scientific analysis have been completely revolutionized.

As with any new development that brings about significant technological change, there is the problem of communicating the essential basics of the fast Fourier transform (FFT). A unified presentation which relates this technique to one's formal education and practical experience is dictated. The central aim of this book is to provide the student and the practicing professional a readable and meaningful treatment of the FFT and its basic application.

The book communicates with the reader not by the introduction of the topics but rather in the manner by which the topics are presented. Every major concept is developed by a three stage sequential process. First, the concept is introduced by an intuitive development which is usually pictorial in nature. Second, a non-sophisticated (but theoretically sound) mathematical treatment is developed to support the intuitive arguments. The third stage consists of practical examples designed to review and expand the concept being discussed. It is felt that this three step procedure gives *meaning* as well as mathematical substance to the basic properties of the FFT.

The book should serve equally well to senior or first year graduate students and to the practicing scientific professional. As a text, the material covered can be easily introduced into course curriculums including linear systems, transform theory, systems analysis, signal processing, simulation, communication theory, optics, and numerical analysis. To the practicing engineer the book offers a readable introduction to the FFT as well as providing a unified reference. All major developments and computing procedures are tabled for ease of reference.

Apart from an introductory chapter which introduces the Fourier transform concept and presents a historical review of the FFT, the book is essentially divided into four subject areas:

1. The Fourier Transform

In Chapters 2 through 6 we lay the foundation for the entire book. We investigate the Fourier transform, its inversion formula, and its basic properties; graphical explanations of each discussion lends physical insight to the concept. Because of their extreme importance in FFT applications the transform properties of the convolution and correlation integrals are explored in detail: Numerous examples are presented to aid in interpreting the concepts. For reference in later chapters the concept of Fourier series and waveform sampling are developed in terms of Fourier transform theory.

2. The Discrete Fourier Transform

Chapters 6 through 9 develop the discrete Fourier transform. A graphical presentation develops the discrete transform from the continuous Fourier transform. This graphical presentation is substantiated by a theoretical development. The relationship between the discrete and continuous Fourier transform is explored in detail; numerous waveform classes are considered by illustrative examples. Discrete convolution and correlation are defined and compared with continuous equivalents by illustrative examples. Following a discussion of discrete Fourier transform properties, a series of examples is used to illustrate techniques for applying the discrete Fourier transform.

3. The Fast Fourier Transform

In Chapters 10 through 12 we develop the FFT algorithm. A simplified explanation of why the FFT is efficient is presented. We follow with the development of a signal flow graph, a graphical procedure for examining the FFT. Based on this flow graph we describe sufficient generalities to develop a computer flow chart and FORTRAN and ALGOL computer programs. The remainder of this subject area is devoted toward theoretical development of the FFT algorithm in its various forms.

4. Basic Application of the FFT

Chapter 13 investigates the basic application of the FFT, computing discrete convolution and correlation integrals. In general, applications of the FFT (systems analysis, digital filtering, simulation, power spectrum analysis, optics, communication theory, etc.) are based on a specific implementation of the discrete convolution or correlation integral. For this reason we describe in detail the procedures for applying the FFT to these discrete integrals.

A full set of problems chosen specifically to enhance and extend the presentation is included for all chapters.

I would like to take this opportunity to thank the many people who have contributed to this book. David E. Thouin, Jack R. Grisham, Kenneth W. Daniel, and Frank W. Goss assisted by reading various portions of the manuscript and offering constructive comments. Barry H. Rosenberg contributed the computer programs in Chapter 10 and W. A. J. Sippel was responsible for all computer results. Joanne Spiessbach compiled the bibliography. To each of these people I express my sincere appreciation.

A special note of gratitude goes to my wife, Vangee, who typed the entire manuscript through its many iterations. Her patience, understanding and encouragement made this book possible.

E. O. BRIGHAM, JR.

THE FAST FOURIER TRANSFORM

1

INTRODUCTION

In this chapter we describe briefly and tutorially the concept of transform analysis. The Fourier transform is related to this basic concept and is examined with respect to its basic analysis properties. A survey of the scientific fields which utilize the Fourier transform as a principal analysis tool is included. The requirement for discrete Fourier transforms and the historical development of the fast Fourier transform (FFT) are presented.

1-1 TRANSFORM ANALYSIS

Every reader has at one time or another used transform analysis techniques to simplify a problem solution.

The reader may question the validity of this statement because the term *transform* is not a familiar analysis description. However, recall that the logarithm is in fact a transform which we have all used.

To more clearly relate the logarithm to transform analysis consider Fig. 1-1. We show a flow diagram which demonstrates the general relationship between conventional and transform analysis procedures. In addition, we illustrate on the diagram a simplified transform example, the logarithm transform. We will use this example as a mechanism for solidifying the meaning of the term *transform* analysis.

From Fig. 1-1 the example problem is to determine the quotient $Y = X/Z$. Let us assume that extremely good accuracy is desired and a computer is not available. Conventional analysis implies that we must determine Y by long-hand division. If we must perform the computation of Y repeatedly,

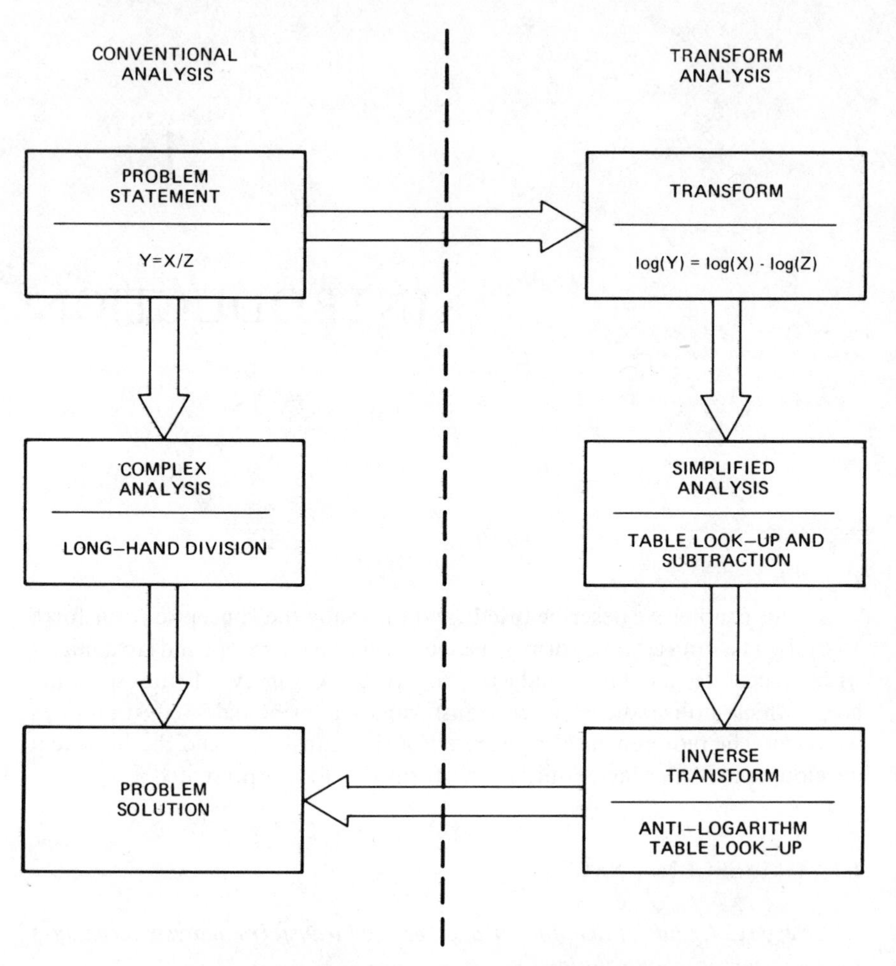

Figure 1-1. Flow diagram relationship of conventional and transform analysis.

then conventional analysis (long-hand division) represents a time consuming process.

The right-hand side of Fig. 1-1 illustrates the basic steps of transform analysis. As shown, the first step is to convert or transform the problem statement. For the example problem, we choose the logarithm to transform division to a subtraction operation.

Because of this simplification, transform analysis then requires only a table look-up of log (X) and log (Z), and a subtraction operation to determine log (Y). From Fig. 1-1, we next find the inverse transform (anti-logarithm) of log (Y) by table look-up and complete the problem solution. We note that by using transform analysis techniques we have reduced the complexity of the example problem.

In general, transforms often result in simplified problem solving analysis. One such transform analysis technique is the Fourier transform. This transform has been found to be especially useful for problem simplification in many fields of scientific endeavor. The Fourier transform is of fundamental concern in this book.

1-2 BASIC FOURIER TRANSFORM ANALYSIS

The logarithm transform considered previously is easily understood because of its single dimensionality; that is, the logarithm function transforms a single value X into the single value $\log(X)$. The Fourier transform is not as easily interpreted because we must now consider functions defined from $-\infty$ to $+\infty$. Hence, contrasted to the logarithm function we must now transform a function of a variable defined from $-\infty$ to $+\infty$ to the function of another variable also defined from $-\infty$ to $+\infty$.

A straightforward interpretation of the Fourier transform is illustrated in Fig. 1-2. As shown, the essence of the Fourier transform of a waveform is to decompose or separate the waveform into a sum of sinusoids of different frequencies. If these sinusoids sum to the original waveform then we have determined the Fourier transform of the waveform. The pictorial representation of the Fourier transform is a diagram which displays the amplitude and frequency of each of the determined sinusoids.

Figure 1-2 also illustrates an example of the Fourier transform of a simple waveform. The Fourier transform of the example waveform is the two sinusoids which add to yield the waveform. As shown, the Fourier transform diagram displays both the amplitude and frequency of each of the sinusoids. We have followed the usual convention and displayed both positive and negative frequency sinusoids for each frequency; the amplitude has been halved accordingly. The Fourier transform then decomposes the example waveform into its two individual sinusoidal components.

The Fourier transform identifies or distinguishes the different frequency sinusoids (and their respective amplitudes) which combine to form an arbitrary waveform. Mathematically, this relationship is stated as

$$S(f) = \int_{-\infty}^{\infty} s(t)e^{-j2\pi ft}\,dt \tag{1-1}$$

where $s(t)$ is the waveform to be decomposed into a sum of sinusoids, $S(f)$ is the Fourier transform of $s(t)$, and $j = \sqrt{-1}$. An example of the Fourier transform of a square wave function is illustrated in Fig. 1-3(a). An intuitive justification that a square waveform can be decomposed into the set of sinusoids determined by the Fourier transform is shown in Fig. 1-3(b).

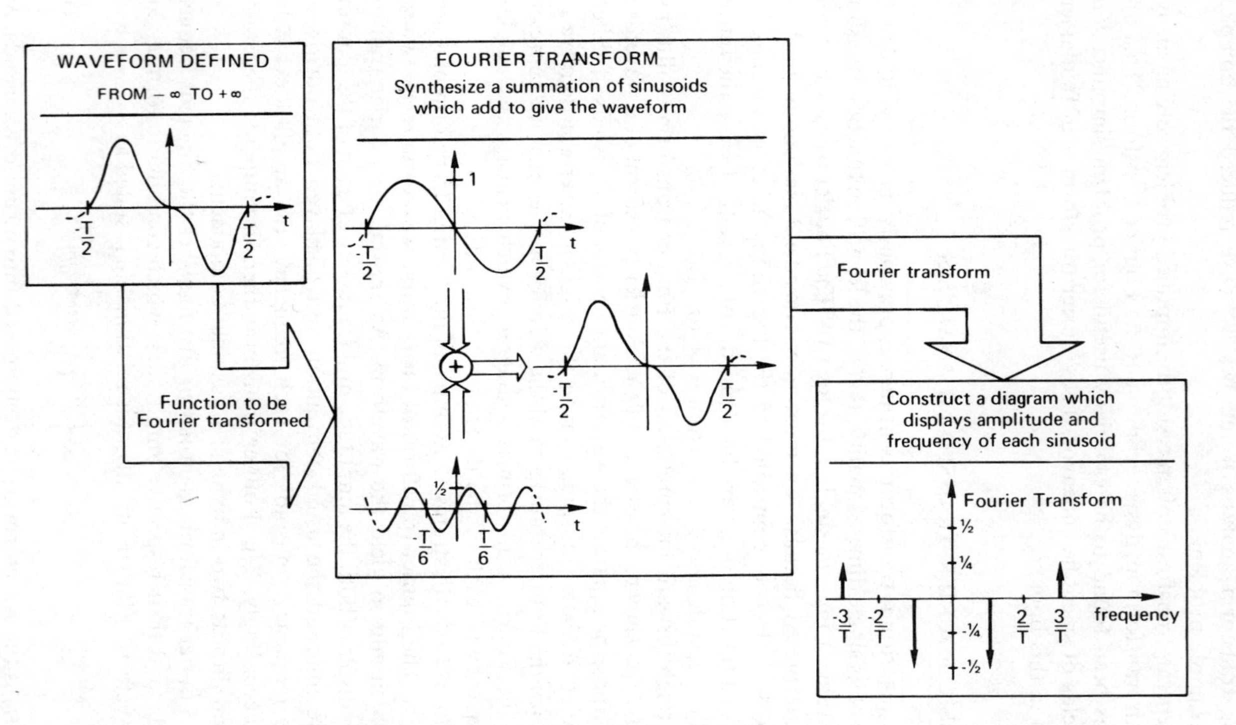

Figure 1-2. Interpretation of the Fourier transform.

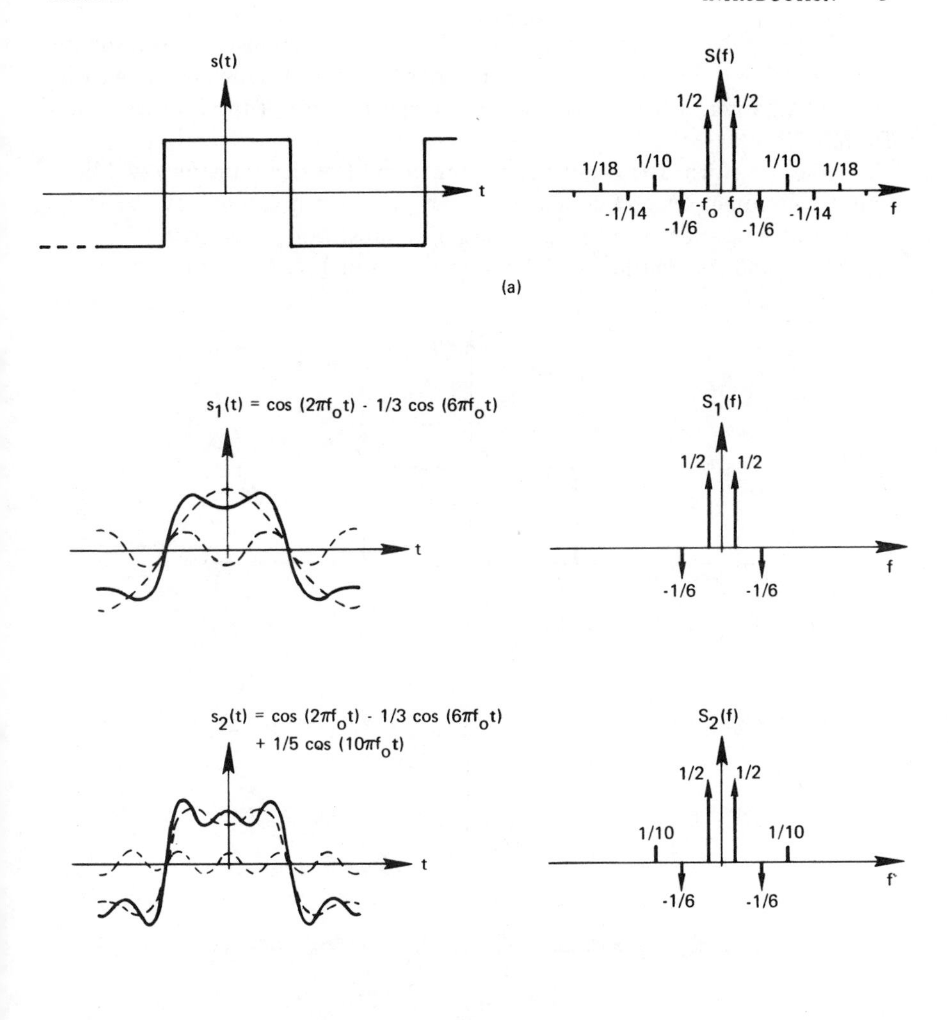

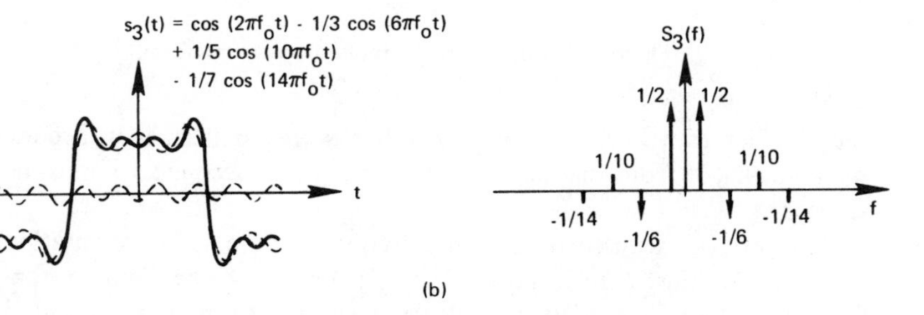

Figure 1-3. Fourier transform of a square wave function.

We normally associate the analysis of periodic functions such as a square wave with Fourier series rather than Fourier transforms. However, as we will show in Chapter 5, the Fourier series is a special case of the Fourier transform.

If the waveform $s(t)$ is not periodic then the Fourier transform will be a continuous function of frequency; that is, $s(t)$ is represented by the summation of sinusoids of all frequencies. For illustration, consider the pulse waveform and its Fourier transform as shown in Fig. 1-4. In this example

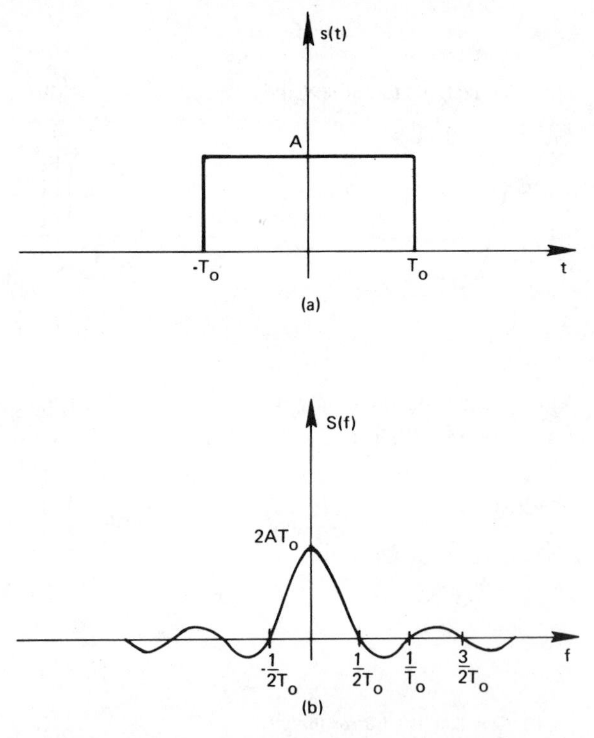

Figure 1-4. Fourier transform of a pulse waveform.

the Fourier transform indicates that one sinusoid frequency becomes indistinguishable from the next and, as a result, all frequencies must be considered.

The Fourier transform is then a frequency domain representation of a function. As illustrated in both Figs. 1-3(a) and 1-4, the Fourier transform frequency domain contains exactly the same information as that of the original function; they differ only in the manner of presentation of the information. Fourier analysis allows one to examine a function from another point of view, the transform domain. As we will see in the discussions to follow, the method of Fourier transform analysis, employed as illustrated in Fig. 1-1, is often the key to problem solving success.

1-3 THE UBIQUITOUS FOURIER TRANSFORM

The term ubiquitous means to be everywhere at the same time. Because of the great variety of seemingly unrelated topics which can be effectively dealt with using the Fourier transform, the modifier ubiquitous is certainly appropriate. One can easily carry over the Fourier analysis techniques developed in one field to many diverse areas. Typical application areas of the Fourier transform include:

Linear Systems. The Fourier transform of the output of a linear system is given by the product of the system transfer function and the Fourier transform of the input signal [1].

Antennas. The field pattern of an antenna is given by the Fourier transform of the antenna current illumination [2].

Optics. Optical systems have the property that a Fourier transform relation exists between the light amplitude distribution at the front and back focal planes of a converging lens [3].

Random Process. The power density spectrum of a random process is given by the Fourier transform of the auto-correlation function of the process [4].

Probability. The characteristic function of a random variable is defined as the Fourier transform of the probability density function of the random variable [5].

Quantum Physics. The *uncertainty principle* in quantum theory is fundamentally associated with the Fourier transform since particle momentum and position are essentially related through the Fourier transform [6].

Boundary-Value Problems. The solution of partial differential equations can be obtained by means of the Fourier transform [7].

Although these application areas are extremely diverse, they are united by the common entity, the Fourier transform. In an age where it is impossible to stay abreast with technology across the spectrum, it is stimulating to find a theory and technique which enables one to invade an unfamiliar field with familiar tools.

1-4 DIGITAL COMPUTER FOURIER ANALYSIS

Because of the wide range of problems which are susceptible to attack by the Fourier transform, we would expect the logical extension of Fourier transform analysis to the digital computer. Numerical integration of Eq. (1-1) implies the relationship

$$S(f_k) = \sum_{i=0}^{N-1} s(t_i)e^{-j2\pi f_k t_i}(t_{i+1} - t_i) \qquad k = 0, 1, \ldots, N-1 \qquad (1\text{-}2)$$

For those problems which do not yield to a closed-form Fourier transform solution, the discrete Fourier transform (1-2) offers a potential method of attack. However, careful inspection of (1-2) reveals that if there are N data points of the function $s(t_i)$ and if we desire to determine the amplitude of N separate sinusoids, then computation time is proportional to N^2, the number of multiplications. Even with high speed computers, computation of the discrete Fourier transform requires excessive machine time for large N.

An obvious requirement existed for the development of techniques to reduce the computing time of the discrete Fourier transform; however, the scientific community met with little success. Then in 1965 Cooley and Tukey published their mathematical algorithm [8] which has become known as the "fast Fourier transform." The fast Fourier transform (FFT) is a computational algorithm which reduces the computing time of Eq. (1-2) to a time proportional to $N \log_2 N$. This increase in computing speed has completely revolutionized many facets of scientific analysis. A historical review of the discovery of the FFT illustrates that this important development was almost ignored.

1-5 HISTORICAL SUMMARY OF THE FAST FOURIER TRANSFORM

During a meeting of the President's Scientific Advisory Committee, Richard L. Garwin noted that John W. Tukey was writing Fourier transforms [9]. Garwin, who in his own research was in desperate need of a fast means to compute the Fourier transform, questioned Tukey as to his knowledge of techniques to compute the Fourier transform. Tukey outlined to Garwin essentially what has led to the famous Cooley-Tukey algorithm.

Garwin went to the computing center at IBM Research in Yorktown Heights to have the technique programmed. James W. Cooley was a relatively new member of the staff at IBM Research and by his own admission was given the problem to work on because he was the only one with nothing important to do [9]. At Garwin's insistence, Cooley quickly worked out a computer program and returned to his own work with the expectation that this project was over and could be forgotten. However, requests for copies of the program and a writeup began accumulating, and Cooley was asked to write a paper on the algorithm. In 1965 Cooley and Tukey published the now famous "An Algorithm for the Machine Calculation of Complex Fourier Series" in the *Mathematics of Computation* [8].

Without the tenacity of Garwin, it is possible that the fast Fourier transform would still be relatively unknown today. The term relative is used because after Cooley and Tukey published their findings, reports of other people using similar techniques began to become known [10]. P. Rudnick [11] of Oceanic Institution in La Jolla, California reported that he was using a similar technique and that he had gotten his idea from a paper published in 1942 by Danielson and Lanczos [12]. This paper in turn referenced Runge [13] and [14] for the source of their methods. These two papers, together with the lecture notes of Runge and König [15], describe essentially the computational advantages of the FFT algorithm as we know it today.

L. H. Thomas of IBM Watson Laboratory also was using a technique [16] very similar to that published by Cooley and Tukey. He implied that he simply had gone to the library and looked up a method to do Fourier series calculations, a book by Stumpff [17]. Thomas generalized the concepts presented in Stumpff and derived a similar technique to what is now known as the fast Fourier transform.

Another line of development also led to an algorithm equivalent to that of Thomas. In 1937, Yates [18] developed an algorithm to compute the interaction of 2^n factorial experiments. Good [19] extended this avenue of approach and outlined a procedure for the computation of N-point Fourier transforms which was essentially equivalent to that of Thomas.

The fast Fourier transform algorithm has had a long and interesting history. Unfortunately, not until recently did the contributions of those involved in its early history become known.

REFERENCES

1. GUPTA, S. C., *Transform and State Variable Methods in Linear Systems.* New York: Wiley, 1966.

2. KRAUS, J. O., *Antennas.* New York: McGraw-Hill, 1950.

3. BORN, M., and E. WOLF, *Principles of Optics.* New York: Pergamon Press, 1959.

4. LEE, Y. W., *Statistical Theory of Communication.* New York: Wiley, 1960.

5. PAPOULIS, A., *Probability, Random Variables, and Stochastic Processes.* New York: McGraw-Hill, 1965.

6. FRENCH, A. P., *Principles of Modern Physics.* New York: Wiley, 1961.

7. BRACEWELL, RON, *The Fourier Transform and Its Applications.* New York: McGraw-Hill, 1965.

8. COOLEY, J. W., and J. W. TUKEY, "An algorithm for the machine calculation of complex Fourier series," *Mathematics of Computation* (1965), Vol. 19, No. 90, pp. 297–301.

9. Cooley, J. W., R. L. Garwin, C. M. Rader, B. P. Bogert, and T. C. Stockham, "The 1968 Arden House Workshop on fast Fourier transform processing," *IEEE Trans. on Audio and Electroacoustics* (June 1969), Vol. AU-17, No. 2.

10. Cooley, J. W., P. W. Lewis, and P. D. Welch, "Historical Notes on the fast Fourier transform," *IEEE Trans. on Audio and Electroacoustics* (June 1967), Vol. AU–15, No. 2, pp. 76–79.

11. Rudnick, P., "Notes on the Calculation of Fourier Series," *Mathematics of Computation* (June 1966), Vol. 20, pp. 429–430.

12. Danielson, G. C., and C. Lanczos, "Some Improvements in Practical Fourier Analysis and Their Application to X-Ray Scattering from Liquids," *J. Franklin Institute* (April 1942), Vol. 233, pp. 365–380, 435–452.

13. Runge C., *Zeit für Math und Physik* (1903), Vol. 48, pp. 433.

14. Runge C., *Zeit für Math und Physik* (1905), Vol. 53, pp. 117.

15. Runge C., and H. König, "Die Grundlehren der Mathematischen Wissenschaften," *Vorlesungen über Numerischen Rechnen*, Vol. 11. Berlin: Julius Springer, 1964.

16. Thomas, L. H., "Using a Computer to Solve Problems in Physics," *Applications of Digital Computers*. Boston, Mass.: Guin, 1963.

17. Stumpff, K., *Tafeln und Aufgaben für Harmonischen Analyse und Periodogrammrechnung*. Berlin: Julius Springer, 1939.

18. Yates, F., "The Design and Analysis of Factorial Experiments," Commonwealth Agriculture Bureaux. Burks, England: Farnam Royal, 1937.

19. Good, I. J., "The Interaction Algorithm and Practical Fourier Analysis," *J. Royal Statistical Society* (1968), Ser. B., Vol. 20, pp. 361–372.

2

THE FOURIER TRANSFORM

A principal analysis tool in many of today's scientific challenges is the Fourier transform. Possibly the most well-known application of this mathematical technique is the analysis of linear time-invariant systems. But as emphasized in Chapter 1, the Fourier transform is essentially a universal problem solving technique. Its importance is based on the fundamental property that one can examine a particular relationship from an entirely different viewpoint. Simultaneous visualization of a function and its Fourier transform is often the key to successful problem solving.

2-1 THE FOURIER INTEGRAL

The Fourier integral is defined by the expression

$$H(f) = \int_{-\infty}^{\infty} h(t)e^{-j2\pi ft}\,dt \tag{2-1}$$

If the integral exists for every value of the parameter f then Eq. (2-1) defines $H(f)$, the Fourier transform of $h(t)$. Typically $h(t)$ is termed a function of the variable time and $H(f)$ is termed a function of the variable frequency. We will use this terminology throughout the book; ***t is time*** and ***f is frequency***. Further, a lower case symbol will represent a function of time; the Fourier transform of this time function will be represented by the same upper case symbol as a function of frequency.

In general the Fourier transform is a complex quantity:

$$H(f) = R(f) + jI(f) = |H(f)|\, e^{j\theta(f)} \tag{2-2}$$

where $R(f)$ is the real part of the Fourier transform,

$I(f)$ is the imaginary part of the Fourier transform,

$|H(f)|$ is the *amplitude* or *Fourier spectrum* of $h(t)$ and is given by $\sqrt{R^2(f) + I^2(f)}$,

$\theta(f)$ is the *phase angle* of the Fourier transform and is given by $\tan^{-1}[I(f)/R(f)]$.

EXAMPLE 2-1

To illustrate the various defining terms of the Fourier transform consider the function of time

$$\begin{aligned} h(t) &= \beta e^{-\alpha t} \qquad t > 0 \\ &= 0 \qquad\quad\;\; t < 0 \end{aligned} \tag{2-3}$$

From Eq. (2-1)

$$\begin{aligned} H(f) &= \int_0^\infty \beta e^{-\alpha t} e^{-j2\pi f t}\, dt = \beta \int_0^\infty e^{-(\alpha + j2\pi f)t}\, dt \\ &= \frac{-\beta}{\alpha + j2\pi f} e^{-(\alpha + j2\pi f)t} \Big|_0^\infty = \frac{\beta}{\alpha + j2\pi f} \\ &= \frac{\beta\alpha}{\alpha^2 + (2\pi f)^2} - j\frac{2\pi f \beta}{\alpha^2 + (2\pi f)^2} \\ &= \frac{\beta}{\sqrt{\alpha^2 + (2\pi f)^2}} e^{j \tan^{-1}[-2\pi f/\alpha]} \end{aligned} \tag{2-4}$$

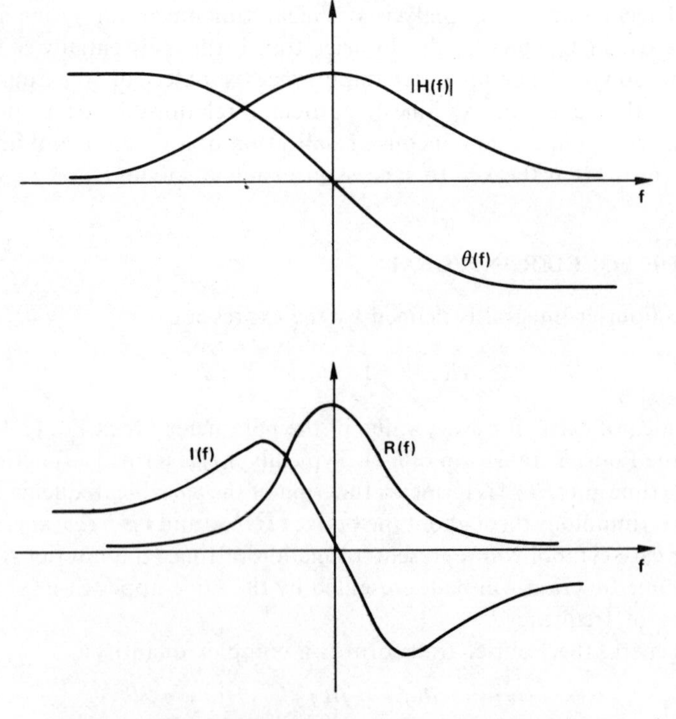

Figure 2-1. Real, imaginary, magnitude, and phase presentations of the Fourier transform.

Hence

$$R(f) = \frac{\beta\alpha}{\alpha^2 + (2\pi f)^2}$$

$$I(f) = \frac{-2\pi f \beta}{\alpha^2 + (2\pi f)^2}$$

$$|H(f)| = \frac{\beta}{\sqrt{\alpha^2 + (2\pi f)^2}}$$

$$\theta(f) = \tan^{-1}\left[\frac{-2\pi f}{\alpha}\right]$$

Each of these functions is plotted in Fig. 2-1 to illustrate the various forms of Fourier transform presentation.

2-2 THE INVERSE FOURIER TRANSFORM

The inverse Fourier transform is defined as

$$h(t) = \int_{-\infty}^{\infty} H(f)e^{j2\pi ft}\, df \tag{2-5}$$

Inversion transformation (2-5) allows the determination of a function of time from its Fourier transform. If the functions $h(t)$ and $H(f)$ are related by Eqs. (2-1) and (2-5), the two functions are termed a *Fourier transform pair*, and we indicate this relationship by the notation

$$h(t) \quad \Longleftrightarrow \quad H(f) \tag{2-6}$$

EXAMPLE 2-2

Consider the frequency function determined in the previous example

$$H(f) = \frac{\beta}{\alpha + j2\pi f} = \frac{\beta\alpha}{\alpha^2 + (2\pi f)^2} - j\frac{2\pi f \beta}{\alpha^2 + (2\pi f)^2}$$

From Eq. (2-5)

$$h(t) = \int_{-\infty}^{\infty} \left[\frac{\beta\alpha}{\alpha^2 + (2\pi f)^2} - j\frac{2\pi f \beta}{\alpha^2 + (2\pi f)^2}\right] e^{j2\pi ft}\, df$$

Since $e^{j2\pi ft} = \cos(2\pi ft) + j\sin(2\pi ft)$, then

$$h(t) = \int_{-\infty}^{\infty} \left[\frac{\beta\alpha\cos(2\pi ft)}{\alpha^2 + (2\pi f)^2} + \frac{2\pi f\beta\sin(2\pi ft)}{\alpha^2 + (2\pi f)^2}\right] df$$

$$+ j\int_{-\infty}^{\infty} \left[\frac{\beta\alpha\sin(2\pi ft)}{\alpha^2 + (2\pi f)^2} - \frac{2\pi f\beta\cos(2\pi ft)}{\alpha^2 + (2\pi f)^2}\right] df \tag{2-7}$$

The second integral of Eq. (2-7) is zero since each integrand term is an odd function. This point is clarified by examination of Fig. 2-2; the first integrand term in the second integral of Eq. (2-7) is illustrated. Note that the function is odd; that is, $g(t) = -g(-t)$. Consequently, the area under the function from $-f_0$ to f_0 is zero. Therefore, in the limit as f_0 approaches infinity, the integral of the function remains zero; the infinite integral of any odd function is zero.

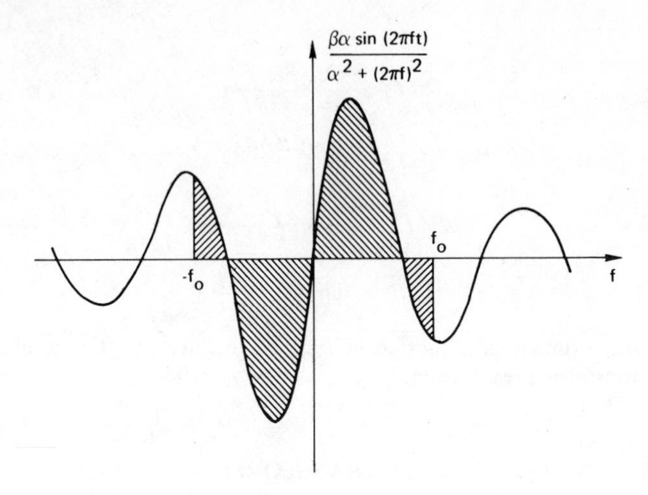

Figure 2-2. Integration of an odd function.

Eq. (2-7) becomes

$$h(t) = \frac{\beta\alpha}{(2\pi)^2} \int_{-\infty}^{\infty} \frac{\cos(2\pi tf)}{(\alpha/2\pi)^2 + f^2} df + \frac{2\pi\beta}{(2\pi)^2} \int_{-\infty}^{\infty} \frac{f \sin(2\pi tf)}{(\alpha/2\pi)^2 + f^2} df \tag{2-8}$$

From a standard table of integrals [4]:

$$\int_{-\infty}^{\infty} \frac{\cos ax}{b^2 + x^2} dx = \frac{\pi}{b} e^{-ab} \qquad a > 0$$

$$\int_{-\infty}^{\infty} \frac{x \sin ax}{b^2 + x^2} dx = \pi e^{-ab} \qquad a > 0$$

Hence Eq. (2-8) can be written as

$$\begin{aligned} h(t) &= \frac{\beta\alpha}{(2\pi)^2}\left[\frac{\pi}{(\alpha/2\pi)} e^{-(2\pi t)(\alpha/2\pi)}\right] + \frac{2\pi\beta}{(2\pi)^2}[\pi e^{-(2\pi t)(\alpha/2\pi)}] \\ &= \frac{\beta}{2} e^{-\alpha t} + \frac{\beta}{2} e^{-\alpha t} = \beta e^{-\alpha t} \qquad t > 0 \end{aligned} \tag{2-9}$$

The time function

$$h(t) = \beta e^{-\alpha t} \qquad t > 0$$

and the frequency function

$$H(f) = \frac{\beta}{\alpha + j(2\pi f)}$$

are related by both Eqs. (2-1) and (2-5) and hence are a Fourier transform pair;

$$\beta e^{-\alpha t} \quad t > 0 \quad \Longleftrightarrow \quad \frac{\beta}{\alpha + j(2\pi f)} \tag{2-10}$$

2-3 EXISTENCE OF THE FOURIER INTEGRAL

To this point we have not considered the validity of Eqs. (2-1) and (2-5); the integral equations have been assumed to be well defined for all functions. In general, for most functions encountered in practical scientific analysis, the Fourier transform and its inverse are well defined. We do not intend to present a highly theoretical discussion of the existence of the Fourier transform but rather to point out conditions for its existence and to give examples of these conditions. Our discussion follows that of Papoulis [5].

Condition 1. If $h(t)$ is integrable in the sense

$$\int_{-\infty}^{\infty} |h(t)|\, dt < \infty \tag{2-11}$$

then its Fourier transform $H(f)$ exists and satisfies the inverse Fourier transform (2-5).

It is important to note that Condition 1 is a sufficient but not a necessary condition for existence of a Fourier transform. There are functions which do not satisfy Condition 1 but have a transform satisfying (2-5). This class of functions will be covered by Condition 2.

EXAMPLE 2-3

To illustrate Condition 1 consider the pulse time waveform

$$\begin{aligned} h(t) &= A && |t| < T_0 \\ &= \frac{A}{2} && t = \pm T_0 \\ &= 0 && |t| > T_0 \end{aligned} \tag{2-12}$$

which is shown in Fig. 2-3. Equation (2-11) is satisfied for this function; therefore, the Fourier transform exists and is given by

$$\begin{aligned} H(f) &= \int_{-T_0}^{T_0} Ae^{-j2\pi ft}\, dt \\ &= A\int_{-T_0}^{T_0} \cos(2\pi ft)\, dt - jA\int_{-T_0}^{T_0} \sin(2\pi ft)\, dt \end{aligned}$$

The second integral is equal to zero since the integrand is odd;

$$\begin{aligned} H(f) &= \frac{A}{2\pi f} \sin(2\pi ft)\Big|_{-T_0}^{T_0} \\ &= 2AT_0 \frac{\sin(2\pi T_0 f)}{2\pi T_0 f} \end{aligned} \tag{2-13}$$

Those terms which obviously can be canceled are retained to emphasize the $[\sin(af)/(af)]$ characteristic of the Fourier transform of a pulse waveform (Fig. 2-3).

Because this example satisfies Condition 1 then $H(f)$ as given by (2-13) must satisfy Eq. (2-5).

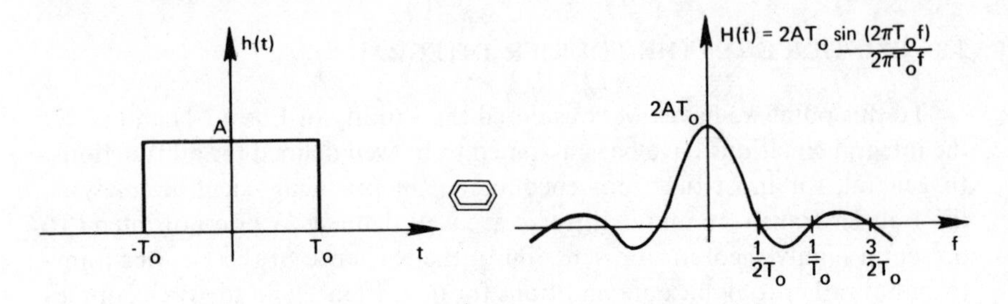

Figure 2-3. Fourier transform of a pulse waveform.

$$h(t) = \int_{-\infty}^{\infty} 2AT_0 \frac{\sin(2\pi T_0 f)}{2\pi T_0 f} e^{j2\pi ft}\, df$$

$$= 2AT_0 \int_{-\infty}^{\infty} \frac{\sin(2\pi T_0 f)}{2\pi T_0 f} [\cos(2\pi ft) + j\sin(2\pi ft)]\, df \tag{2-14}$$

The imaginary integrand term is odd; therefore

$$h(t) = \frac{A}{\pi} \int_{-\infty}^{\infty} \frac{\sin(2\pi T_0 f)\cos(2\pi ft)}{f}\, df \tag{2-15}$$

From the trigonometric identity

$$\sin(x)\cos(y) = \tfrac{1}{2}[\sin(x+y) + \sin(x-y)] \tag{2-16}$$

$h(t)$ becomes

$$h(t) = \frac{A}{2\pi} \int_{-\infty}^{\infty} \frac{\sin[2\pi f(T_0 + t)]}{f}\, df + \frac{A}{2\pi} \int_{-\infty}^{\infty} \frac{\sin[2\pi f(T_0 - t)]}{f}\, df$$

and can be rewritten as

$$h(t) = A(T_0 + t) \int_{-\infty}^{\infty} \frac{\sin[2\pi f(T_0 + t)]}{2\pi f(T_0 + t)}\, df$$

$$+ A(T_0 - t) \int_{-\infty}^{\infty} \frac{\sin[2\pi f(T_0 - t)]}{2\pi f(T_0 - t)}\, df \tag{2-17}$$

Since (| | denotes magnitude or absolute value)

$$\int_{-\infty}^{\infty} \frac{\sin(2\pi ax)}{2\pi ax}\, dx = \frac{1}{2|a|} \tag{2-18}$$

then

$$h(t) = \frac{A}{2} \frac{T_0 + t}{|T_0 + t|} + \frac{A}{2} \frac{T_0 - t}{|T_0 - t|} \tag{2-19}$$

Each term of Eq. (2-19) is illustrated in Fig. 2-4; by inspection these terms add to yield

$$\begin{aligned} h(t) &= A & |t| < T_0 \\ &= \frac{A}{2} & t = \pm T_0 \\ &= 0 & |t| > T_0 \end{aligned} \tag{2-20}$$

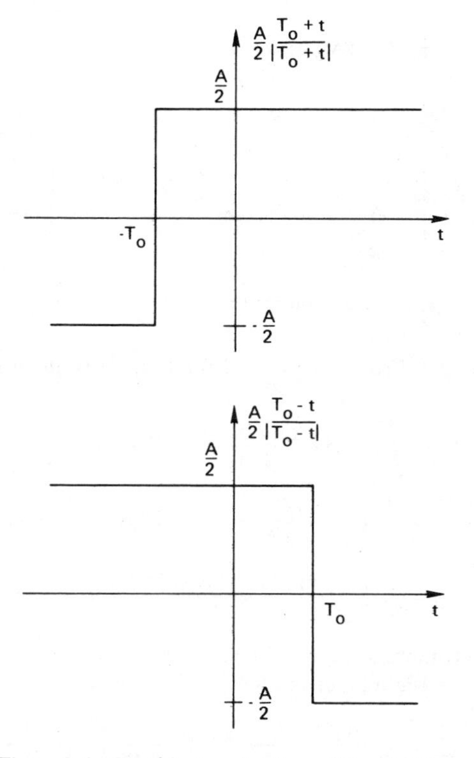

Figure 2-4. Graphical evaluation of Eq. (2-19).

The existence of the Fourier transform and the inverse Fourier transform has been demonstrated for a function satisfying Condition 1. We have established the Fourier transform pair (Fig. 2-3)

$$h(t) = A \quad |t| < T_0 \quad \Longleftrightarrow \quad 2AT_0 \frac{\sin(2\pi T_0 f)}{2\pi T_0 f} \tag{2-21}$$

Condition 2. If $h(t) = \beta(t) \sin(2\pi ft + \alpha)$ (f and α are arbitrary constants), if $\beta(t + k) < \beta(t)$, and if for $|t| > \lambda > 0$, the function $h(t)/t$ is absolutely integrable in the sense of Eq. (2-11) then $H(f)$ exists and satisfies the inverse Fourier transform Eq. (2-5).

An important example is the function $[\sin(af)/(af)]$ which does not satisfy the integrability requirements of Condition 1.

EXAMPLE 2-4

Consider the function

$$h(t) = 2Af_0 \frac{\sin(2\pi f_0 t)}{2\pi f_0 t} \tag{2-22}$$

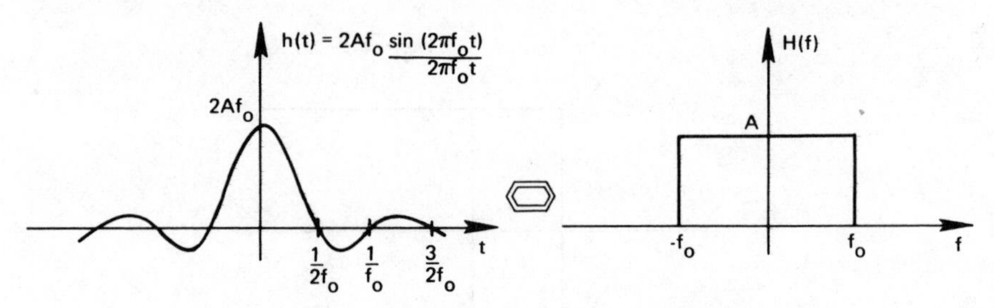

Figure 2-5. Fourier transform of $A \sin (at)/(at)$.

illustrated in Fig. 2-5. From Condition 2 the Fourier transform of $h(t)$ exists and is given by

$$
\begin{aligned}
H(f) &= \int_{-\infty}^{\infty} 2Af_0 \frac{\sin (2\pi f_0 t)}{2\pi f_0 t} e^{-j2\pi ft}\, dt \\
&= \frac{A}{\pi} \int_{-\infty}^{\infty} \frac{\sin (2\pi f_0 t)}{t} [\cos (2\pi ft) - j \sin (2\pi ft)]\, dt \\
&= \frac{A}{\pi} \int_{-\infty}^{\infty} \frac{\sin (2\pi f_0 t) \cos (2\pi ft)}{t}\, dt
\end{aligned} \tag{2-23}
$$

The imaginary term integrates to zero since the integrand term is an odd function. Substitution of the trigonometric identity (2-16) gives

$$
\begin{aligned}
H(f) &= \frac{A}{2\pi} \int_{-\infty}^{\infty} \frac{\sin [2\pi t(f_0 + f)]}{t}\, dt + \frac{A}{2\pi} \int_{-\infty}^{\infty} \frac{\sin [2\pi t(f_0 - f)]}{t}\, dt \\
&= A(f_0 + f) \int_{-\infty}^{\infty} \frac{\sin [2\pi t(f_0 + f)]}{2\pi t(f_0 + f)}\, dt \\
&\quad + A(f_0 - f) \int_{-\infty}^{\infty} \frac{\sin [2\pi t(f_0 - f)]}{2\pi t(f_0 - f)}\, dt
\end{aligned} \tag{2-24}
$$

Equation (2-24) is of the same form as Eq. (2-17); identical analysis techniques yield

$$
\begin{aligned}
H(f) &= A \qquad |f| < f_0 \\
&= \frac{A}{2} \qquad f = \pm f_0 \\
&= 0 \qquad |f| > f_0
\end{aligned} \tag{2-25}
$$

Because this example satisfies Condition 2, $H(f)$ [Eq. (2-25)], must satisfy the inverse Fourier transform relationship (2-5)

$$
\begin{aligned}
h(t) &= \int_{-f_0}^{f_0} Ae^{j2\pi ft}\, df \\
&= A \int_{-f_0}^{f_0} \cos (2\pi ft)\, df = A \frac{\sin (2\pi ft)}{2\pi t} \bigg|_{-f_0}^{f_0} \\
&= 2Af_0 \frac{\sin (2\pi f_0 t)}{2\pi f_0 t}
\end{aligned} \tag{2-26}
$$

By means of Condition 2, the Fourier transform pair

$$2Af_0 \frac{\sin(2\pi f_0 t)}{2\pi f_0 t} \quad \Leftrightarrow \quad H(f) = A \qquad |f| < f_0 \tag{2-27}$$

has been established and is illustrated in Fig. 2-5.

Condition 3. Although not specifically stated, all functions for which Conditions 1 and 2 hold are assumed to be of *bounded variation*; that is, they can be represented by a curve of finite length in any finite time interval. By means of Condition 3 we extend the theory to include singular (impulse) functions.

If $h(t)$ is a periodic or impulse function, then $H(f)$ exists only if one introduces the theory of distributions. Appendix A is an elementary discussion of distribution theory; with the aid of this development the Fourier transform of singular functions can be defined. It is important to develop the Fourier transform of impulse functions because their use greatly simplifies the derivation of many transform pairs.

Impulse function $\delta(t)$ is defined as [Eq. (A-8)]

$$\int_{-\infty}^{\infty} \delta(t - t_0)\, x(t)\, dt = x(t_0) \tag{2-28}$$

where $x(t)$ is an arbitrary function continuous at t_0. Application of the definition (2-28) yields straightforwardly the Fourier transform of many important functions.

EXAMPLE 2-5

Consider the function

$$h(t) = K\delta(t) \tag{2-29}$$

The Fourier transform of $h(t)$ is easily derived using the definition (2-28):

$$H(f) = \int_{-\infty}^{\infty} K\delta(t) e^{-j2\pi ft}\, dt = Ke^0 = K \tag{2-30}$$

The inverse Fourier transform of $H(f)$ is given by

$$h(t) = \int_{-\infty}^{\infty} [K] e^{j2\pi ft}\, df = \int_{-\infty}^{\infty} K \cos(2\pi ft)\, df + j \int_{-\infty}^{\infty} K \sin(2\pi ft)\, df \tag{2-31}$$

Because the integrand of the second integral is an odd function, the integral is zero; the first integral is meaningless unless it is interpreted in the sense of distribution theory. From Eq. (A-21), Eq. (2-31) exists and can be rewritten as

$$h(t) = K \int_{-\infty}^{\infty} e^{j2\pi ft}\, df = K \int_{-\infty}^{\infty} \cos(2\pi ft)\, df = K\, \delta(t) \tag{2-32}$$

These results establish the Fourier transform pair

$$K\delta(t) \quad \Leftrightarrow \quad H(f) = K \tag{2-33}$$

which is illustrated in Fig. 2-6.

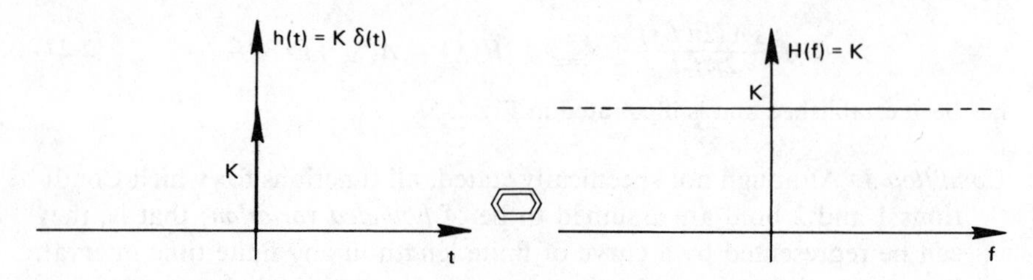

Figure 2-6. Fourier transform of an impulse function.

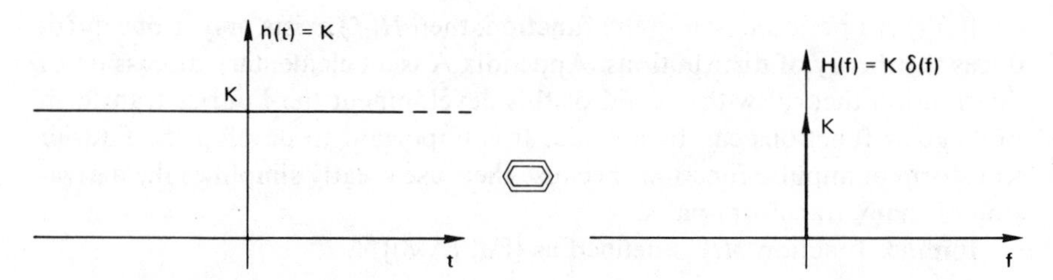

Figure 2-7. Fourier transform of a constant amplitude waveform.

Similarly the Fourier transform pair (Fig. 2-7)

$$h(t) = K \quad \Longleftrightarrow \quad K\delta(f) \tag{2-34}$$

can be established where the reasoning process concerning existence is exactly as argued previously.

EXAMPLE 2-6

To illustrate the Fourier transform of periodic functions consider

$$h(t) = A \cos (2\pi f_0 t) \tag{2-35}$$

The Fourier transform is given by

$$\begin{aligned} H(f) &= \int_{-\infty}^{\infty} A \cos (2\pi f_0 t) e^{-j2\pi f t}\, dt \\ &= \frac{A}{2} \int_{-\infty}^{\infty} [e^{j2\pi f_0 t} + e^{-j2\pi f_0 t}] e^{-j2\pi f t}\, dt \\ &= \frac{A}{2} \int_{-\infty}^{\infty} [e^{-j2\pi t(f-f_0)} + e^{-j2\pi t(f_0+f)}]\, dt \\ &= \frac{A}{2}\delta(f - f_0) + \frac{A}{2}\delta(f + f_0) \end{aligned} \tag{2-36}$$

where arguments identical to those leading to Eq. (2-32) have been employed. The inversion formula yields

$$h(t) = \int_{-\infty}^{\infty} \left[\frac{A}{2}\delta(f - f_0) + \frac{A}{2}\delta(f + f_0)\right] e^{j2\pi ft}\, df$$
$$= \frac{A}{2} e^{j2\pi f_0 t} + \frac{A}{2} e^{-j2\pi f_0 t}$$
$$= A \cos(2\pi f_0 t) \tag{2-37}$$

Fourier transform pair

$$A \cos(2\pi f_0 t) \quad ⬭ \quad \frac{A}{2}\delta(f - f_0) + \frac{A}{2}\delta(f + f_0) \tag{2-38}$$

is illustrated in Fig. 2-8.

Similarly the Fourier transform pair (Fig. 2-9)

$$A \sin(2\pi f_0 t) \quad ⬭ \quad j\frac{A}{2}\delta(f + f_0) - j\frac{A}{2}\delta(f - f_0) \tag{2-39}$$

can be established. Note that the Fourier transform is imaginary.

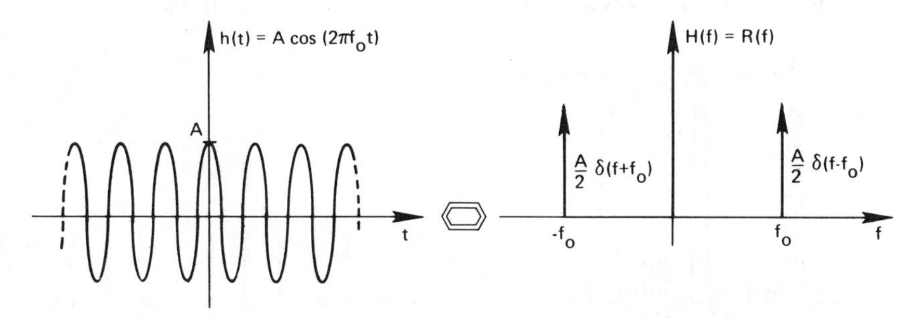

Figure 2-8. Fourier transform of A cos (at).

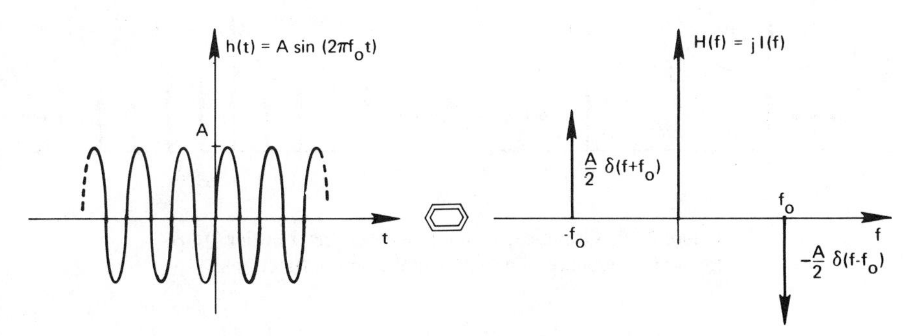

Figure 2-9. Fourier transform of A sin (at).

EXAMPLE 2-7

Without proof†, the Fourier transform of a sequence of equal distant impulse functions is another sequence of equal distant impulses;

†A Papoulis, *The Fourier Integral and Its Applications* (New York: McGraw-Hill, 1962), p. 44.

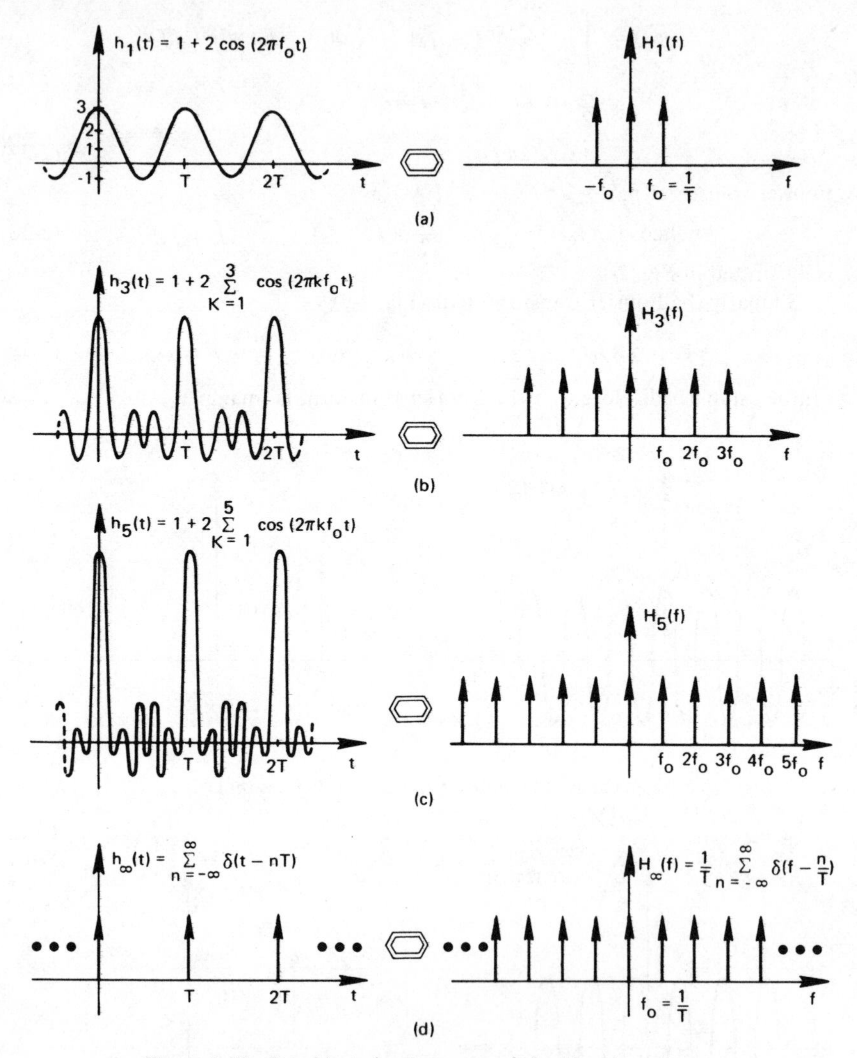

Figure 2-10. Graphical development of the Fourier transform of a sequence of equal distant impulse functions.

$$h(t) = \sum_{n=-\infty}^{\infty} \delta(t - nT) \quad \text{⬭} \quad H(f) = \frac{1}{T} \sum_{n=-\infty}^{\infty} \delta\left(f - \frac{n}{T}\right) \tag{2-40}$$

A graphical development of this Fourier transform pair is illustrated in Fig. 2-10. The importance of Fourier transform pair (2-40) will become obvious in future discussions of discrete Fourier transforms.

Inversion Formula Proof

By means of distribution theory concepts it is possible to derive a simple formal proof of the inversion formula (2-5).

Substitution of $H(f)$ [Eq. (2-1)] into the inverse Fourier transform (2-5) yields

$$\int_{-\infty}^{\infty} H(f)e^{j2\pi ft}\,df = \int_{-\infty}^{\infty} e^{j2\pi ft}\,df \int_{-\infty}^{\infty} h(x)e^{-j2\pi fx}\,dx \tag{2-41}$$

Since [Eq. (A-21)]

$$\int_{-\infty}^{\infty} e^{j2\pi ft}\,dt = \delta(t)$$

then an interchange of integration in (2-41) gives

$$\begin{aligned}\int_{-\infty}^{\infty} H(f)e^{j2\pi ft}\,dt &= \int_{-\infty}^{\infty} h(x)\,dx \int_{-\infty}^{\infty} e^{j2\pi f(t-x)}\,df \\ &= \int_{-\infty}^{\infty} h(x)\,\delta(t-x)\,dx \end{aligned} \tag{2-42}$$

But by the definition of the impulse function (2-28), Eq. (2-42) simply equals $h(x)$. This statement is valid only if $h(t)$ is continuous.† However if it is assumed that

$$h(t) = \frac{h(t^+) + h(t^-)}{2} \tag{2-43}$$

that is, if $h(t)$ is defined as the mid-value at a discontinuity, then the inversion formula still holds. Note that in the previous examples we carefully defined each discontinuous function consistent with Eq. (2-43).

2-4 ALTERNATE FOURIER TRANSFORM DEFINITIONS

It is a well established fact that the Fourier transform is a universally accepted tool of modern analysis. Yet to this day there is not a common definition of the Fourier integral and its inversion formula. To be specific the Fourier transform pair is defined as

$$H(\omega) = a_1 \int_{-\infty}^{\infty} h(t)e^{-j\omega t}\,dt \qquad \omega = 2\pi f \tag{2-44}$$

$$h(t) = a_2 \int_{-\infty}^{\infty} H(\omega)e^{j\omega t}\,d\omega \tag{2-45}$$

where the coefficients a_1 and a_2 assume different values depending on the user. Some set $a_1 = 1$; $a_2 = 1/2\pi$; others set $a_1 = a_2 = 1/\sqrt{2\pi}$, or set $a_1 = 1/2\pi$; $a_2 = 1$. Eqs. (2-44) and (2-45) impose the requirement that $a_1 a_2 = 1/2\pi$. Various users are then concerned with the splitting of the product $a_1 a_2$.

To resolve this question, we must define the relationship desired between the Fourier transform and the Laplace transform and the definition we wish to assume for the relationship between the total energy computed in the time

†See Appendix A. The definition of the impulse response is based on continuity of the testing function, $h(x)$.

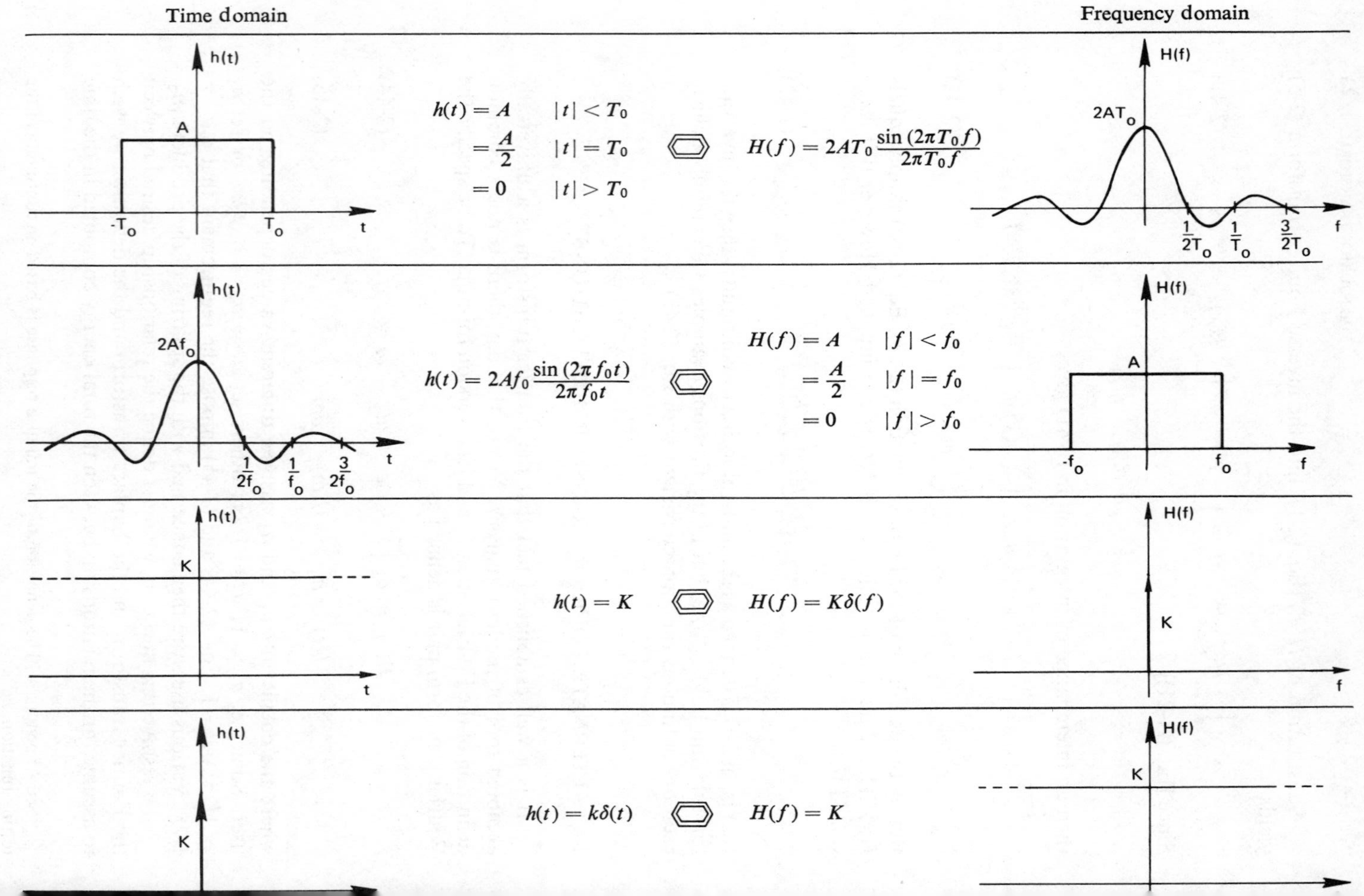
Time domain
Frequency domain
h(t)
A
-T₀
T₀
t
$h(t) = A \quad |t| < T_0$
$= \frac{A}{2} \quad |t| = T_0$
$= 0 \quad |t| > T_0$
$H(f) = 2AT_0 \frac{\sin(2\pi T_0 f)}{2\pi T_0 f}$
H(f)
2AT₀
1/2T₀
1/T₀
3/2T₀
f
2Af₀
1/2f₀
1/f₀
3/2f₀
$h(t) = 2Af_0 \frac{\sin(2\pi f_0 t)}{2\pi f_0 t}$
$H(f) = A \quad |f| < f_0$
$= \frac{A}{2} \quad |f| = f_0$
$= 0 \quad |f| > f_0$
-f₀
f₀
K
$h(t) = K$
$H(f) = K\delta(f)$
$h(t) = k\delta(t)$
$H(f) = K$

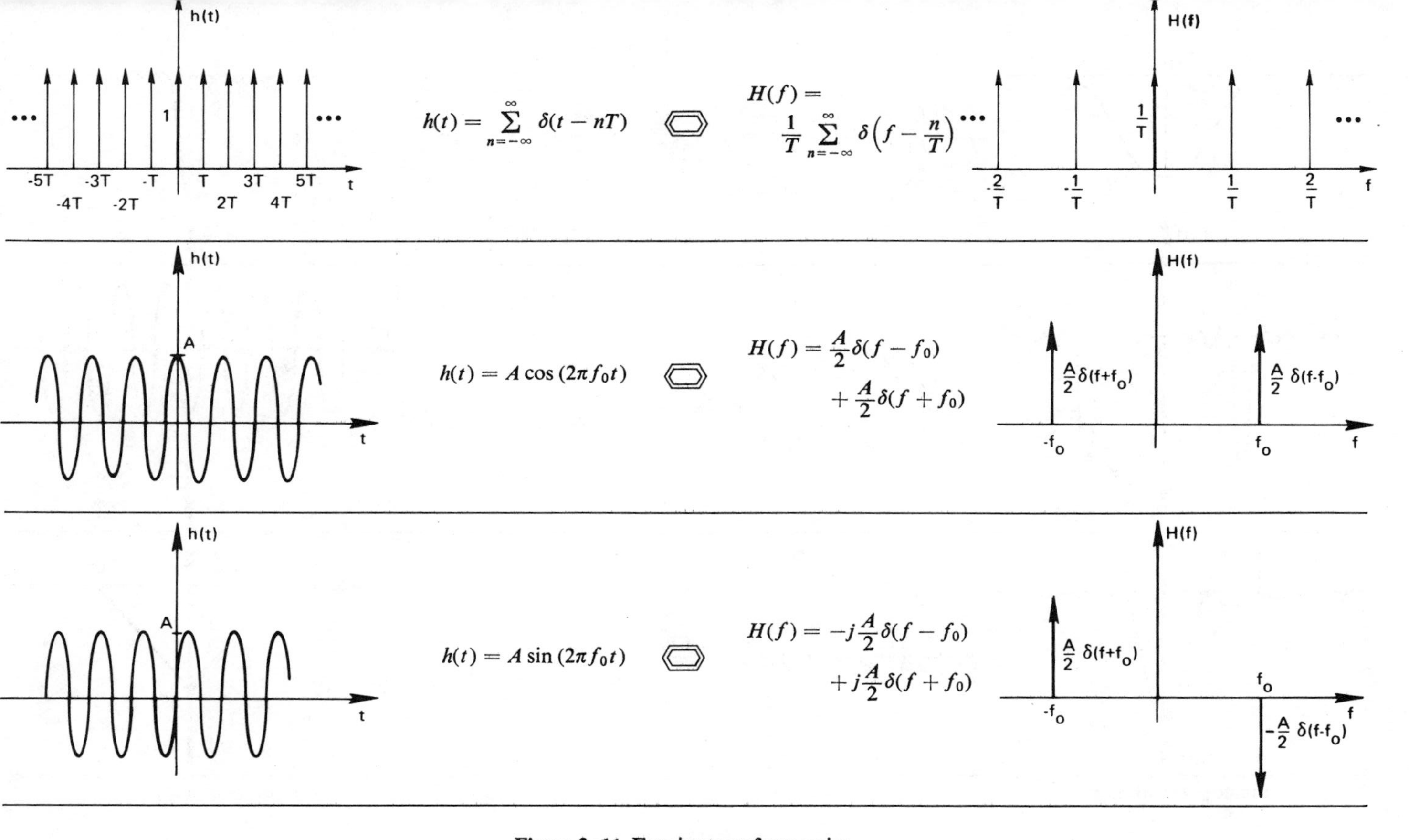

Figure 2-11. Fourier transform pairs.

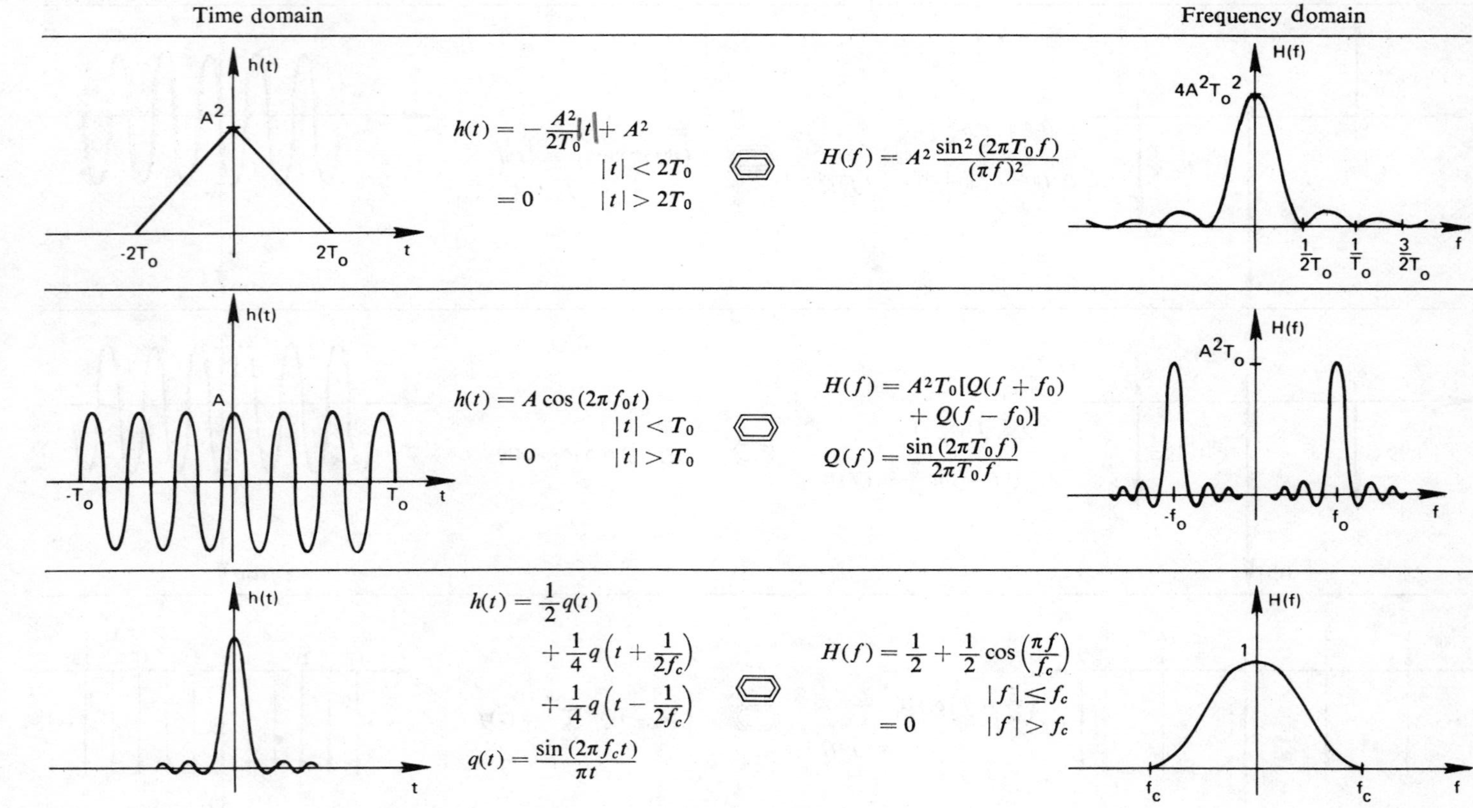

Time domain
Frequency domain
h(t)
A²
-2T₀
2T₀
t
h(t) = −(A²/2T₀)|t| + A² |t| < 2T₀
= 0 |t| > 2T₀
H(f) = A² sin²(2πT₀f)/(πf)²
H(f)
4A²T₀²
1/2T₀
1/T₀
3/2T₀
f
h(t)
A
-T₀
T₀
t
h(t) = A cos(2πf₀t) |t| < T₀
= 0 |t| > T₀
H(f) = A²T₀[Q(f + f₀) + Q(f − f₀)]
Q(f) = sin(2πT₀f)/2πT₀f
H(f)
A²T₀
-f₀
f₀
f
h(t)
t
h(t) = ½q(t) + ¼q(t + 1/2f_c) + ¼q(t − 1/2f_c)
q(t) = sin(2πf_ct)/πt
H(f) = ½ + ½ cos(πf/f_c) |f| ≤ f_c
= 0 |f| > f_c
H(f)
1
f_c
f_c
f

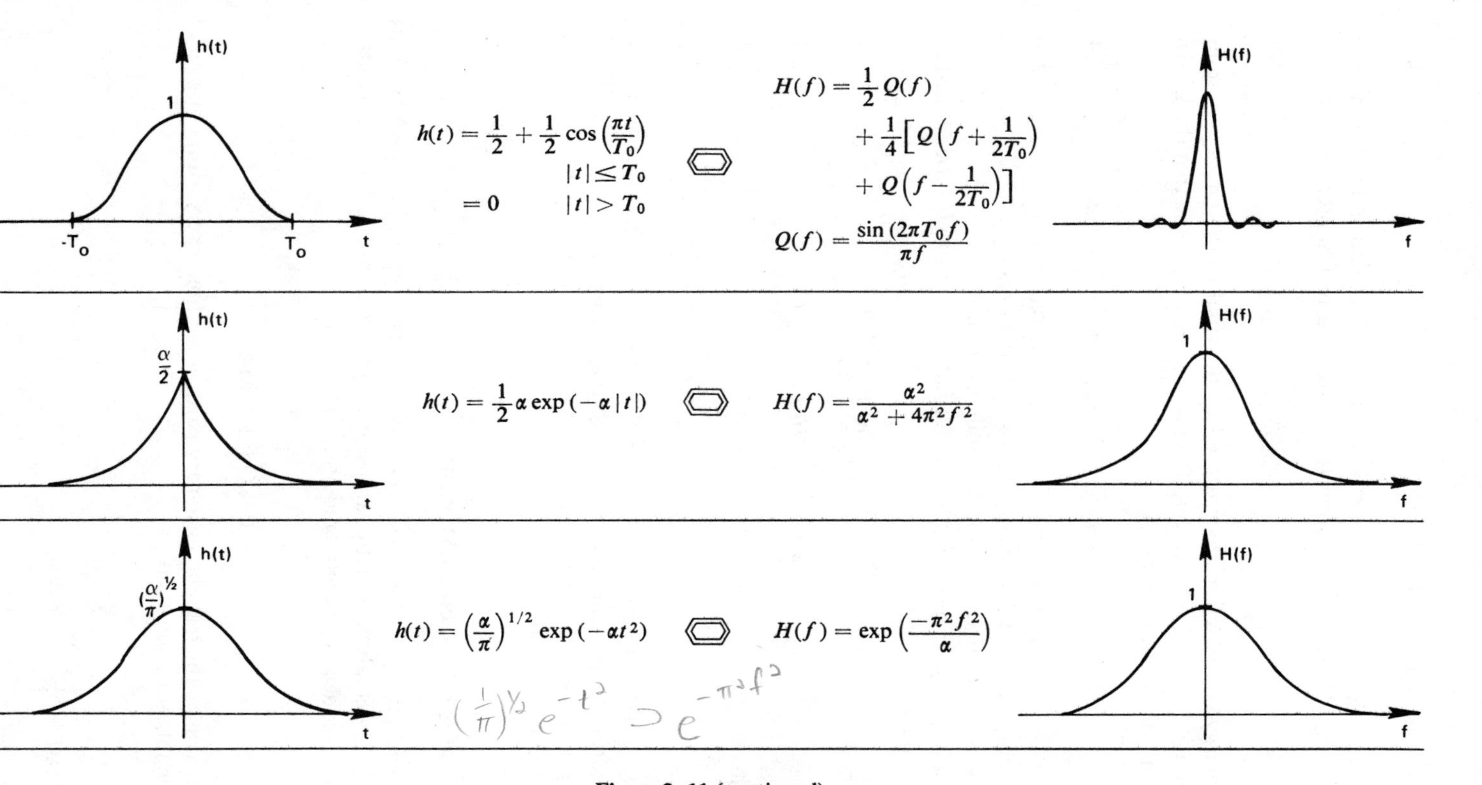

Figure 2-11 (continued).

domain and the total energy computed in ω, the radian frequency domain. For example, Parseval's Theorem (to be derived in Chapter 4) states:

$$\int_{-\infty}^{\infty} h^2(t)\,dt = 2\pi a_1^2 \int_{-\infty}^{\infty} |H(\omega)|^2\,d\omega \tag{2-46}$$

If the energy computed in t is required to be equal to the energy computed in ω, then $a_1 = 1/\sqrt{2\pi}$. However, if the requirement is made that the Laplace transform, universally defined as

$$L[h(t)] = \int_{-\infty}^{\infty} h(t)e^{-st}\,dt = \int_{-\infty}^{\infty} h(t)e^{-(\alpha+j\omega)t}\,dt \tag{2-47}$$

shall reduce to the Fourier transform when the real part of s is set to zero, then a comparison of Eqs. (2-44) and (2-47) requires $a_2 = 1$, i.e., $a_1 = 1/2\pi$ which is in contradiction to the previous hypothesis.

A logical way to resolve this conflict is to define the Fourier transform pair as follows:

$$H(f) = \int_{-\infty}^{\infty} h(t)e^{-j2\pi ft}\,dt \tag{2-48}$$

$$h(t) = \int_{-\infty}^{\infty} H(f)e^{j2\pi ft}\,df \tag{2-49}$$

With this definition Parseval's Theorem becomes

$$\int_{-\infty}^{\infty} h^2(t)\,dt = \int_{-\infty}^{\infty} |H(f)|^2\,df$$

and Eq. (2-48) is consistent with the definition of the Laplace transform. Note that as long as integration is with respect to f, the scale factor $1/2\pi$ never appears. For this reason, the latter definition of the Fourier transform pair was chosen for this book.

2-5 FOURIER TRANSFORM PAIRS

A pictorial table of Fourier transform pairs is given in Fig. 2-11. This graphical and analytical dictionary is by no means complete but does contain the most frequently encountered transform pairs.

PROBLEMS

2-1. Determine the real and imaginary parts of the Fourier transform of each of the following functions:

a. $h(t) = e^{-a|t|} \qquad -\infty < t < \infty$

b. $h(t) = \begin{cases} k & t > 0 \\ \dfrac{k}{2} & t = 0 \\ 0 & t < 0 \end{cases}$

c. $h(t) = \begin{cases} -A & t < 0 \\ 0 & t = 0 \\ A & t > 0 \end{cases}$

d. $h(t) = \begin{cases} A\cos(2\pi f_0 t) & t > 0 \\ \frac{A}{2} & t = 0 \\ 0 & t < 0 \end{cases}$

e. $h(t) = \begin{cases} A & a < t < b;\ a, b > 0 \\ \frac{A}{2} & t = a;\ t = b \\ 0 & \text{elsewhere} \end{cases}$

f. $h(t) = \begin{cases} Ae^{-\alpha t}\sin(2\pi f_0 t) & t \geq 0 \\ 0 & t < 0 \end{cases}$

g. $h(t) = \frac{1}{2}\left[\delta(t+a) + \delta(t-a) + \delta\left(t + \frac{a}{2}\right) + \delta\left(t - \frac{a}{2}\right)\right]$

2-2. Determine the amplitude spectrum $|H(f)|$ and phase $\theta(f)$ of the Fourier transform of $h(t)$:

a. $h(t) = \frac{1}{|t|} \qquad -\infty < t < \infty$

b. $h(t) = e^{-\pi t^2} \qquad -\infty < t < \infty$

c. $h(t) = A\sin(2\pi f_0 t) \qquad 0 \leq t < \infty$

d. $h(t) = Ae^{-\alpha t}\cos(2\pi f_0 t) \qquad 0 \leq t < \infty$

2-3. Determine the inverse Fourier transform of each of the following:

a. $H(f) = \frac{\alpha^2}{\alpha^2 + (2\pi f)^2}$

b. $H(f) = \frac{\sin(2\pi fT)\cos(2\pi fT)}{(2\pi f)}$

c. $H(f) = \frac{1}{(j2\pi f + \alpha)^2}$

d. $H(f) = \frac{1}{(j2\pi f + \alpha)^3}$

e. $H(f) = \frac{\beta}{(\alpha + j2\pi f)^2 + \beta^2}$

f. $H(f) = (1 - f^2)^2 \qquad |f| < 1$
$\quad = 0 \qquad \text{otherwise}$

g. $H(f) = \frac{f^3}{f^4 + \alpha}$

h. $H(f) = \frac{f^2 + \alpha}{f^4 + 2\alpha}$

i. $H(f) = \frac{f}{(f^2 + \alpha)(f^2 + 4\alpha)}$

REFERENCES

1. ARASC, J., *Fourier Transforms and the Theory of Distributions.* Englewood Cliffs, N. J.: Prentice-Hall, 1966.

2. BRACEWELL, R., *The Fourier Transform and Its Applications.* New York: McGraw-Hill, 1965.

3. CAMPBELL, G. A., and R. M. FOSTER, *Fourier Integrals for Practical Applications.* New York: Van Nostrand Reinhold, 1948.

4. ERDILYI, A., *Tables of Integral Transforms*, Vol. 1. New York: McGraw-Hill, 1954.

5. PAPOULIS, A., *The Fourier Integral and Its Applications.* New York: McGraw-Hill, 1962.

3

FOURIER TRANSFORM PROPERTIES

In dealing with Fourier transforms there are a few properties which are basic to a thorough understanding. A visual interpretation of these fundamental properties is of equal importance to knowledge of their mathematical relationships. The purpose of this chapter is to develop not only the theoretical concepts of the basic Fourier transform pairs, but also the *meaning* of these properties. For this reason we use ample analytical and graphical examples.

3-1 LINEARITY

If $x(t)$ and $y(t)$ have the Fourier transforms $X(f)$ and $Y(f)$, respectively, then the sum $x(t) + y(t)$ has the Fourier transform $X(f) + Y(f)$. This property is established as follows:

$$\int_{-\infty}^{\infty} [x(t) + y(t)]e^{-j2\pi ft}\,dt = \int_{-\infty}^{\infty} x(t)e^{-j2\pi ft}\,dt + \int_{-\infty}^{\infty} y(t)e^{-j2\pi ft}\,dt$$
$$= X(f) + Y(f) \tag{3-1}$$

Fourier transform pair

$$x(t) + y(t) \quad \Longleftrightarrow \quad X(f) + Y(f) \tag{3-2}$$

is of considerable importance because it reflects the applicability of the Fourier transform to linear system analysis.

EXAMPLE 3-1

To illustrate the linearity property, consider the Fourier transform pairs

$$x(t) = K \quad \Longleftrightarrow \quad X(f) = K\delta(f) \tag{3-3}$$

$$y(t) = A \cos(2\pi f_0 t) \quad \Longleftrightarrow \quad Y(f) = \frac{A}{2}\delta(f - f_0) + \frac{A}{2}\delta(f + f_0) \tag{3-4}$$

By the linearity theorem

$$x(t) + y(t) = K + A\cos(2\pi f_0 t) \quad \Longleftrightarrow \quad X(f) + Y(f) = K\delta(f) + \frac{A}{2}\delta(f - f_0) + \frac{A}{2}\delta(f + f_0) \tag{3-5}$$

Figures 3-1(a), (b), and (c), illustrate each of the Fourier transform pairs, respectively.

3-2 SYMMETRY

If $h(t)$ and $H(f)$ are a Fourier transform pair then

$$H(t) \quad \Longleftrightarrow \quad h(-f) \tag{3-6}$$

Fourier transform pair (3-6) is established by rewriting Eq. (2-5)

$$h(-t) = \int_{-\infty}^{\infty} H(f) e^{-j2\pi f t}\, df \tag{3-7}$$

and by interchanging the parameters t and f

$$h(-f) = \int_{-\infty}^{\infty} H(t) e^{-j2\pi f t}\, dt \tag{3-8}$$

EXAMPLE 3-2

To illustrate this property consider the Fourier transform pair

$$h(t) = A \qquad |t| < T_0 \quad \Longleftrightarrow \quad \frac{2AT_0 \sin(2\pi T_0 f)}{2\pi T_0 f} \tag{3-9}$$

illustrated previously in Fig. 2-3. By the symmetry theorem

$$2AT_0 \frac{\sin(2\pi T_0 t)}{2\pi T_0 t} \quad \Longleftrightarrow \quad h(-f) = h(f) = A \qquad |f| < T_0 \tag{3-10}$$

which is identical to the Fourier transform pair (2-27) illustrated in Fig. 2-5. Utilization of the symmetry theorem can eliminate many complicated mathematical developments; a case in point is the development of the Fourier transform pair (2-27).

3-3 TIME SCALING

If the Fourier transform of $h(t)$ is $H(f)$, then the Fourier transform of $h(kt)$ where k is a real constant greater than zero is determined by substituting $t' = kt$ in the Fourier integral equation;

$$\int_{-\infty}^{\infty} h(kt) e^{-j2\pi f t}\, dt = \int_{-\infty}^{\infty} h(t') e^{-j2\pi t'(f/k)} \frac{dt'}{k} = \frac{1}{k} H\left(\frac{f}{k}\right) \tag{3-11}$$

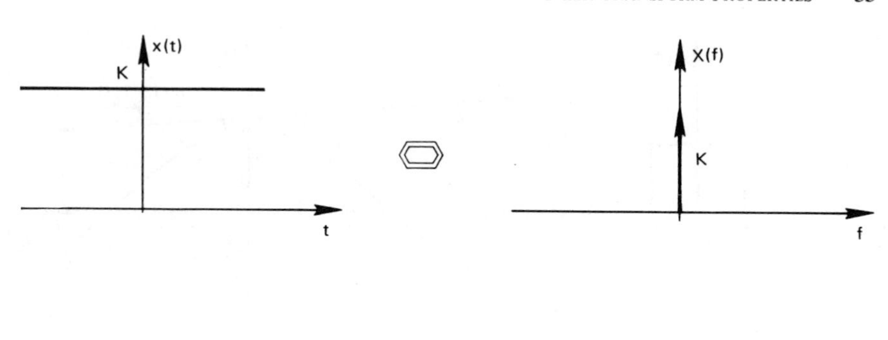

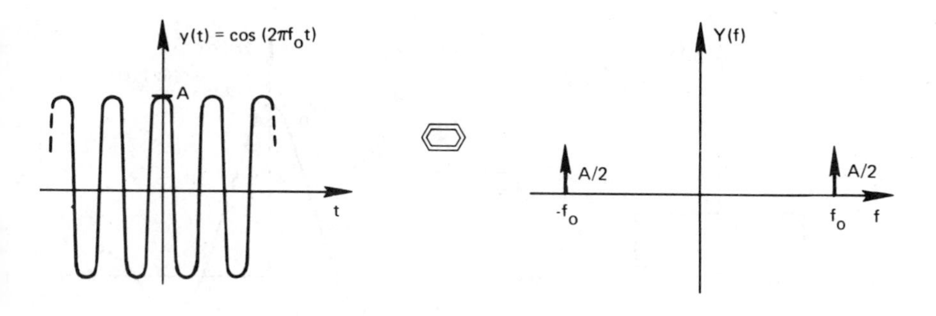

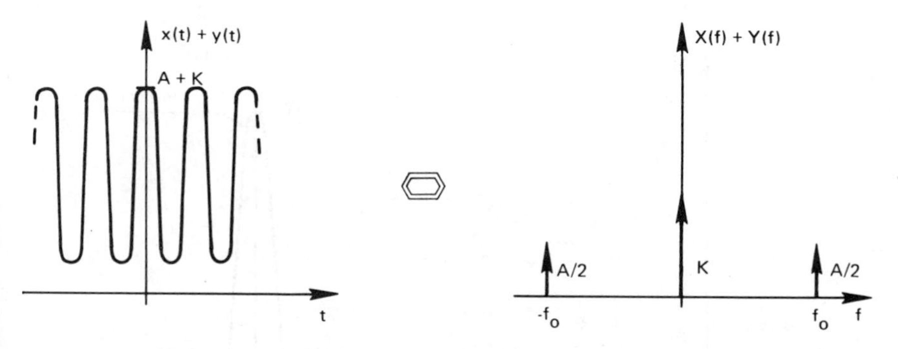

Figure 3-1. The linearity property.

For k negative, the term on the right-hand side changes sign because the limits of integration are interchanged. Therefore, time scaling results in the Fourier transform pair

$$h(kt) \quad \Leftrightarrow \quad \frac{1}{|k|} H\left(\frac{f}{k}\right) \tag{3-12}$$

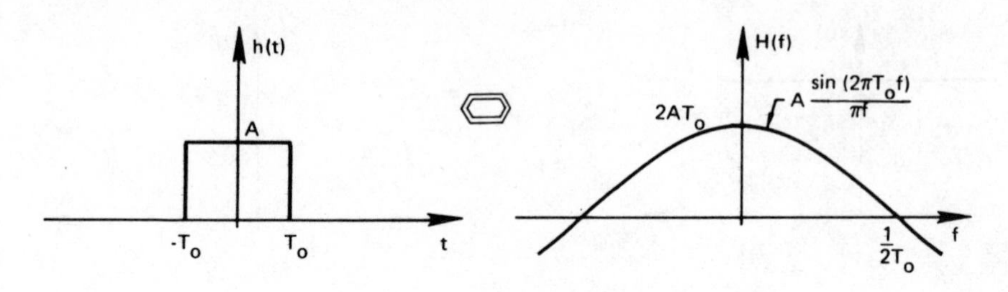

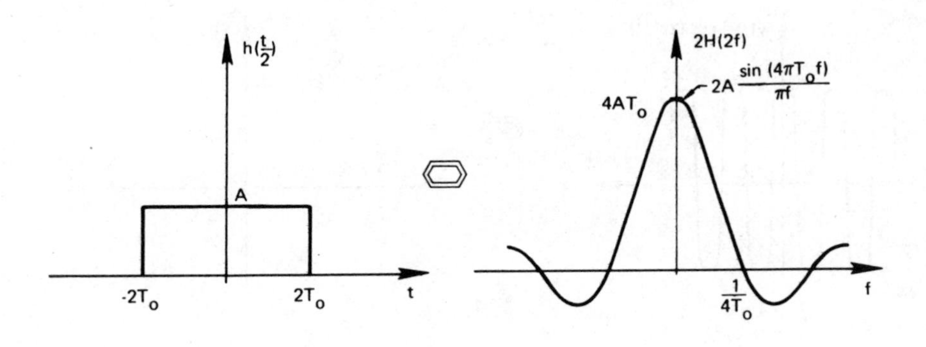

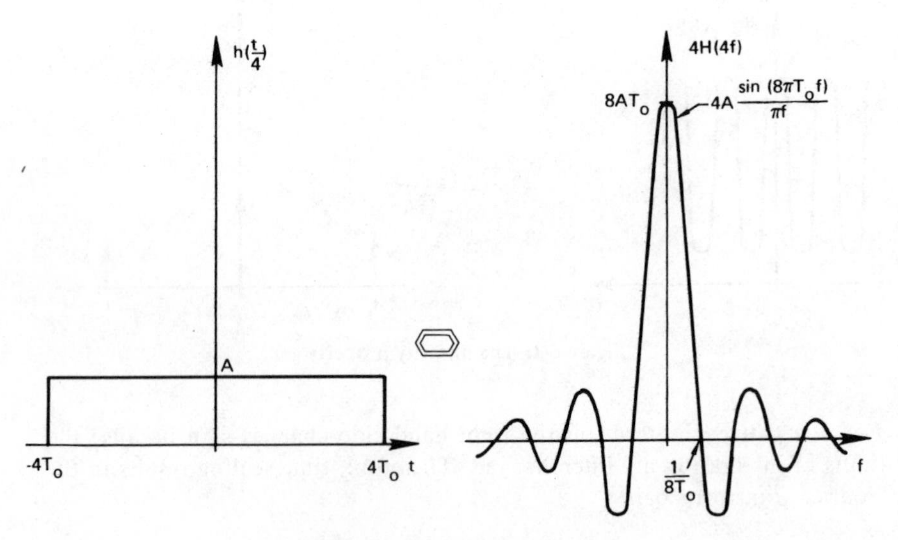

Figure 3-2. Time scaling property.

When dealing with time scaling of impulses, extra care must be exercised; from Eq. (A-10)

$$\delta(at) = \frac{1}{|a|}\delta(t) \tag{3-13}$$

EXAMPLE 3-3

The time scaling Fourier transform property is well-known in many fields of scientific endeavor. As shown in Fig. 3-2 time scale expansion corresponds to frequency scale compression. Note that as the time scale expands, the frequency scale not only contracts but the amplitude increases vertically in such a way as to keep the area constant. This is a well-known concept in radar and antenna theory.

3-4 FREQUENCY SCALING

If the inverse Fourier transform of $H(f)$ is $h(t)$, the inverse Fourier transform of $H(kf)$; k a real constant; is given by the Fourier transform pair

$$\frac{1}{|k|}h\left(\frac{t}{k}\right) \quad \Longleftrightarrow \quad H(kf) \tag{3-14}$$

Relationship (3-14) is established by substituting $f' = kf$ into the inversion formula;

$$\int_{-\infty}^{\infty} H(kf)e^{j2\pi ft}\,df = \int_{-\infty}^{\infty} H(f')e^{j2\pi f'(t/k)}\frac{df'}{k} = \frac{1}{|k|}h\left(\frac{t}{k}\right) \tag{3-15}$$

Frequency scaling of impulse functions is given by

$$\delta(af) = \frac{1}{|a|}\delta(f) \tag{3-16}$$

EXAMPLE 3-4

Analogous to time scaling, frequency scale expansion results in a contraction of the time scale. This effect is illustrated in Fig. 3-3. Note that as the frequency scale expands, the amplitude of the time function increases. This is simply a reflection of the symmetry property (3-6) and the time scaling relationship (3-12).

EXAMPLE 3-5

Many texts state Fourier transform pairs in terms of the radian frequency ω. For example, Papoulis [2, page 44] gives

$$h(t) = \sum_{n=-\infty}^{\infty} \delta(t - nT) \quad \Longleftrightarrow \quad H(\omega) = \frac{2\pi}{T}\sum_{n=-\infty}^{\infty} \delta\left(\omega - \frac{2n\pi}{T}\right) \tag{3-17}$$

By the frequency scaling relationship (3-16) we know that

$$\frac{2\pi}{T}\sum_{n=-\infty}^{\infty} \delta\left[2\pi\left(f - \frac{n}{T}\right)\right] = \frac{1}{T}\sum_{n=-\infty}^{\infty} \delta\left(f - \frac{n}{T}\right) \tag{3-18}$$

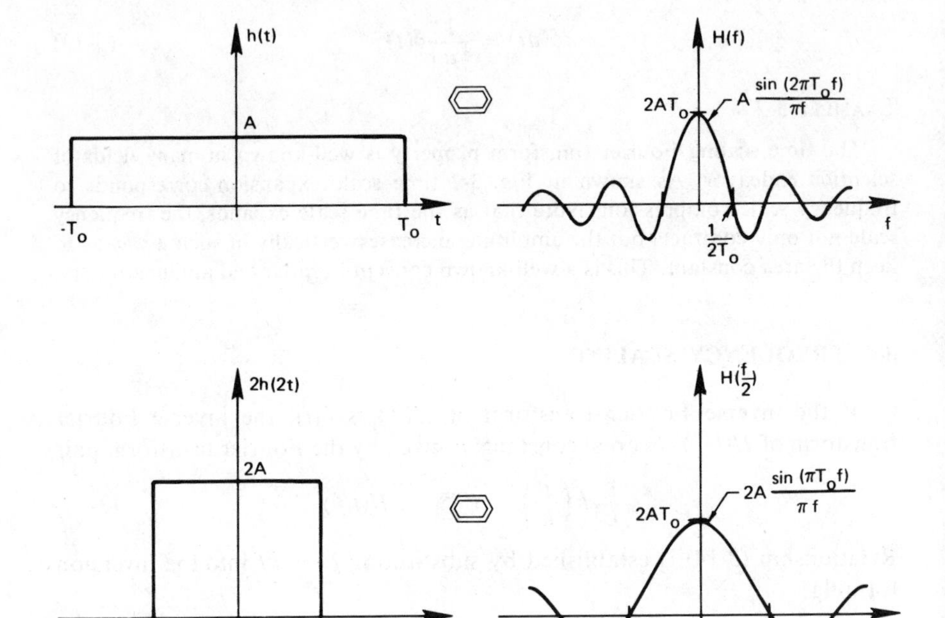

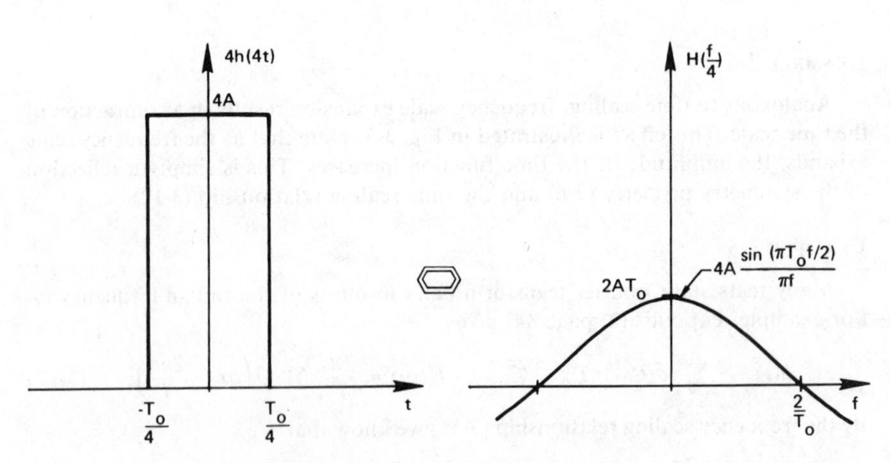

Figure 3-3. Frequency scaling property.

and (3-17) can be rewritten in terms of the frequency variable f

$$h(t) = \sum_{n=-\infty}^{\infty} \delta(t - nT) \quad \Longleftrightarrow \quad H(f) = \frac{1}{T} \sum_{n=-\infty}^{\infty} \delta\left(f - \frac{n}{T}\right) \tag{3-19}$$

which is Eq. (2-40).

3-5 TIME-SHIFTING

If $h(t)$ is shifted by a constant t_0 then by substituting $s = t - t_0$ the Fourier transform becomes

$$\begin{aligned}\int_{-\infty}^{\infty} h(t - t_0)e^{-j2\pi ft}\,dt &= \int_{-\infty}^{\infty} h(s)e^{-j2\pi f(s+t_0)}\,ds \\ &= e^{-j2\pi ft_0}\int_{-\infty}^{\infty} h(s)e^{-j2\pi fs}\,ds \\ &= e^{-j2\pi ft_0}H(f)\end{aligned} \tag{3-20}$$

The time-shifted Fourier transform pair is

$$h(t - t_0) \quad \Longleftrightarrow \quad H(f)e^{-j2\pi ft_0} \tag{3-21}$$

EXAMPLE 3-6

A pictorial description of this pair is illustrated in Fig. 3-4. As shown, time-shifting results in a change in the phase angle $\theta(f) = \tan^{-1}[I(f)/R(f)]$. Note that time-shifting does not alter the magnitude of the Fourier transform. This follows since

$$H(f)e^{-j2\pi ft_0} = H(f)[\cos(2\pi ft_0) - j\sin(2\pi ft_0)]$$

and hence the magnitude is given by

$$|H(f)e^{-j2\pi ft_0}| = \sqrt{H^2(f)[\cos^2(2\pi ft_0) + \sin^2(2\pi ft_0)]} = \sqrt{H^2(f)} \tag{3-22}$$

where $H(f)$ has been assumed to be real for simplicity. These results are easily extended to the case of $H(f)$, a complex function.

3-6 FREQUENCY SHIFTING

If $H(f)$ is shifted by a constant f_0, its inverse transform is multiplied by $e^{j2\pi tf_0}$

$$h(t)e^{j2\pi tf_0} \quad \Longleftrightarrow \quad H(f - f_0) \tag{3-23}$$

This Fourier transform pair is established by substituting $s = f - f_0$ into the inverse Fourier transform-defining relationship

$$\begin{aligned}\int_{-\infty}^{\infty} H(f - f_0)e^{j2\pi ft}\,dt &= \int_{-\infty}^{\infty} H(s)e^{j2\pi t(s+f_0)}\,ds \\ &= e^{j2\pi tf_0}\int_{-\infty}^{\infty} H(s)e^{j2\pi st}\,ds \\ &= e^{j2\pi tf_0}\,h(t)\end{aligned} \tag{3-24}$$

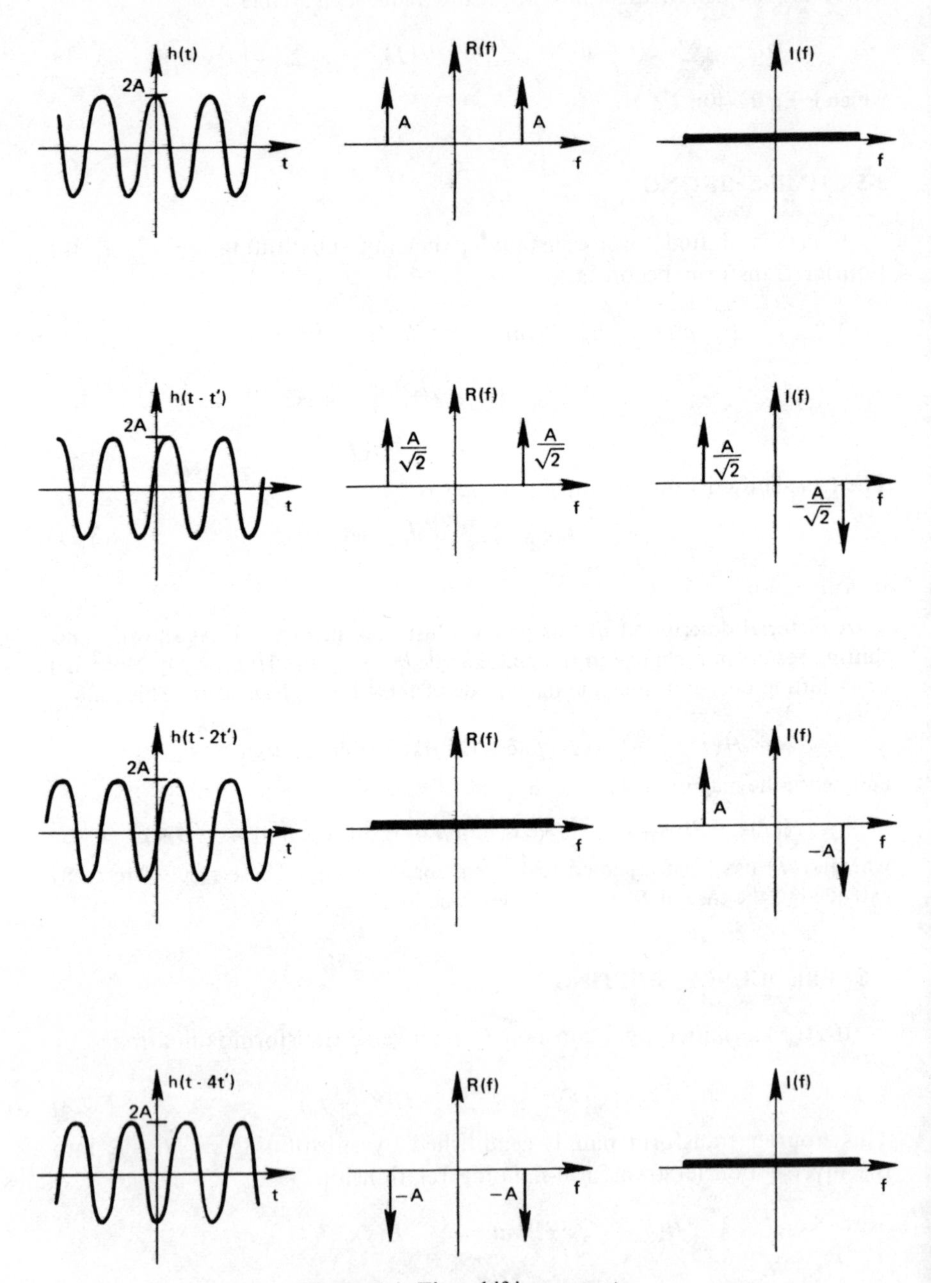

Figure 3-4. Time shifting property.

EXAMPLE 3-7

To illustrate the effect of frequency-shifting let us assume that the frequency function $H(f)$ is real. For this case, frequency-shifting results in a multiplication of the time function $h(t)$ by a cosine whose frequency is determined by the frequency shift f_0 (Fig. 3-5). This process is commonly known as modulation.

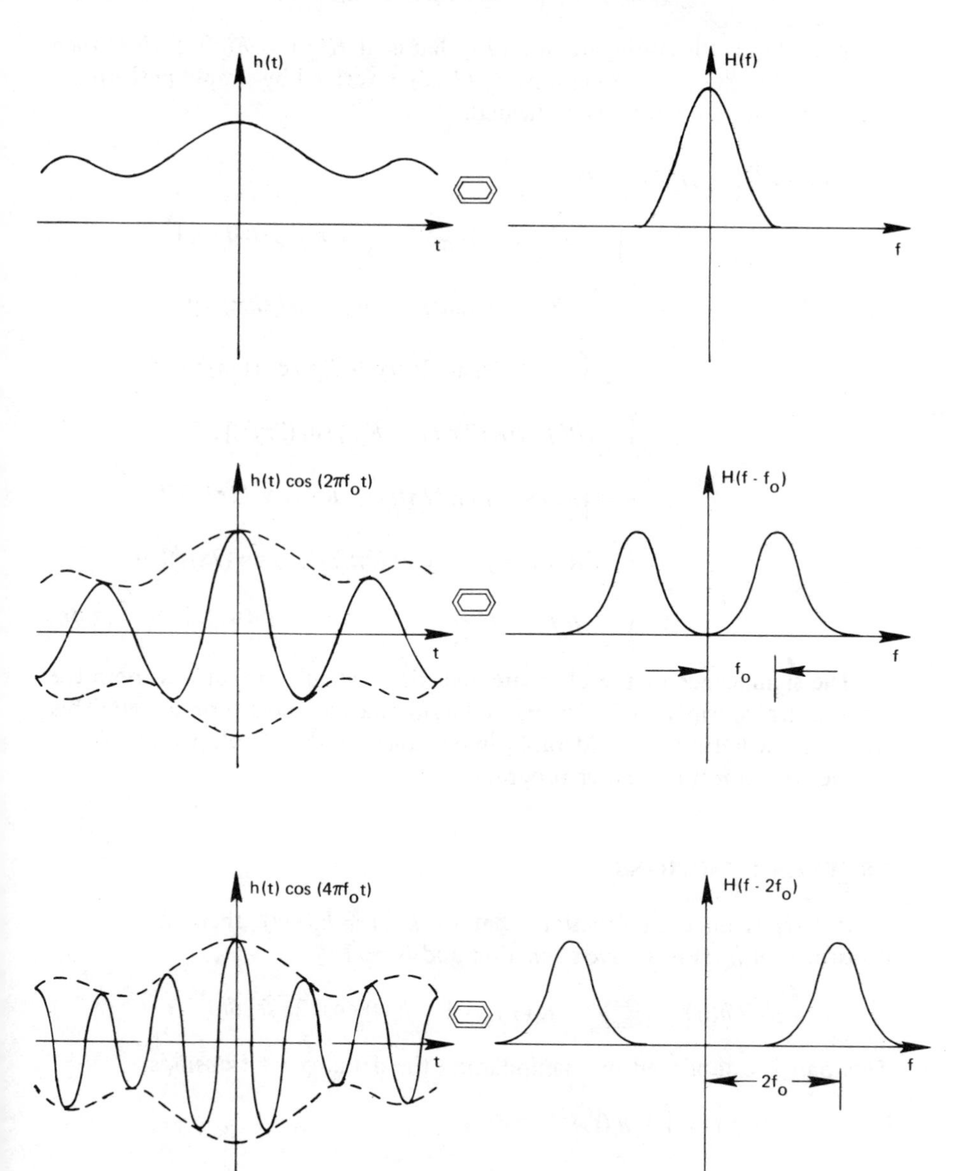

Figure 3-5. Frequency shifting property.

3-7 ALTERNATE INVERSION FORMULA

The inversion formula (2-5) may also be written as

$$h(t) = \left[\int_{-\infty}^{\infty} H^*(f)e^{-j2\pi ft}\,df\right]^* \tag{3-25}$$

where $H^*(f)$ is the conjugate of $H(f)$; that is, if $H(f) = R(f) + jI(f)$ then $H^*(f) = R(f) - jI(f)$. Relationship (3-25) is verified by simply performing the conjugation operations indicated.

$$\begin{aligned}
h(t) &= \left[\int_{-\infty}^{\infty} H^*(f)e^{-j2\pi ft}\,df\right]^* \\
&= \left[\int_{-\infty}^{\infty} R(f)e^{-j2\pi ft}\,df - j\int_{-\infty}^{\infty} I(f)e^{-j2\pi ft}\,df\right]^* \\
&= \left[\int_{-\infty}^{\infty} [R(f)\cos(2\pi ft) - I(f)\sin(2\pi ft)]\,df \right.\\
&\qquad \left. - j\int_{-\infty}^{\infty} [R(f)\sin(2\pi ft) + I(f)\cos(2\pi ft)]\,df\right]^* \\
&= \int_{-\infty}^{\infty} [R(f)\cos(2\pi ft) - I(f)\sin(2\pi ft)]\,df \\
&\qquad + j\int_{-\infty}^{\infty} [R(f)\sin(2\pi ft) + I(f)\cos(2\pi ft)]\,df \\
&= \int_{-\infty}^{\infty} [R(f) + jI(f)][\cos(2\pi ft) + j\sin(2\pi ft)]\,df \\
&= \int_{-\infty}^{\infty} H(f)e^{j2\pi ft}\,df
\end{aligned} \tag{3-26}$$

The significance of the alternate inversion formula is that now both the Fourier transform and its inverse contain the common term $e^{-j2\pi ft}$. This similarity will be of considerable importance in the development of fast Fourier transform computer programs.

3-8 EVEN FUNCTIONS

If $h_e(t)$ is an even function, that is, $h_e(t) = h_e(-t)$, then the Fourier transform of $h_e(t)$ is an even function and is real;

$$h_e(t) \quad \Longleftrightarrow \quad R_e(f) = \int_{-\infty}^{\infty} h_e(t)\cos(2\pi ft)\,dt \tag{3-27}$$

This pair is established by manipulating the defining relationships;

$$\begin{aligned}
H(f) &= \int_{-\infty}^{\infty} h_e(t)e^{-j2\pi ft}\,dt \\
&= \int_{-\infty}^{\infty} h_e(t)\cos(2\pi ft)\,dt - j\int_{-\infty}^{\infty} h_e(t)\sin(2\pi ft)\,dt \\
&= \int_{-\infty}^{\infty} h_e(t)\cos(2\pi ft)\,dt = R_e(f)
\end{aligned} \tag{3-28}$$

The imaginary term is zero since the integrand is an odd function. Since $\cos(2\pi ft)$ is an even function then $h_e(t)\cos(2\pi ft) = h_e(t)\cos[2\pi(-f)t]$ and $H_e(f) = H_e(-f)$; the frequency function is even. Similarly, if $H(f)$ is given as a real and even frequency function, the inversion formula yields

$$\begin{aligned} h(t) &= \int_{-\infty}^{\infty} H_e(f)e^{j2\pi ft}\,dt = \int_{-\infty}^{\infty} R_e(f)e^{j2\pi ft}\,df \\ &= \int_{-\infty}^{\infty} R_e(f)\cos(2\pi ft)\,df + j\int_{-\infty}^{\infty} R_e(f)\sin(2\pi ft)\,df \\ &= \int_{-\infty}^{\infty} R_e(f)\cos(2\pi ft)\,df = h_e(t) \end{aligned} \tag{3-29}$$

EXAMPLE 3-8

As shown in Fig. 3-6 the Fourier transform of an even time function is a real and even frequency function; conversely, the inverse Fourier transform of a real and even frequency function is an even function of time.

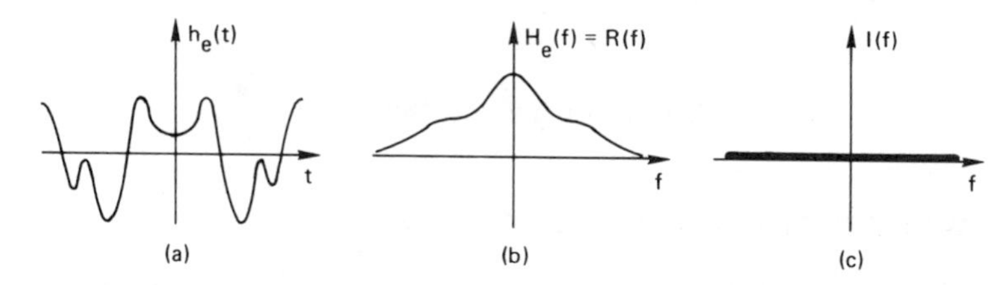

Figure 3-6. Fourier transform of an even function.

3-9 ODD FUNCTIONS

If $h_0(t) = -h_0(-t)$, then $h_0(t)$ is an odd function, and its Fourier transform is an odd and imaginary function,

$$\begin{aligned} H(f) &= \int_{-\infty}^{\infty} h_0(t)e^{-j2\pi ft}\,dt \\ &= \int_{-\infty}^{\infty} h_0(t)\cos(2\pi ft)\,dt - j\int_{-\infty}^{\infty} h_0(t)\sin(2\pi ft)\,dt \\ &= -j\int_{-\infty}^{\infty} h_0(t)\sin(2\pi ft)\,dt = jI_0(f) \end{aligned} \tag{3-30}$$

The real integral is zero since the multiplication of an odd and an even function is an odd function. Since $\sin(2\pi ft)$ is an odd function, then $h_0(t)\sin(2\pi ft) = -h_0(t)\sin[2\pi(-f)t]$ and $H_0(f) = -H_0(-f)$; the frequency function is odd. For $H(f)$ given as an odd and imaginary function, then

$$\begin{aligned} h(t) &= \int_{-\infty}^{\infty} H(f)e^{j2\pi ft}\,dt = j\int_{-\infty}^{\infty} I_0(f)e^{j2\pi ft}\,df \\ &= j\int_{-\infty}^{\infty} I_0(f)\cos(2\pi ft)\,df + j\int_{-\infty}^{\infty} I_0(f)\sin(2\pi ft)\,df \\ &= j\int_{-\infty}^{\infty} I_0(f)\sin(2\pi ft)\,df = h_0(t) \end{aligned} \tag{3-31}$$

and the resulting $h_0(t)$ is an odd function. The Fourier transform pair is th
established:

$$h_0(t) \Leftrightarrow jI_0(f) = -j\int_{-\infty}^{\infty} h_0(t) \sin(2\pi ft)\, dt \qquad (3\text{-}$$

EXAMPLE 3-9

An illustrative example of this transform pair is shown in Fig. 3-7. The funct
$h(t)$ depicted is odd; therefore, the Fourier transform is an odd and imagin
function of frequency. If a frequency function is odd and imaginary then its inve
transform is an odd function of time.

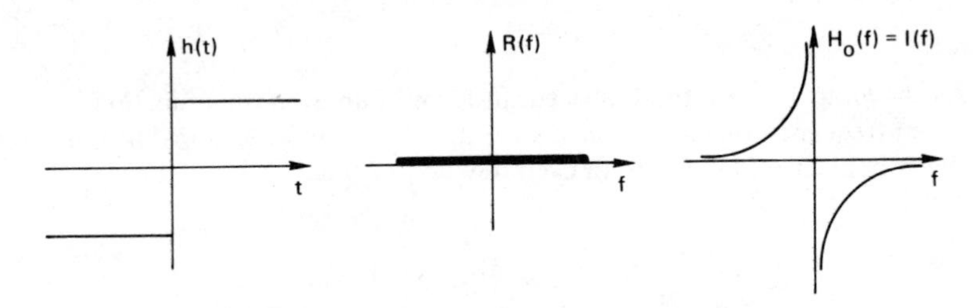

Figure 3-7. Fourier transform of an odd function.

3-10 WAVEFORM DECOMPOSITION

An arbitrary function can always be decomposed or separated into
sum of an even and an odd function;

$$\begin{aligned} h(t) &= \frac{h(t)}{2} + \frac{h(t)}{2} \\ &= \left[\frac{h(t)}{2} + \frac{h(-t)}{2}\right] + \left[\frac{h(t)}{2} - \frac{h(-t)}{2}\right] \\ &= h_e(t) + h_0(t) \end{aligned} \qquad (3\text{-}$$

The terms in brackets satisfy the definition of an even and an odd fu
tion, respectively. From Eqs. (3-27) and (3-32) the Fourier transform
(3-33) is

$$H(f) = R(f) + jI(f) = H_e(f) + H_0(f) \qquad (3\text{-}$$

where $H_e(f) = R(f)$ and $H_0(f) = jI(f)$. We will show in Chapter 10 t
decomposition can be utilized to increase the speed of computation of
discrete Fourier transform.

EXAMPLE 3-10

To demonstrate the concept of waveform decomposition consider the expon
tial function [Fig. 3-8(a)]

$$h(t) = e^{-at} \qquad t \geq 0 \qquad (3\text{-}$$

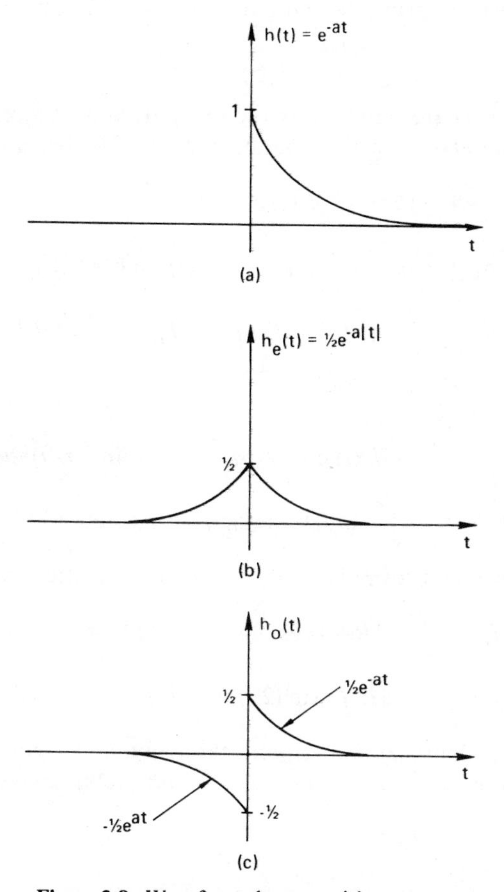

Figure 3-8. Waveform decomposition property.

Following the developments leading to (3-33) we obtain

$$\begin{aligned} h(t) &= \left[\frac{e^{-at}}{2}\right] + \left[\frac{e^{-at}}{2}\right] \\ &= \left\{\left[\frac{e^{-at}}{2}\right]_{t\geq 0} + \left[\frac{e^{at}}{2}\right]_{t\leq 0}\right\} + \left\{\left[\frac{e^{-at}}{2}\right]_{t\geq 0} - \left[\frac{e^{at}}{2}\right]_{t\leq 0}\right\} \\ &= \{e^{-a|t|}\} + \left\{\left[\frac{e^{-at}}{2}\right]_{t\geq 0} - \left[\frac{e^{at}}{2}\right]_{t\leq 0}\right\} \\ &= \{h_e(t)\} + \{h_0(t)\} \end{aligned} \tag{3-36}$$

Figures 3-8(b) and (c) illustrate the even and odd decomposition, respectively.

3-11 COMPLEX TIME FUNCTIONS

For ease of presentation we have to this point considered only real functions of time. The Fourier transform (2-1), the inversion integral (2-5), and the Fourier transform properties hold for the case of $h(t)$, a complex function of time. If

$$h(t) = h_r(t) + jh_i(t) \tag{3-37}$$

where $h_r(t)$ and $h_i(t)$ are respectively the real part and imaginary part of the complex function $h(t)$, then the Fourier integral (2-1) becomes

$$\begin{aligned} H(f) &= \int_{-\infty}^{\infty} [h_r(t) + jh_i(t)]e^{-j2\pi ft}\,dt \\ &= \int_{-\infty}^{\infty} [h_r(t)\cos(2\pi ft) + h_i(t)\sin(2\pi ft)]\,dt \\ &\quad -j\int_{-\infty}^{\infty} [h_r(t)\sin(2\pi ft) - h_i(t)\cos(2\pi ft)]\,dt \\ &= R(f) + jI(f) \end{aligned} \tag{3-38}$$

Therefore

$$R(f) = \int_{-\infty}^{\infty} [h_r(t)\cos(2\pi ft) + h_i(t)\sin(2\pi ft)]\,dt \tag{3-39}$$

$$I(f) = -\int_{-\infty}^{\infty} [h_r(t)\sin(2\pi ft) - h_i(t)\cos(2\pi ft)]\,dt \tag{3-40}$$

Similarly the inversion formula (2-5) for complex functions yields

$$h_r(t) = \int_{-\infty}^{\infty} [R(f)\cos(2\pi ft) - I(f)\sin(2\pi ft)]\,df \tag{3-41}$$

$$h_i(t) = \int_{-\infty}^{\infty} [R(f)\sin(2\pi ft) + I(f)\cos(2\pi ft)]\,df \tag{3-42}$$

If $h(t)$ is real, then $h(t) = h_r(t)$, and the real and imaginary parts of the Fourier transform are given by Eqs. (3-39) and (3-40), respectively,

$$R_e(f) = \int_{-\infty}^{\infty} h_r(t)\cos(2\pi ft)\,dt \tag{3-43}$$

$$I_0(f) = -\int_{-\infty}^{\infty} h_r(t)\sin(2\pi ft)\,dt \tag{3-44}$$

$R_e(f)$ is an even function, since $R_e(f) = R_e(-f)$. Similarly $I_0(-f) = -I_0(f)$ and $I_0(f)$ is odd.

For $h(t)$ purely imaginary, $h(t) = jh_i(t)$ and

$$R_0(f) = \int_{-\infty}^{\infty} h_i(t)\sin(2\pi ft)\,dt \tag{3-45}$$

$$I_e(f) = \int_{-\infty}^{\infty} h_i(t)\cos(2\pi ft)\,dt \tag{3-46}$$

$R_0(f)$ is an odd function and $I_e(f)$ is an even function. Table 3-1 lists various complex time functions and their respective Fourier transforms.

EXAMPLE 3-11

We can employ relationships (3-43), (3-44), (3-45), and (3-46) to simultaneously determine the Fourier transform of two real functions. To illustrate this point,

TABLE 3-1 PROPERTIES OF THE FOURIER TRANSFORM FOR COMPLEX FUNCTIONS

Time domain $h(t)$	Frequency domain $H(f)$
Real	Real part even Imaginary part odd
Imaginary	Real part odd Imaginary part even
Real even, imaginary odd	Real
Real odd, imaginary even	Imaginary
Real and even	Real and even
Real and odd	Imaginary and odd
Imaginary and even	Imaginary and even
Imaginary and odd	Real and odd
Complex and even	Complex and even
Complex and odd	Complex and odd

recall the linearity property (3-2);

$$x(t) + y(t) \quad \Longleftrightarrow \quad X(f) + Y(f) \tag{3-47}$$

Let $x(t) = h(t)$ and $y(t) = jg(t)$ where both $h(t)$ and $g(t)$ are real functions. It follows that $X(f) = H(f)$ and $Y(f) = jG(f)$. Since $x(t)$ is real then from (3-43) and (3-44)

$$x(t) = h(t) \quad \Longleftrightarrow \quad X(f) = H(f) = R_e(f) + jI_0(f) \tag{3-48}$$

Similarly, since $y(t)$ is imaginary then from (3-45) and (3-46)

$$y(t) = jg(t) \quad \Longleftrightarrow \quad Y(f) = jG(f) = R_0(f) + jI_e(f) \tag{3-49}$$

Hence

$$h(t) + jg(t) \quad \Longleftrightarrow \quad H(f) + jG(f) \tag{3-50}$$

where

$$H(f) = R_e(f) + jI_0(f) \tag{3-51}$$

$$G(f) = I_e(f) - jR_0(f) \tag{3-52}$$

Thus if

$$z(t) = h(t) + jg(t) \tag{3-53}$$

then the Fourier transform of $z(t)$ can be expressed as

$$\begin{aligned} Z(f) &= R(f) + jI(f) \\ &= \left[\frac{R(f)}{2} + \frac{R(-f)}{2}\right] + \left[\frac{R(f)}{2} - \frac{R(-f)}{2}\right] \\ &\quad + j\left[\frac{I(f)}{2} + \frac{I(-f)}{2}\right] + j\left[\frac{I(f)}{2} - \frac{I(-f)}{2}\right] \end{aligned} \tag{3-54}$$

and from (3-51) and (3-52)

$$H(f) = \left[\frac{R(f)}{2} + \frac{R(-f)}{2}\right] + j\left[\frac{I(f)}{2} - \frac{I(-f)}{2}\right] \tag{3-55}$$

$$G(f) = \left[\frac{I(f)}{2} + \frac{I(-f)}{2}\right] - j\left[\frac{R(f)}{2} - \frac{R(-f)}{2}\right] \tag{3-56}$$

Thus it is possible to separate the frequency function $Z(f)$ into the Fourier transforms of $h(t)$ and $g(t)$, respectively. As will be demonstrated in Chapter 10, this technique can be used advantageously to increase the speed of computation of the discrete Fourier transform.

3-12 SUMMARY OF PROPERTIES

For future reference the basic properties of the Fourier transform are summarized in Table 3-2. These relationships will be of considerable importance throughout the remainder of this book.

TABLE 3-2 PROPERTIES OF FOURIER TRANSFORMS

Time domain	Equation no.	Frequency domain
Linear addition $x(t) + y(t)$	(3-2)	Linear addition $X(f) + Y(f)$
Symmetry $H(t)$	(3-6)	Symmetry $h(-f)$
Time scaling $h(kt)$	(3-12)	Inverse scale change $\frac{1}{k}H\left(\frac{f}{k}\right)$
Inverse scale change $\frac{1}{k}h\left(\frac{t}{k}\right)$	(3-14)	Frequency scaling $H(kf)$
Time shifting $h(t - t_0)$	(3-21)	Phase shift $H(f)e^{-j2\pi f t_0}$
Modulation $h(t)e^{j2\pi t f_0}$	(3-23)	Frequency shifting $H(f - f_0)$
Even function $h_e(t)$	(3-27)	Real function $H_e(f) = R_e(f)$
Odd function $h_0(t)$	(3-30)	Imaginary $H_0(f) = jI_0(f)$
Real function $h(t) = h_r(t)$	(3-43) (3-44)	Real part even Imaginary part odd $H(f) = R_e(f) + jI_0(f)$
Imaginary function $h(t) = jh_i(t)$	(3-45) (3-46)	Real part odd Imaginary part even $H(f) = R_0(f) + jI_e(f)$

PROBLEMS

3-1. Let

$$h(t) = \begin{cases} A & |t| < 2 \\ \frac{A}{2} & t = \pm 2 \\ 0 & |t| > 2 \end{cases}$$

$$x(t) = \begin{cases} -A & |t| < 1 \\ -\frac{A}{2} & t = \pm 1 \\ 0 & |t| > 1 \end{cases}$$

Sketch $h(t)$, $x(t)$, and $[h(t) - x(t)]$. Use Fourier transform pair (2-21) and the linearity theorem to find the Fourier transform of $[h(t) - x(t)]$.

3-2. Consider the functions $h(t)$ illustrated in Fig. 3-9. Use the linearity property to derive the Fourier transform of $h(t)$.

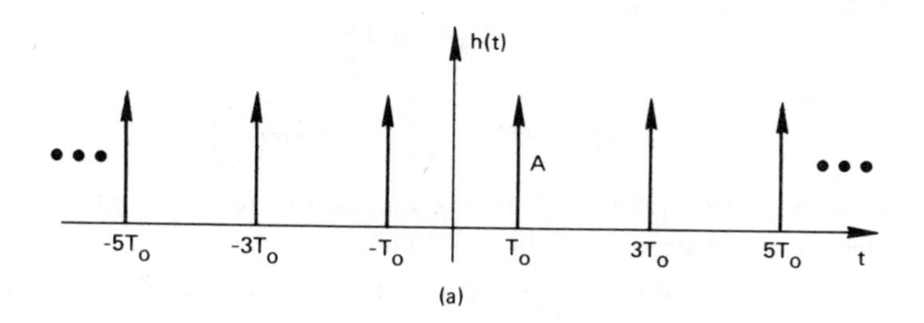

(a)

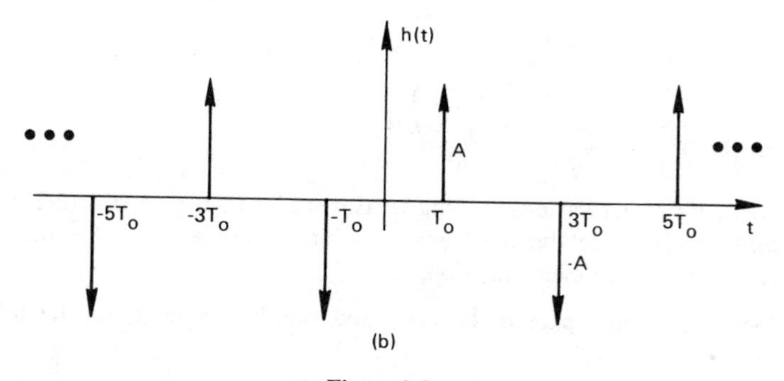

(b)

Figure 3-9.

3-3. Use the symmetry theorem and the Fourier transform pairs of Fig. 2-11 to determine the Fourier transform of the following:

a. $h(t) = \dfrac{A^2 \sin^2 (2\pi T_0 t)}{(\pi t)^2}$

b. $h(t) = \dfrac{\alpha^2}{(\alpha^2 + 4\pi^2 t^2)}$

c. $h(t) = \exp\left(\dfrac{-\pi^2 t^2}{\alpha}\right)$

3-4. Derive the frequency scaling property from the time scaling property by means of the symmetry theorem.

3-5. Consider

$$h(t) = \begin{cases} A^2 - \dfrac{A^2|t|}{2T_0} & |t| < 2T_0 \\ 0 & |t| > 2T_0 \end{cases}$$

Sketch the Fourier transform of $h(2t)$, $h(4t)$, and $h(8t)$. (The Fourier transform of $h(t)$ is given in Fig. 2-11.)

3-6. Derive the time scaling property for the case k negative.

3-7. By means of the shifting theorem find the Fourier transform of the following functions:

a. $h(t) = \dfrac{A \sin [2\pi f_0(t - t_0)]}{\pi(t - t_0)}$

b. $h(t) = K\delta(t - t_0)$

c. $h(t) = \begin{cases} A^2 - \dfrac{A^2}{2T_0}|t - t_0| & |t - t_0| < 2T_0 \\ 0 & |t - t_0| > 2T_0 \end{cases}$

3-8. Show that

$$h(\alpha t - \beta) \quad \Longleftrightarrow \quad \frac{1}{|\alpha|} e^{j2\pi\beta f/\alpha} H\left(\frac{f}{\alpha}\right)$$

3-9. Show that $|H(f)| = |e^{-j2\pi f t_0} H(f)|$; that is, the magnitude of a frequency function is independent of the time delay.

3-10. Find the inverse Fourier transform of the following functions by using the frequency shifting theorem:

a. $H(f) = \dfrac{A \sin [2\pi T_0(f - f_0)]}{\pi(f - f_0)}$

b. $H(f) = \dfrac{\alpha^2}{[\alpha^2 + 4\pi^2(f + f_0)^2]}$

c. $H(f) = \dfrac{A^2 \sin^2 [2\pi T_0(f - f_0)]}{[\pi(f - f_0)]^2}$

3-11. Review the derivations leading to Eqs. (2-9), (2-13), (2-20), (2-25), (2-26), and (2-32). Note the mathematics which result are real for the Fourier transform of an even function.

3-12. Decompose and sketch the even and odd components of the following functions:

a. $h(t) = \begin{cases} 1 & 1 < t < 2 \\ 0 & \text{otherwise} \end{cases}$

b. $h(t) = \dfrac{1}{[2 - (t - 2)^2]}$

c. $h(t) = \begin{cases} -t + 1 & 0 < t \leq 1 \\ 0 & \text{otherwise} \end{cases}$

3-13. Prove each of the properties listed in Table 3-1.

3-14. If $h(t)$ is real, show that $|H(f)|$ is an even function.

3-15. By making a substitution of variable in Eq. (2-28) show that

$$\int_{-\infty}^{\infty} x(t)\delta(at - t_0)\,dt = \frac{1}{a}x\left(\frac{t_0}{a}\right)$$

3-16. Prove the following Fourier transform pairs:

a. $\dfrac{dh(t)}{dt} \Longleftrightarrow j2\pi f H(f)$

b. $[-j2\pi t]s(t) \Longleftrightarrow \dfrac{dH(f)}{df}$

3-17. Use the derivative relationship of Problem 3-16(a) to find the Fourier transform of a pulse waveform given the Fourier transform of a triangular waveform.

REFERENCES

1. BRACEWELL, R., *The Fourier Transform and Its Applications.* New York: McGraw-Hill, 1965.

2. PAPOULIS, A., *The Fourier Integral and Its Applications.* New York: McGraw-Hill, 1962.

4

CONVOLUTION AND CORRELATION

In the previous chapter we investigated those properties which are fundamental to the Fourier transform. However, there exists a class of Fourier transform relationships whose importance far outranks those previously considered. These properties are the convolution and correlation theorems which are to be discussed at length in this chapter.

4-1 CONVOLUTION INTEGRAL

Convolution of two functions is a significant physical concept in many diverse scientific fields. However, as in the case of many important mathematical relationships, the convolution integral does not readily *unveil* itself as to its true implications. To be more specific, the convolution integral is given by

$$y(t) = \int_{-\infty}^{\infty} x(\tau)h(t - \tau)\, d\tau = x(t) * h(t) \tag{4-1}$$

Function $y(t)$ is said to be the convolution of the functions $x(t)$ and $h(t)$. Note that it is extremely difficult to *visualize* the mathematical operation of Eq. (4-1). We will develop the true meaning of convolution by graphical analysis.

4-2 GRAPHICAL EVALUATION OF THE CONVOLUTION INTEGRAL

Let $x(t)$ and $h(t)$ be two time functions given by graphs as represented in Fig. 4-1(a) and (b), respectively. To evaluate Eq. (4-1), functions $x(\tau)$ and $h(t - \tau)$ are required. $x(\tau)$ and $h(\tau)$ are simply $x(t)$ and $h(t)$, respectively,

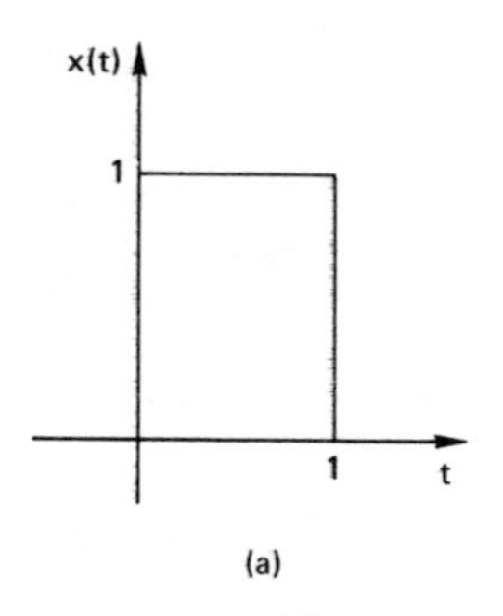

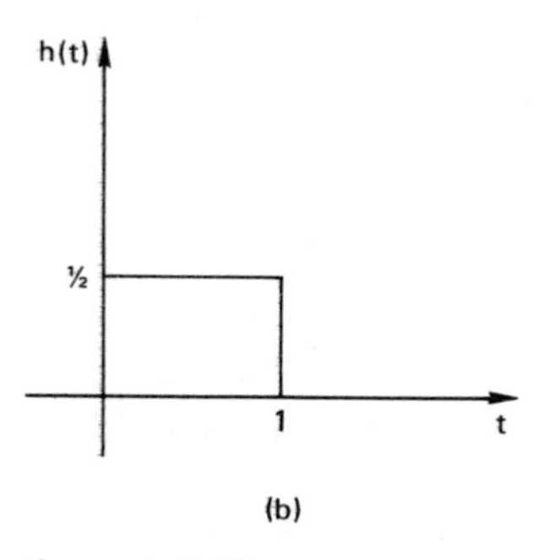

Figure 4-1. Example waveforms for convolution.

where the variable t has been replaced by the variable τ. $h(-\tau)$ is the image of $h(\tau)$ about the ordinate axis and $h(t - \tau)$ is simply the function $h(-\tau)$ shifted by the quantity t. Functions $x(\tau)$, $h(-\tau)$, and $h(t - \tau)$ are shown in Fig. 4-2. To compute the integral Eq. (4-1), it is necessary to multiply and integrate the functions $x(\tau)$ [Fig. 4-2(a)] and $h(t - \tau)$ [Fig. 4-2(c)] for each value of t from $-\infty$ to $+\infty$. As illustrated in Figs. 4-3(a) and (*h*), this product is zero for the choice of the parameter $t = -t_1$. The product remains zero until t is reduced to zero. As illustrated in Figs. 4-3(c) and (*h*), the product of $x(\tau)$ and $h(t_1 - \tau)$ is the function emphasized by shading. The integral of this function is simply the shaded area beneath the curve. As t is increased to $2t_1$ and further to $3t_1$, Figs. 4-3(d), (e), and (*h*) illustrate the relationships of the functions to be multiplied as well as the resulting integrations. For $t = 4t_1$, the product again becomes zero as shown by Figs. 4-3(*f*) and (*h*). This product remains zero for all t greater than $4t_1$[Figs. 4-3(g) and (h)]. If t is allowed to be a continuum of values, then the convolution of $x(t)$ and $h(t)$ is the triangular function illustrated in Fig. 4-3(h).

The procedure described is a convenient graphical technique for evaluating convolution integrals. Summarizing the steps:

1. *Folding*. Take the mirror image of $h(\tau)$ about ordinate axis.
2. *Displacement*. Shift $h(-\tau)$ by the amount t.
3. *Multiplication*. Multiply the shifted function $h(t - \tau)$ by $x(\tau)$.
4. *Integration*. Area under the product of $h(t - \tau)$ and $x(\tau)$ is the value of the convolution at time t.

Example 4-1

To illustrate further the rules for graphical evaluation of the convolution integral, convolve the functions illustrated in Figs. 4-4(a) and (b). First, fold $h(\tau)$ to obtain $h(-\tau)$ as illustrated in Fig. 4-4(c). Next, displace or shift $h(-\tau)$ by the amount t as shown in Fig. 4-4(d). Then, multiply $h(t - \tau)$ by $x(\tau)$ [Fig. 4-4(e)] and finally, integrate to obtain the convolution result for time t' [Fig. 4-4(*f*)].

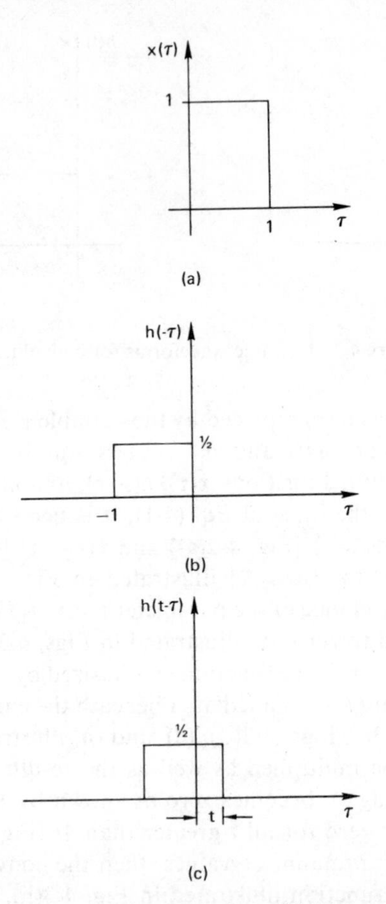

Figure 4-2. Graphical description of folding operation.

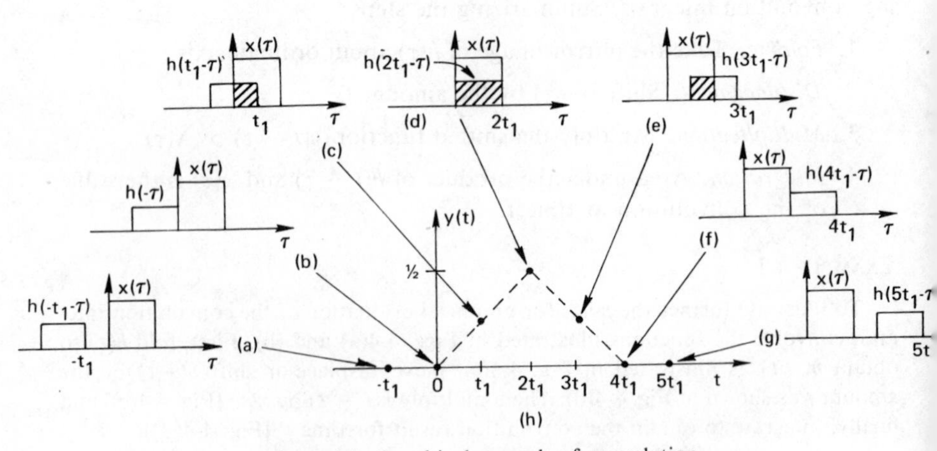

Figure 4-3. Graphical example of convolution.

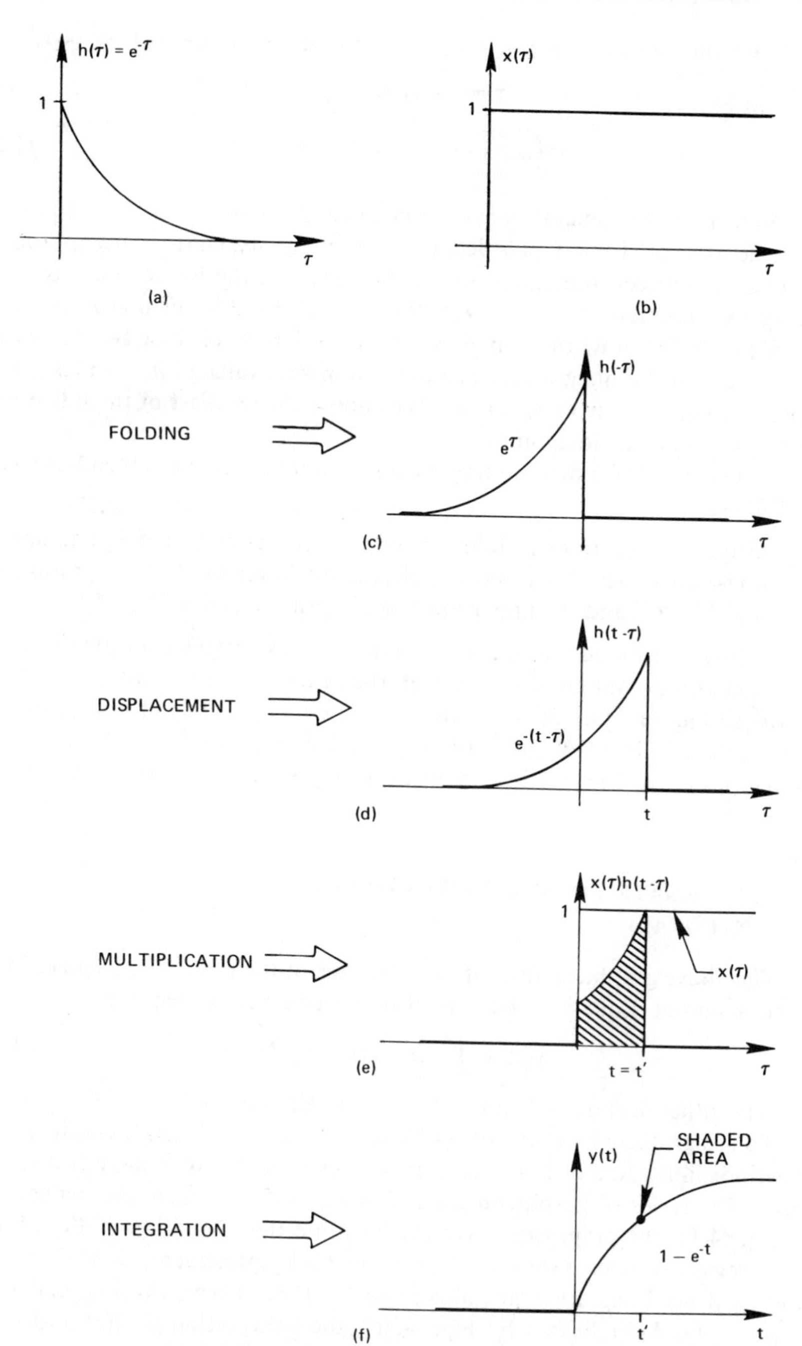

Figure 4-4. Convolution procedure: folding, displacement, multiplication, and integration.

The result illustrated in Fig. 4-4(f) can be determined directly from (4-1)

$$y(t) = \int_{-\infty}^{\infty} x(\tau)h(t-\tau)\,d\tau = \int_{0}^{t} (1)e^{-(t-\tau)}\,d\tau$$

$$= e^{-t}\left(e^{\tau}\Big|_{0}^{t}\right) = e^{-t}[e^{t}-1] = 1 - e^{-t} \qquad (4$$

Note that the general convolution integration limits of $-\infty$ to + become 0 to t for Ex. 4-1. It is desired to develop a straightforward appro to find the correct integration limits. For Ex. 4-1, the lower non-zero v ue of the function $h(t-\tau) = e^{-(t-\tau)}$ is $-\infty$ and the lower non-zero va for $x(\tau)$ is 0. When we integrated, we chose the largest of these two values our lower limit of integration. The upper non-zero value of $h(t-\tau)$ is t; upper non-zero value of $x(\tau)$ is ∞. We choose the smallest of these two our upper limit of integration.

A general rule for determining the limits of integration can then be sta as follows:

> Given two functions with lower non-zero values of L_1 and L_2 and up non-zero values of U_1 and U_2, choose the lower limit of integration max $[L_1, L_2]$ and the upper limit of integration as min $[U_1, U_2]$.

It should be noted that the lower and upper non-zero values for the fi function $x(\tau)$ do not change; however, the lower and upper non-zero val of the sliding function $h(t-\tau)$ change as t changes. Thus, it is possible have different limits of integration for different ranges of t. A graph sketch similar to Fig. 4-4 is also an extremely valuable aid in choosing correct limit of integration.

4-3 ALTERNATE FORM OF THE CONVOLUTION INTEGRAL

The above graphical illustration is but one of the possible interpretati of convolution. Equation (4-1) can also be written equivalently as

$$y(t) = \int_{-\infty}^{\infty} h(\tau)x(t-\tau)\,d\tau \qquad ($$

That is, either $h(\tau)$ or $x(\tau)$ can be folded and shifted.

To see graphically that Eqs. (4-1) and (4-3) are equivalent, consider functions illustrated in Fig. 4-5(a). It is desired to convolve these two fu tions. The series of graphs on the left in Fig. 4-5 illustrates the evalua of Eq. (4-1); the graphs on the right illustrate the evaluation of Eq. (4 The previously defined steps of (1) Folding, (2) Displacement, (3) Multipl tion, and (4) Integration are illustrated by Figs. 4-5(b), (c), (d), and respectively. As indicated by Fig. 4-5(e), the convolution of $x(\tau)$ and is the same irrespective of which function is chosen for folding and displ ment.

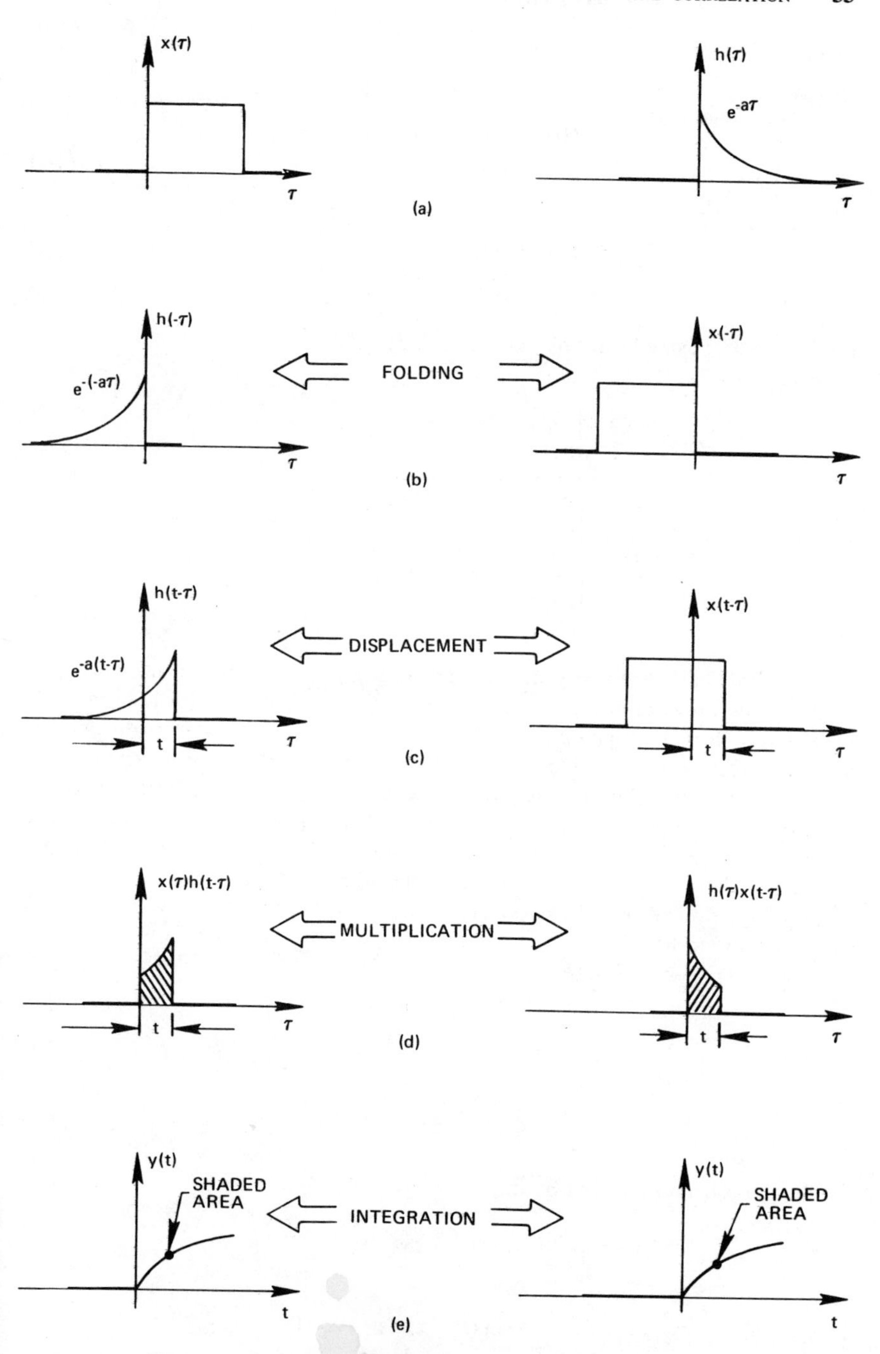

Figure 4-5. Graphical example of convolution by Eqs. (4-1) and (4-3).

EXAMPLE 4-2

Let

$$h(t) = e^{-t} \qquad t \geq 0$$
$$= 0 \qquad t < 0 \tag{4-4}$$

and

$$x(t) = \sin t \qquad 0 \leq t \leq \frac{\pi}{2}$$
$$= 0 \qquad \text{otherwise} \tag{4-5}$$

Find $h(t) * x(t)$ using both Eqs. (4-1) and (4-3).

From (4-1)

$$y(t) = \int_{-\infty}^{\infty} x(\tau)h(t - \tau)\, d\tau$$

$$y(t) = \begin{cases} \displaystyle\int_0^t \sin(\tau)e^{-(t-\tau)}\, d\tau & 0 \leq t \leq \frac{\pi}{2} \\ \displaystyle\int_0^{\pi/2} \sin(\tau)e^{-(t-\tau)}\, d\tau & t \geq \frac{\pi}{2} \\ 0 & t \leq 0 \end{cases} \tag{4-6}$$

The integral limits are easily determined by using the procedure described previously. The lower and upper non-zero value of the function $x(\tau)$ is 0 and $\pi/2$, respectively. For the function $h(t - \tau) = e^{-(t-\tau)}$ the lower non-zero value is $-\infty$ and the upper non-zero value is t. We take the maximum of the lower non-zero values for our lower limit of integration; i.e., 0. The upper limit of integration is a function of t. For $0 \leq t \leq \pi/2$ the minimum of the upper non-zero values is t and hence the upper limit of integration is t. For $t \geq \pi/2$ the minimum of the upper non-zero values is $\pi/2$ and consequently the upper limit of integration for this range of t is $\pi/2$. A graphical sketch of the convolution process will also yield these integration limits.

Evaluating (4-6) we obtain

$$y(t) = \begin{cases} 0 & t \leq 0 \\ \dfrac{1}{2}(\sin t - \cos t + e^{-t}) & 0 < t \leq \dfrac{\pi}{2} \\ \dfrac{e^{-t}}{2}(1 + e^{\pi/2}) & t \geq \dfrac{\pi}{2} \end{cases} \tag{4-7}$$

Similarly from (4-3) we obtain

$$y(t) = \int_{-\infty}^{\infty} h(\tau)\, x(t - \tau)\, d\tau$$

$$y(t) = \begin{cases} \displaystyle\int_0^t e^{-\tau} \sin(t - \tau)\, d\tau & 0 < t < \frac{\pi}{2} \\ \displaystyle\int_{t-\pi/2}^{t} e^{-\tau} \sin(t - \tau)\, d\tau & t \geq \frac{\pi}{2} \\ 0 & t < 0 \end{cases} \tag{4-8}$$

Although Eqs. (4-8) are different from Eqs. (4-6), evaluation yields identical results to (4-7).

4-4 CONVOLUTION INVOLVING IMPULSE FUNCTIONS

The simplest type of convolution integral to evaluate is one in which either $x(t)$ or $h(t)$ is an impulse function. To illustrate this point, let $h(t)$ be

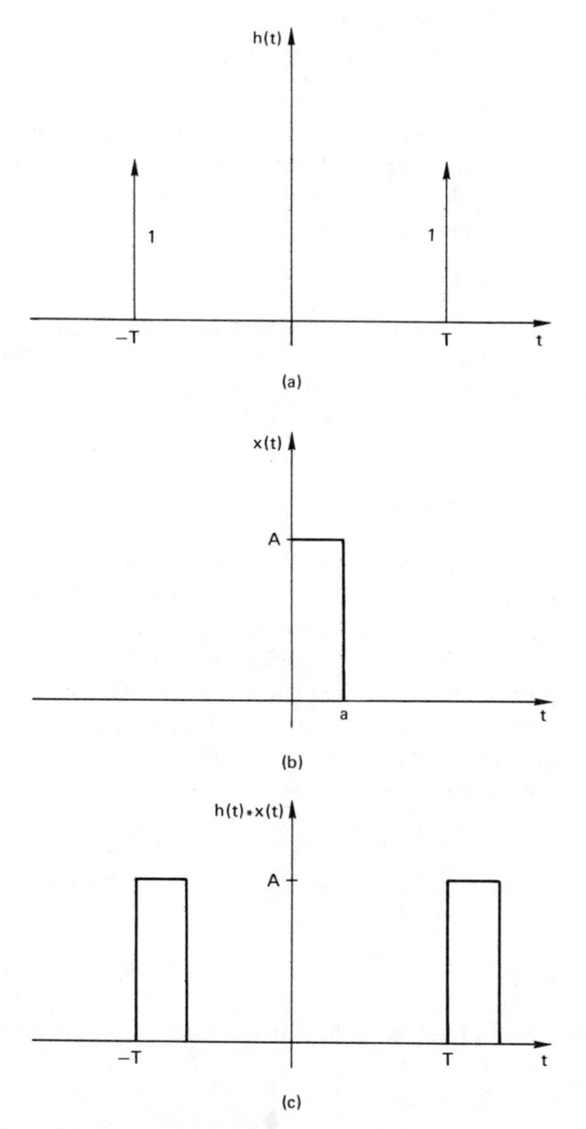

Figure 4-6. Illustration of convolution involving impulse functions.

the singular function shown graphically in Fig. 4-6(a) and let $x(t)$ be the rectangular function shown in Fig. 4-6(b). For these example functions Eq. (4-1) becomes

$$y(t) = \int_{-\infty}^{\infty} [\delta(\tau - T) + \delta(\tau + T)]x(t - \tau)\, d\tau \tag{4-9}$$

Recall from Eq. (2-28) that

$$\int_{-\infty}^{\infty} \delta(\tau - T)\, x(\tau)\, d\tau = x(T)$$

Hence, Eq. (4-9) can be written as

$$y(t) = x(t - T) + x(t + T) \tag{4-10}$$

Function $y(t)$ is illustrated in Fig. 4-6(c). Note that convolution of the function $x(t)$ with an impulse function is evaluated by simply reconstructing $x(t)$ with the position of the impulse function replacing the ordinate of $x(t)$. As we will see in the developments to follow, the ability to visualize convolution involving impulse function is of considerable importance.

EXAMPLE 4-3

Let $h(t)$ be a series of impulse functions as illustrated in Fig. 4-7(a). To evaluate the convolution of $h(t)$ with the rectangular pulse shown in Fig. 4-7(b), we simply reproduce the rectangular pulse at each of the impulse functions. The resulting convolution results are illustrated in Fig. 4-7(c).

4-5 CONVOLUTION THEOREM

Possibly the most important and powerful tool in modern scientific analysis is the relationship between Eq. (4-1) and its Fourier transform. This relationship, known as the convolution theorem, allows one the complete freedom to convolve mathematically (or visually) in the time domain by simple multiplication in the frequency domain. That is, if $h(t)$ has the Fourier transform $H(f)$ and $x(t)$ has the Fourier transform $X(f)$, then $h(t) * x(t)$ has the Fourier transform $H(f)X(f)$. The convolution theorem is thus given by the Fourier transform pair

$$h(t) * x(t) \quad \Longleftrightarrow \quad H(f)X(f) \tag{4-11}$$

To establish this result, first form the Fourier transform of both sides of Eq. (4-1)

$$\int_{-\infty}^{\infty} y(t)e^{-j2\pi ft}\, dt = \int_{-\infty}^{\infty} \left[\int_{-\infty}^{\infty} x(\tau)\, h(t - \tau)\, d\tau\right] e^{-j2\pi ft}\, dt \tag{4-12}$$

which is equivalent to (assuming the order of integration can be changed)

$$Y(f) = \int_{-\infty}^{\infty} x(\tau)\left[\int_{-\infty}^{\infty} h(t - \tau)\, e^{-j2\pi ft}\, dt\right] d\tau \tag{4-13}$$

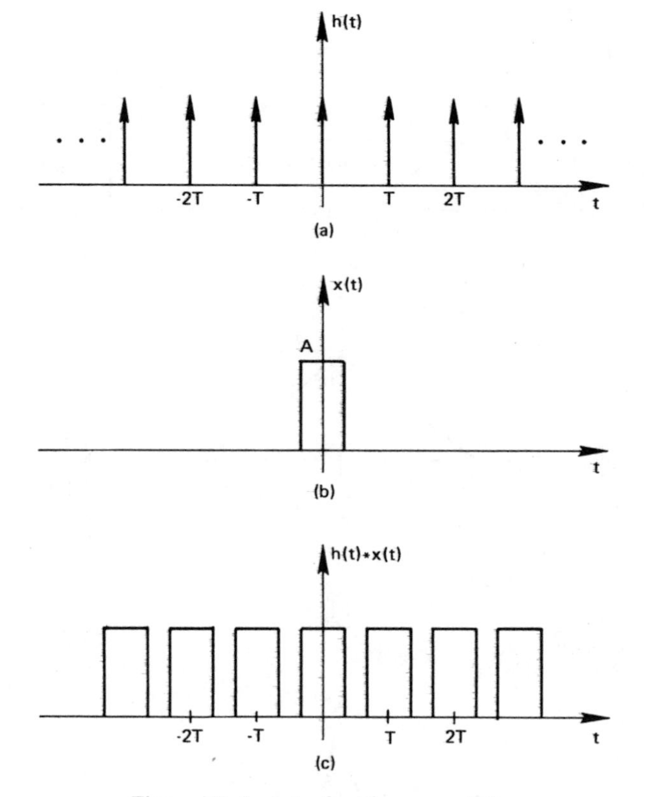

Figure 4-7. Impulse function convolution.

By substituting $\sigma = t - \tau$ the term in the brackets becomes

$$\int_{-\infty}^{\infty} h(\sigma)\, e^{-j2\pi f(\sigma+\tau)}\, d\sigma = e^{-j2\pi f\tau} \int_{-\infty}^{\infty} h(\sigma)\, e^{-j2\pi f\sigma}\, d\sigma$$
$$= e^{-j2\pi f\tau}\, H(f) \qquad (4\text{-}14)$$

Equation (4-13) can then be rewritten as

$$Y(f) = \int_{-\infty}^{\infty} x(\tau)\, e^{-j2\pi f\tau}\, H(f)\, d\tau = H(f)X(f) \qquad (4\text{-}15)$$

The converse is proven similarly.

Example 4-4

To illustrate the application of the convolution theorem, consider the convolution of the two rectangular functions shown in Figs. 4-8(a) and (b). As we have seen previously, the convolution of two rectangular functions is a triangular function as shown in Fig. 4-8(e). Recall from Fourier transform pair (2-21) that the Fourier transform of a rectangular function is the $\sin(f)/f$ function illustrated in Figs. 4-8(c) and (d). The convolution theorem states that convolution in the time domain

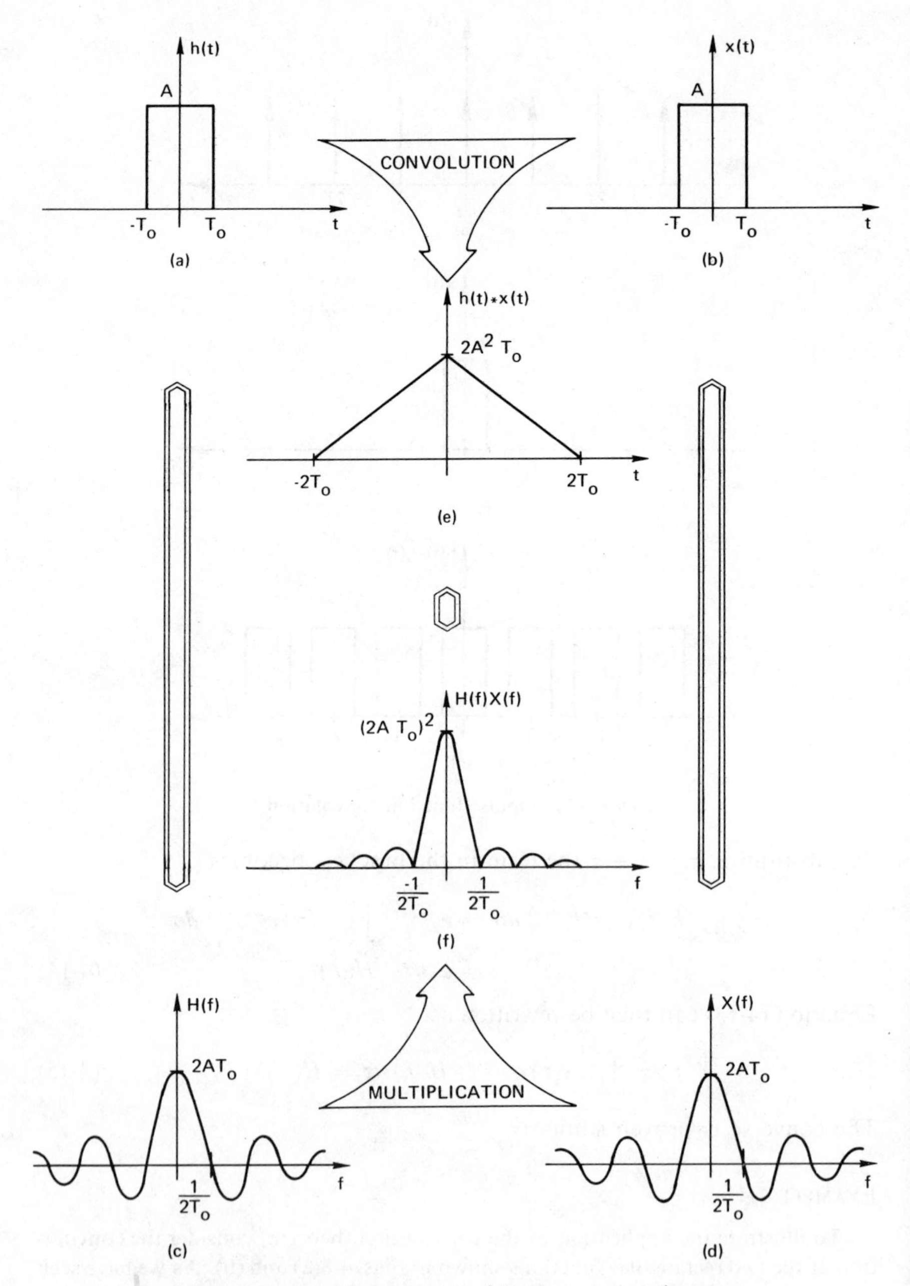

Figure 4-8. Graphical example of the convolution theorem.

corresponds to multiplication in the frequency domain; therefore, the triangular waveform of Fig. 4-8(e) and the $\sin^2 f/f^2$ function of Fig. 4-8(f) are Fourier transform pairs. Thus, we can use the theorem as a convenient tool for developing additional Fourier transform pairs.

EXAMPLE 4-5

One of the most significant contributions of distribution theory results from the fact that the product of a continuous function and an impulse function is well defined (Appendix A); hence, if $h(t)$ is continuous at $t = t_0$ then

$$h(t)\,\delta(t - t_0) = h(t_0)\,\delta(t - t_0) \tag{4-16}$$

This result coupled with the convolution theorem allows one to eliminate the tedious derivation of many Fourier transform pairs. To illustrate, consider the two time functions $h(t)$ and $x(t)$ shown in Figs. 4-9(a) and (b). As described previously, the convolution of these two functions is the infinite pulse train illustrated in Fig. 4-9(e). It is desired to determine the Fourier transform of this infinite sequence of pulses. We simply use the convolution theorem; the Fourier transform of $h(t)$ is the sequence of impulse functions; transform pair (2-40), illustrated in Fig. 4-9(c), and the Fourier transform of a rectangular function is the $\sin(f)/f$ function shown in Fig. 4-9(d). Multiplication of these two frequency functions yields the desired Fourier transform. As illustrated in Fig. 4-9(f), the Fourier transform of a pulse train is a sequence of impulse functions whose amplitude is weighted by a $\sin f/f$ function. This is a well-known result in the field of radar systems. It is to be noted that the multiplication of the two frequency functions must be interpreted in the sense of distribution theory; otherwise the product is meaningless. We can see that the ability to change from a convolution in the time domain to multiplication in the frequency domain often renders unwieldy problems rather straightforward.

4-6 FREQUENCY CONVOLUTION THEOREM

We can equivalently go from convolution in the frequency domain to multiplication in the time domain by using the frequency convolution theorem; the Fourier transform of the product $h(t)x(t)$ is equal to the convolution $H(f) * X(f)$. The frequency convolution theorem is

$$h(t)x(t) \quad \Longleftrightarrow \quad H(f) * X(f) \tag{4-17}$$

This pair is established by simply substituting the Fourier transform pair (4-11) into the symmetry Fourier transform relationship (3-6).

EXAMPLE 4-6

To illustrate the frequency convolution theorem, consider the cosine waveform of Fig. 4-10(a) and the rectangular waveform of Fig. 4-10(b). It is desired to determine the Fourier transform of the product of these two functions [Fig. 4-10(e)]. The Fourier transforms of the cosine and rectangular waveforms are given in Figs.

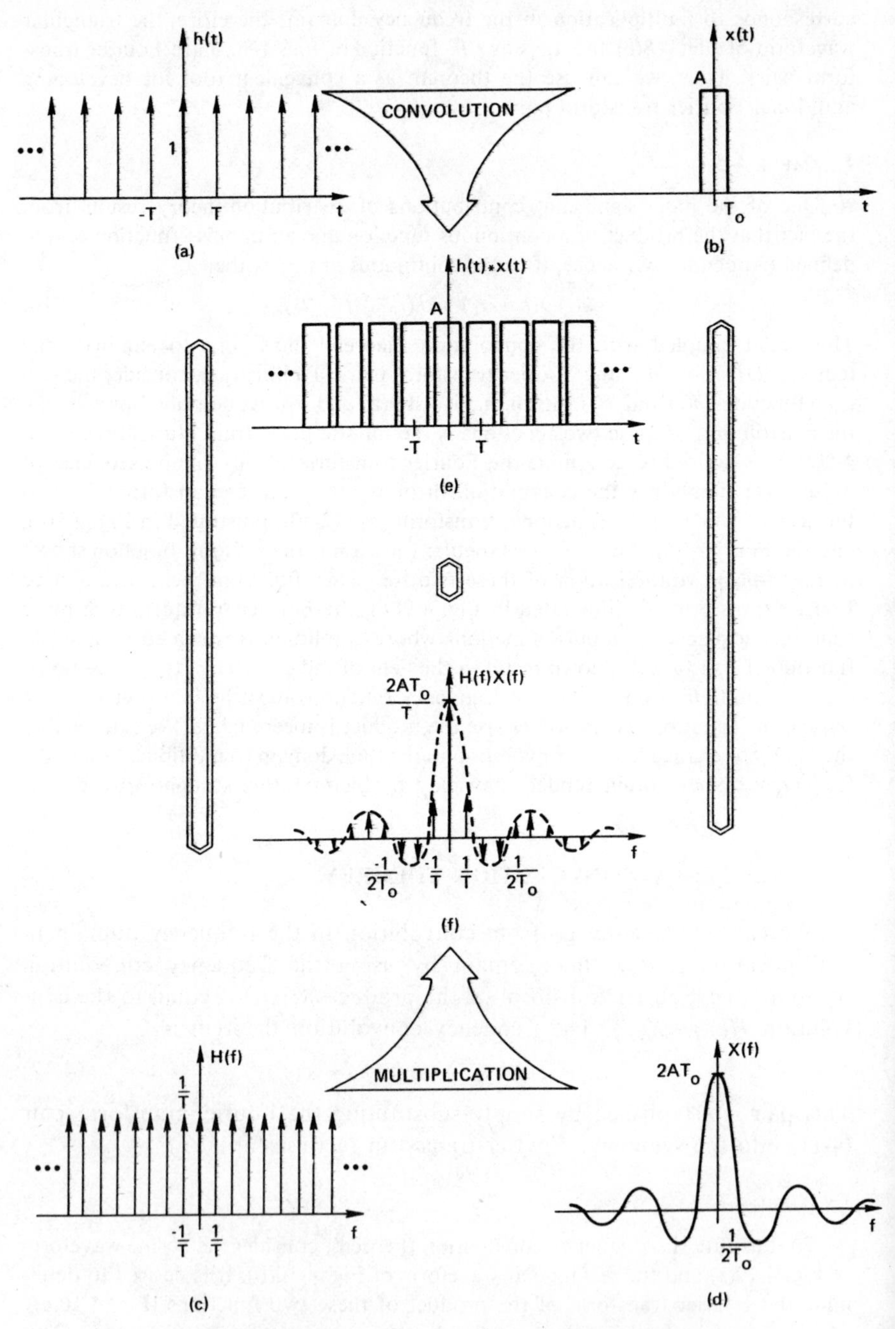

Figure 4-9. Example application of the convolution theorem.

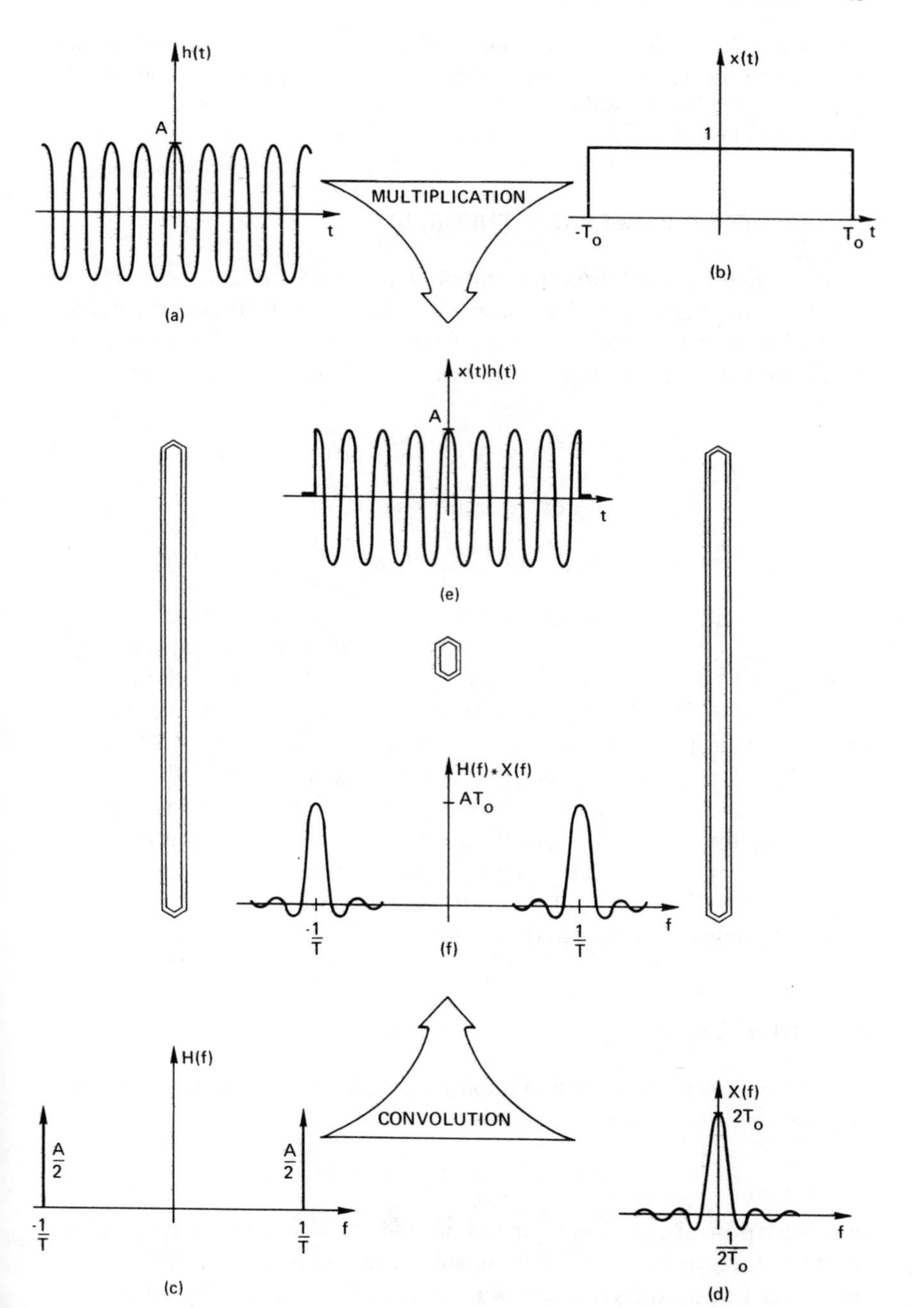

Figure 4-10. Graphical example of the frequency convolution theorem.

4-10(c) and (d), respectively. Convolution of these two frequency functions yields the function shown in Fig. 4-10(f); Figs. 4-10(e) and (f) are thus Fourier transform pairs. This is the well-known Fourier transform pair of a single, frequency modulated pulse.

4-7 PROOF OF PARSEVAL'S THEOREM

Because the convolution theorem often simplifies difficult problems, it is fitting to summarize this discussion by utilizing the theorem to develop a simple proof of Parseval's Theorem. Consider the function $y(t) = h(t) \cdot h(t)$. By the convolution theorem the Fourier transform of $y(t)$ is $H(f) * H(f)$; that is

$$\int_{-\infty}^{\infty} h^2(t)\, e^{-j2\pi\sigma t}\, dt = \int_{-\infty}^{\infty} H(f)\, H(\sigma - f)\, df \tag{4-18}$$

Setting $\sigma = 0$ in the above expression yields

$$\int_{-\infty}^{\infty} h^2(t)\, dt = \int_{-\infty}^{\infty} H(f)\, H(-f)\, df = \int_{-\infty}^{\infty} |H(f)|^2\, df \tag{4-19}$$

The last equality follows since $H(f) = R(f) + jI(f)$ and thus $H(-f) = R(-f) + jI(-f)$. From Eqs. (3-43) and (3-44), $R(f)$ is even and $I(f)$ is odd. Consequently $R(-f) = R(f)$; $I(-f) = -I(f)$; and $H(-f) = R(f) - jI(f)$. The product $H(f)\, H(-f)$ is equal to $R^2(f) + I^2(f)$ which is the square of the Fourier spectrum $|H(f)|$ defined in Eq. (2-2). Equation (4-19) is Parseval's Theorem; it states that the energy in a waveform $h(t)$ computed in the time domain must equal the energy of $H(f)$ as computed in the frequency domain. As shown, the use of the convolution theorem allows us to prove rather simply an important result. The convolution theorem is fundamental to many facets of Fourier transform analysis and, as we shall see, the theorem is of considerable importance in the application of the fast Fourier transform.

4-8 CORRELATION

Another integral equation of importance in both theoretical and practical application is the correlation integral;

$$z(t) = \int_{-\infty}^{\infty} x(\tau)\, h(t + \tau)\, d\tau \tag{4-20}$$

A comparison of the above expression and the convolution integral (4-1) indicates that the two are closely related. The nature of this relationship is best described by the graphical illustrations of Fig. 4-11. The functions to be both *convolved* and *correlated* are shown in Fig. 4-11(a). Illustrations on the left depict the process of convolution as described in the previous section; illustrations on the right graphically portrary the process of correlation. As

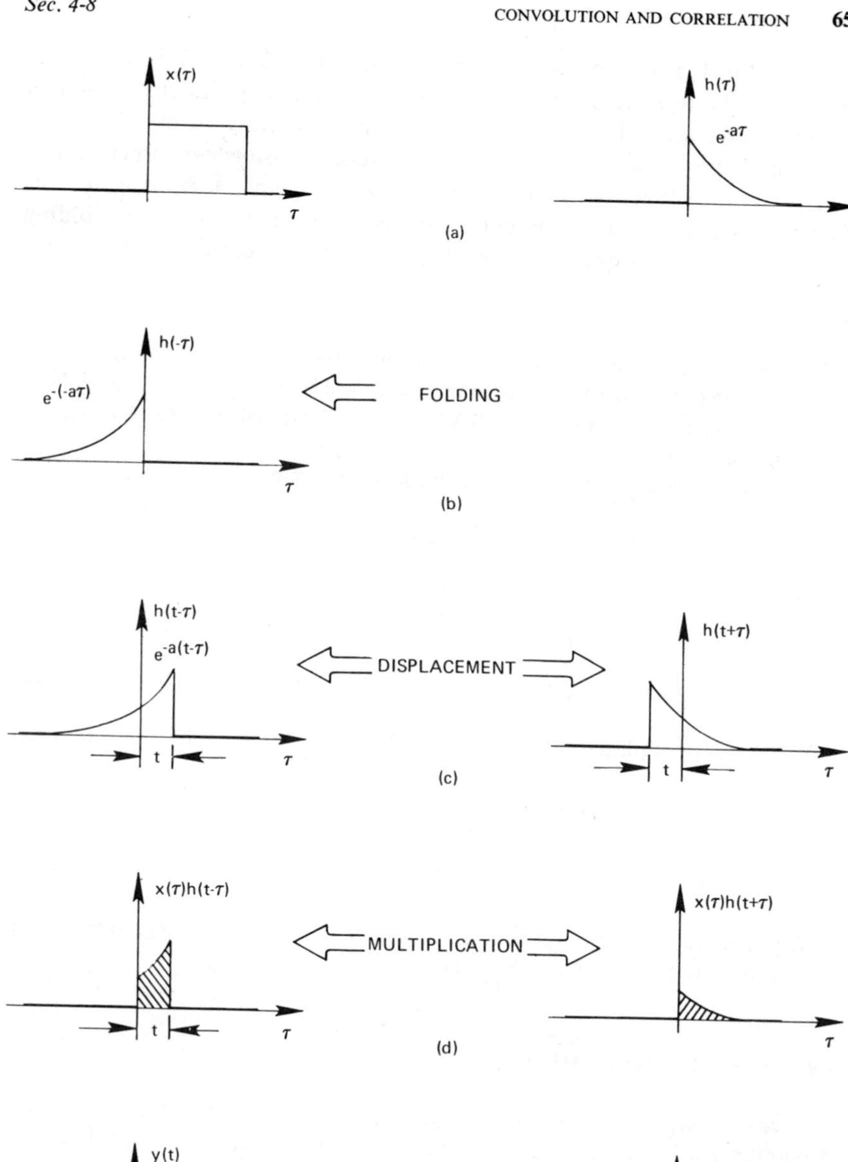

Figure 4-11. Graphical comparison of convolution and correlation.

evidenced in Fig. 4-11(b), the two integrals differ in that there is no folding of one of the integrands in correlation. The previously described rules of displacement, multiplication, and integration are performed identically for both convolution and correlation. For the special case where either $x(t)$ or $h(t)$ is an even function, convolution and correlation are equivalent; this follows since an even function and its image are identical and, thus, folding can be eliminated from the steps in computing the convolution integral.

EXAMPLE 4-7

Correlate graphically and analytically the waveforms illustrated in Fig. 4-12(a).

According to the rules for correlation we displace $h(\tau)$ by the shift t; multiply by $x(\tau)$; and integrate the product $x(\tau)\,h(t+\tau)$ as illustrated in Figs. 4-12(b), (c), and (d), respectively.

From Eq. (4-20), for positive displacement t we obtain

$$
\begin{aligned}
z(t) &= \int_{-\infty}^{\infty} x(\tau)\,h(t+\tau)\,d\tau \\
&= \int_{0}^{a-t} (1)\frac{Q}{a}\tau\,d\tau \\
&= \frac{Q}{2a}\tau^2\Big|_0^{a-t} = \frac{Q}{2a}(a-t)^2 \qquad 0 \le t \le a
\end{aligned}
\tag{4-21}
$$

For negative displacement, see Fig. 4-12(c) to justify the limits of integration.

$$
\begin{aligned}
z(t) &= \int_{t}^{a} (1)\frac{Q}{a}\tau\,d\tau \\
&= \frac{Q}{2a}(a^2 - t^2) \qquad -a \le t \le 0
\end{aligned}
\tag{4-22}
$$

A general rule can be developed for determining the limits of integration for the correlation integral (see Problem 4-15).

4-9 CORRELATION THEOREM

Recall that convolution-multiplication forms a Fourier transform pair. A similar result can be obtained for correlation. To derive this relationship, first evaluate the Fourier transform of Eq. (4-20)

$$
\int_{-\infty}^{\infty} z(t)e^{-j2\pi ft}\,dt = \int_{-\infty}^{\infty}\left[\int_{-\infty}^{\infty} x(\tau)\,h(t+\tau)\,d\tau\right]e^{-j2\pi ft}\,dt \tag{4-23}
$$

or (assuming the order of integration can be interchanged)

$$
Z(f) = \int_{-\infty}^{\infty} x(\tau)\left[\int_{-\infty}^{\infty} h(t+\tau)\,e^{-j2\pi ft}\,dt\right]d\tau \tag{4-24}
$$

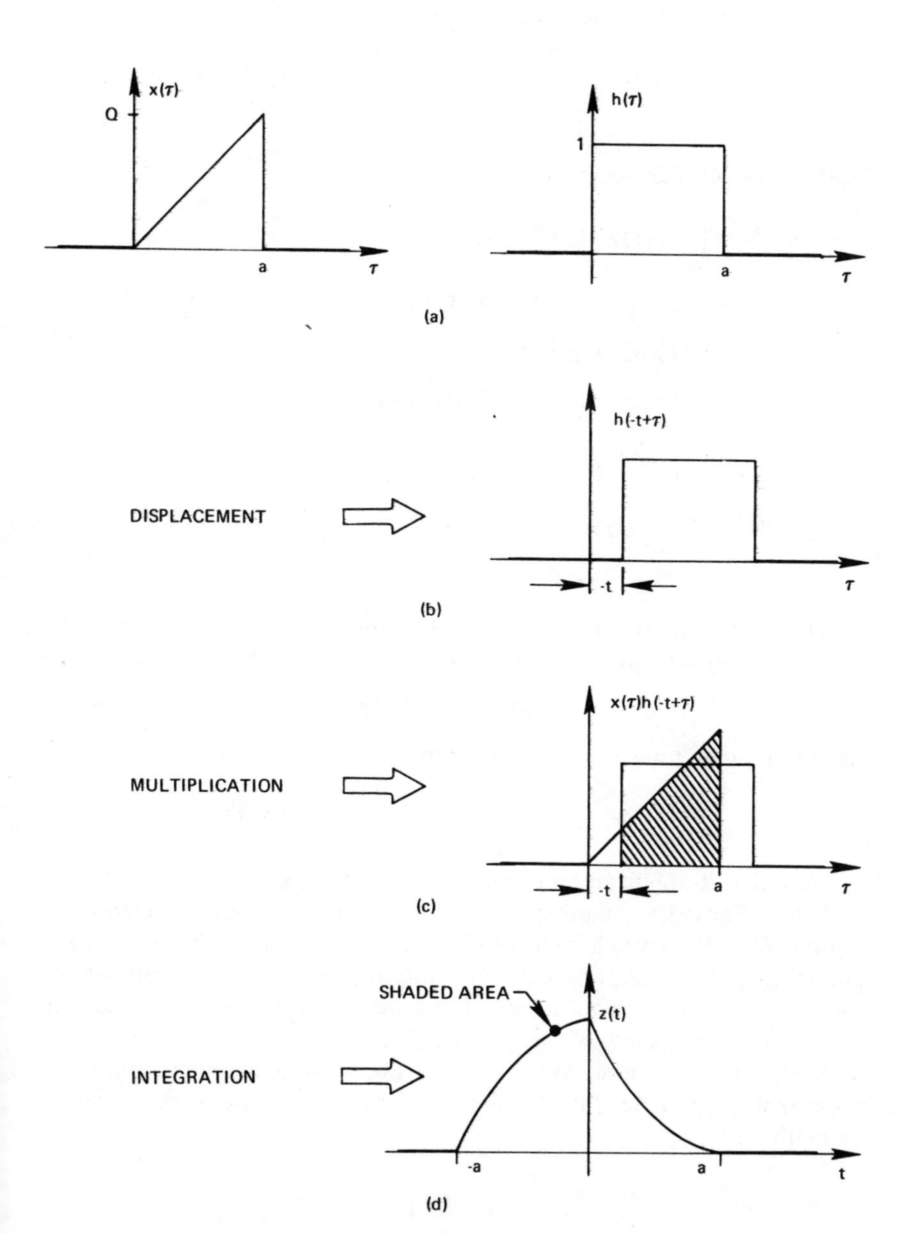

Figure 4-12. Correlation procedure: displacement, multiplication, and integration.

Let $\sigma = t + \tau$ and rewrite the term in brackets as

$$\int_{-\infty}^{\infty} h(\sigma)\, e^{-j2\pi f(\sigma-\tau)}\, d\sigma = e^{j2\pi f\tau} \int_{-\infty}^{\infty} h(\sigma)\, e^{-j2\pi f\sigma}\, d\sigma$$
$$= e^{j2\pi f\tau} H(f) \tag{4-25}$$

Equation (4-24) then becomes

$$Z(f) = \int_{-\infty}^{\infty} x(\tau)\, e^{j2\pi f\tau} H(f)\, d\tau$$
$$= H(f)\left[\int_{-\infty}^{\infty} x(\tau) \cos(2\pi f\tau)\, d\tau + j \int_{-\infty}^{\infty} x(\tau) \sin(2\pi f\tau)\, d\tau\right]$$
$$= H(f)[R(f) + jI(f)] \tag{4-26}$$

Now the Fourier transform of $x(\tau)$ is given by

$$X(f) = \int_{-\infty}^{\infty} x(\tau)\, e^{-j2\pi f\tau}\, d\tau$$
$$= \int_{-\infty}^{\infty} x(\tau) \cos(2\pi f\tau)\, d\tau - j \int_{-\infty}^{\infty} x(\tau) \sin(2\pi f\tau)\, d\tau$$
$$= R(f) - jI(f) \tag{4-27}$$

The bracketed term of (4-26) and the expression on the right in (4-27) are called conjugates [defined in Eq. (3-25)]. Equation (4-26) may be written as

$$Z(f) = H(f)X^*(f) \tag{4-28}$$

and the Fourier transform pair for correlation is

$$\int_{-\infty}^{\infty} h(\tau)\, x(t+\tau)\, d\tau \quad \Longleftrightarrow \quad H(f)X^*(f) \tag{4-29}$$

Note that if $x(t)$ is an even function then $X(f)$ is purely real and $X(f) = X^*(f)$. For these conditions the Fourier transform of the correlation integral is $H(f)X(f)$ which is identical to the Fourier transform of the convolution integral. These arguments for identity of the two integrals are simply the frequency domain equivalents of the previously discussed time domain requirement for equality of the two integrals.

If $x(t)$ and $h(t)$ are the same function, Eq. (4-20) is normally termed the *autocorrelation* function; if $x(t)$ and $h(t)$ differ, the term *crosscorrelation* is normally used.

EXAMPLE 4-8

Determine the autocorrelation function of the waveform

$$h(t) = e^{-at} \qquad t > 0$$
$$= 0 \qquad t < 0 \tag{4-30}$$

From (4-20)

$$\begin{aligned} z(t) &= \int_{-\infty}^{\infty} h(\tau)\, h(t+\tau)\, d\tau \\ &= \int_{0}^{\infty} e^{-a\tau}\, e^{-a(t+\tau)}\, d\tau \qquad t > 0 \\ &= \int_{t}^{\infty} e^{-a\tau} e^{-a(t+\tau)}\, d\tau \qquad t < 0 \\ &= \frac{e^{-a|t|}}{2a} \qquad -\infty < t < \infty \end{aligned} \tag{4-31}$$

PROBLEMS

4-1. Prove the following convolution properties:

a. Convolution is commutative; $(h(t) * x(t)) = (x(t) * h(t))$

b. Convolution is associative; $h(t) * [g(t) * x(t)] = [h(t) * g(t)] * x(t)$

c. Convolution is distributive over addition; $h(t) * [g(t) + x(t)] = h(t) * g(t) + h(t) * x(t)$

4-2. Determine $h(t) * g(t)$ where

a. $h(t) = e^{-at} \quad t > 0$
$\quad = 0 \quad t < 0$
$g(t) = e^{-bt} \quad t > 0$
$\quad = 0 \quad t < 0$

b. $h(t) = te^{-t} \quad t \geq 0$
$\quad = 0 \quad t < 0$
$g(t) = e^{-t} \quad t > 0$
$\quad = 0 \quad t < 0$

c. $h(t) = te^{-t} \quad t \geq 0$
$\quad = 0 \quad t < 0$
$g(t) = e^{t} \quad t < -1$
$\quad = 0 \quad t > -1$

d. $h(t) = 2e^{3t} \quad t > 1$
$\quad = 0 \quad t < 0$
$g(t) = 2e^{t} \quad t < 0$
$\quad = 0 \quad t > 0$

e. $h(t) = \sin(2\pi t) \quad 0 \leq t \leq \frac{1}{2}$
$\quad = 0 \quad$ elsewhere
$g(t) = 1 \quad 0 < t < \frac{1}{8}$
$\quad = 0 \quad t < 0;\ t > \frac{1}{8}$

f. $h(t) = 1 - t \quad 0 < t < 1$
$\quad = 0 \quad t < 0;\ t > 1$
$g(t) = h(t)$

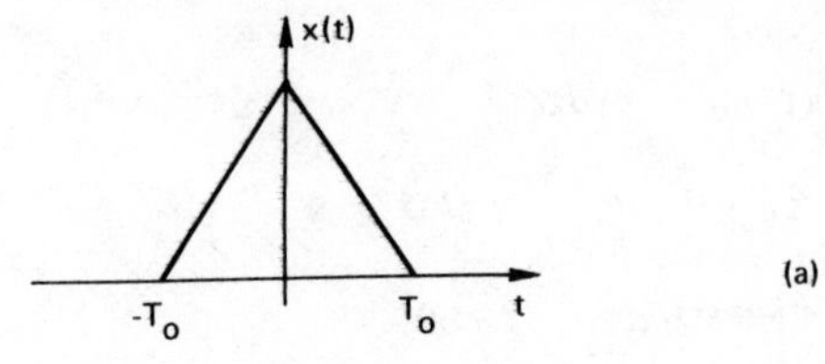

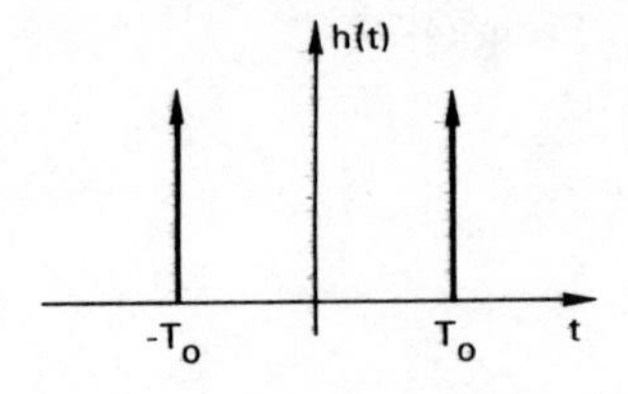

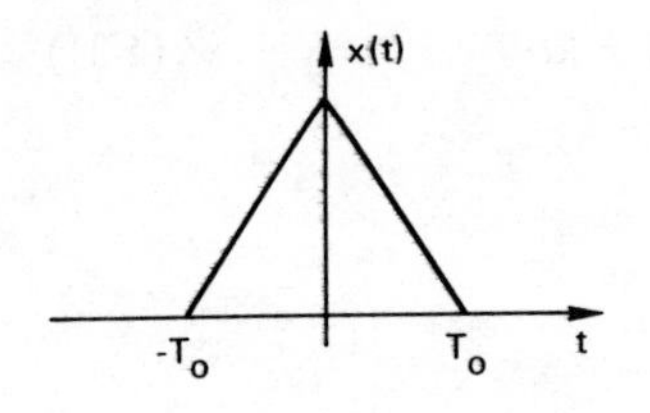

(b)
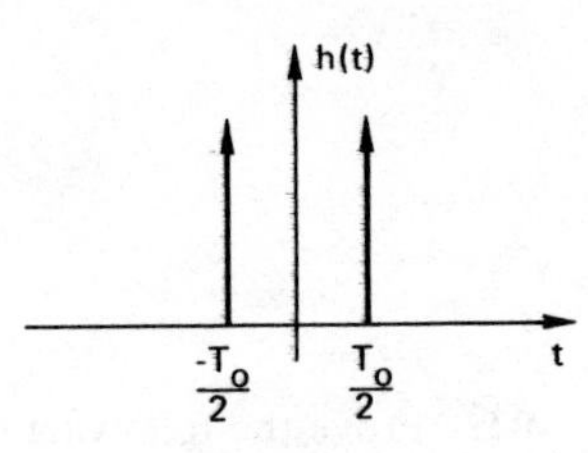

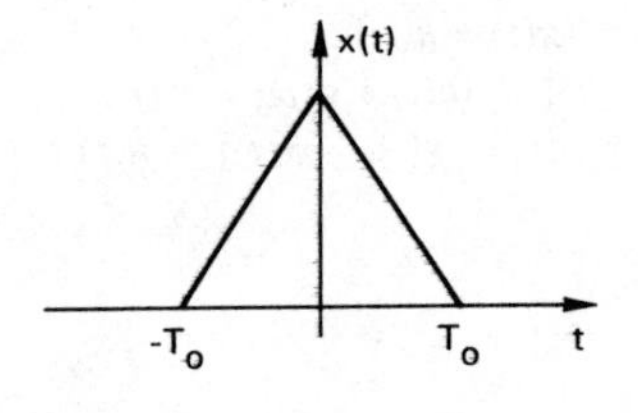

(c)
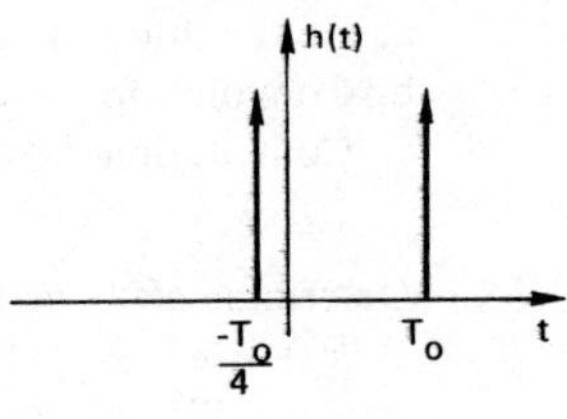

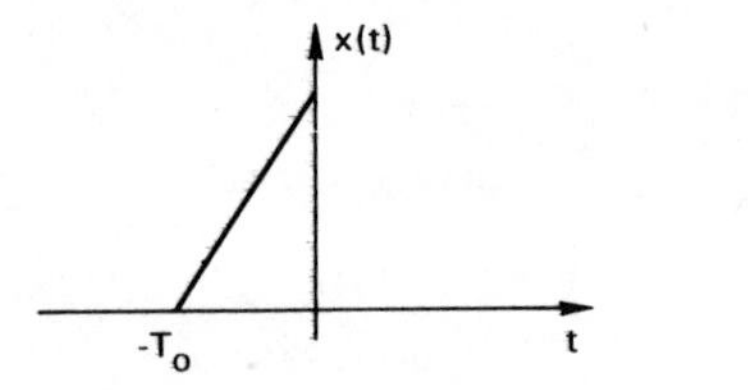

(d)
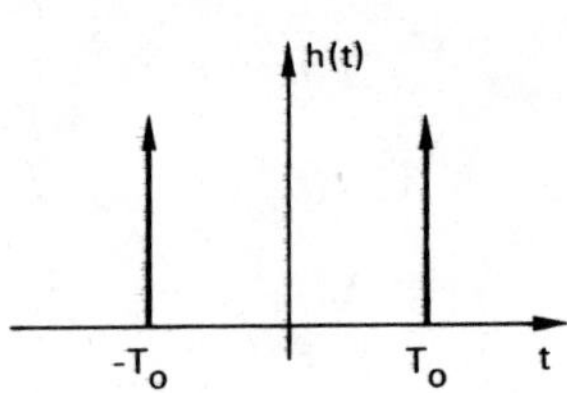

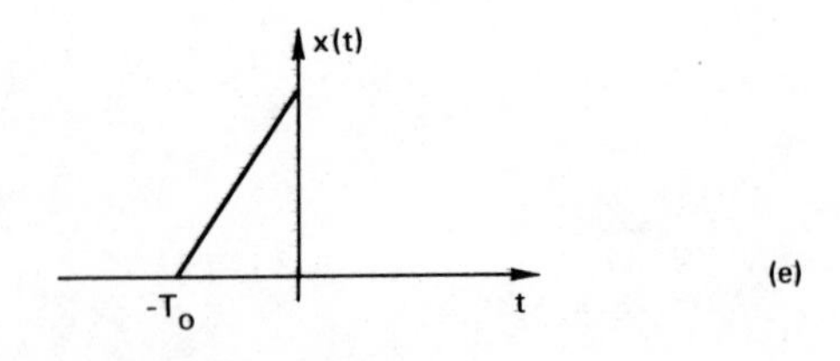

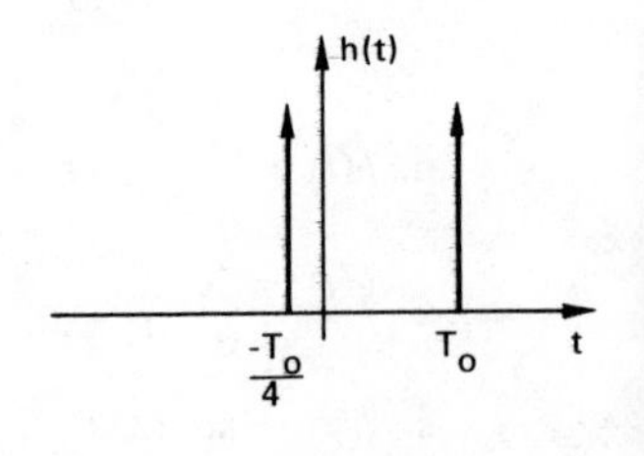

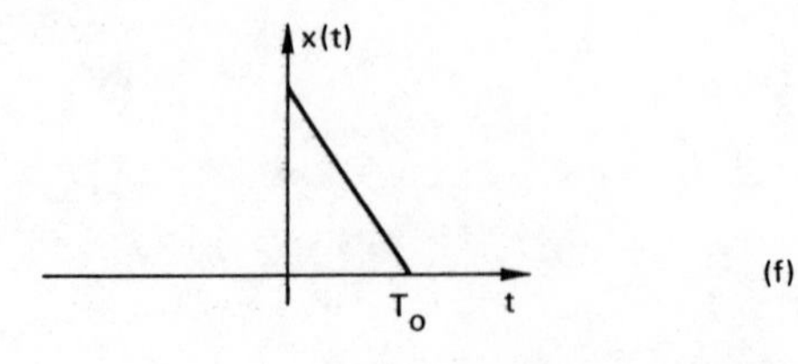

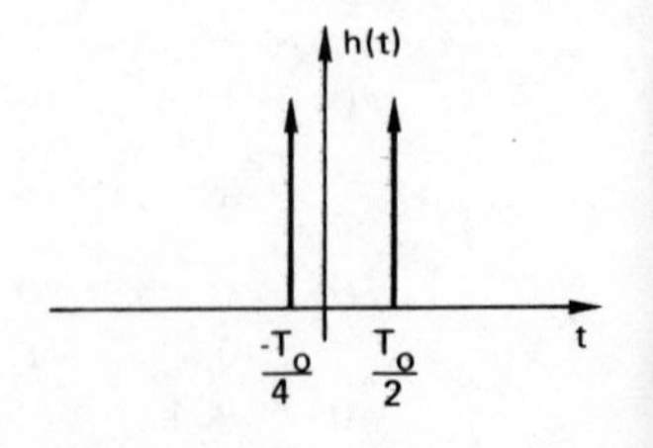

Figure 4-13.

g. $h(t) = (a - |t|)^3 \qquad -a \leq t \leq a$
$\quad = 0 \qquad$ elsewhere
$g(t) = h(t)$

h. $h(t) = e^{-at} \qquad t > 0$
$\quad = 0 \qquad t < 0$
$g(t) = 1 - t \qquad 0 < t < 1$
$\quad = 0 \qquad t < 0; t > 1$

4-3. Graphically sketch the convolution of the functions $x(t)$ and $h(t)$ illustrated in Fig. 4-13.

4-4. Sketch the convolution of the two odd functions $x(t)$ and $h(t)$ illustrated in Fig. 4-14. Show that the convolution of two odd functions is an even function.

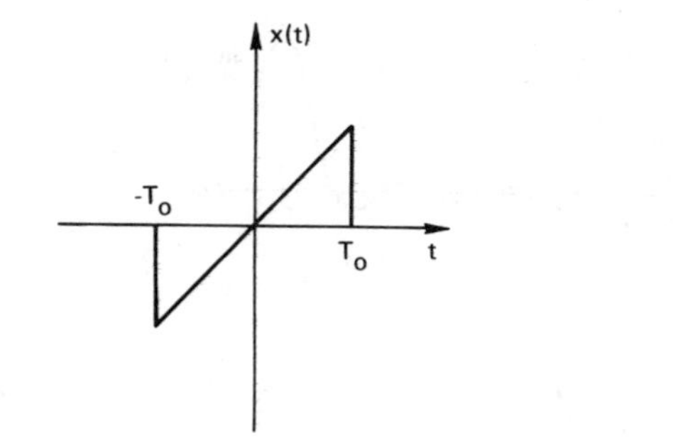

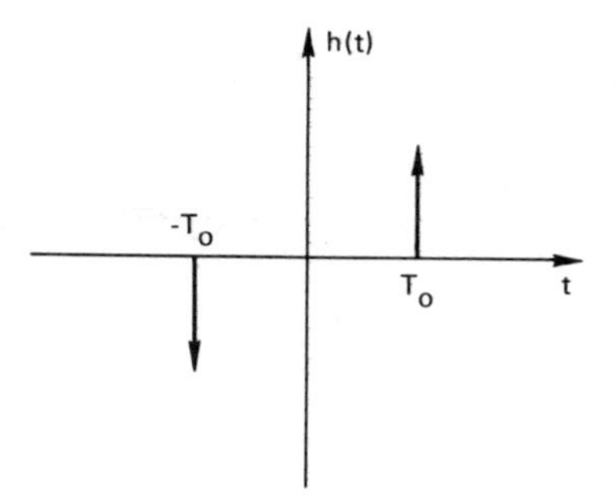

Figure 4-14.

4-5. Use the convolution theorem to graphically determine the Fourier transform of the functions illustrated in Fig. 4-15.

4-6. Analytically determine the Fourier transform of $e^{-\alpha t^2} * e^{-\beta t^2}$. (Hint: Use the convolution theorem.)

4-7. Use the frequency convolution theorem to graphically determine the convolution of the functions $x(t)$ and $h(t)$ illustrated in Fig. 4-16.

4-8. Graphically determine the correlation of the functions $x(t)$ and $h(t)$ illustrated in Fig. 4-13.

4-9. Let $h(t)$ be a time-limited function which is non-zero over the range

$$\frac{-T_0}{2} \leq t \leq \frac{T_0}{2}$$

Show that $h(t) * h(t)$ is non-zero over the range $-T_0 \leq t \leq T_0$; that is, $h(t) * h(t)$ has a "width" twice that of $h(t)$.

4-10. Show that if $h(t) = f(t) * g(t)$ then

$$\frac{dh(t)}{dt} = \frac{df(t)}{dt} * g(t) = f(t) * \frac{dg(t)}{dt}$$

4-11. If $[x(t)]*^3$ implies $[x(t) * x(t) * x(t)]$ how does one evaluate $[x(t)]*^{3/2}$?

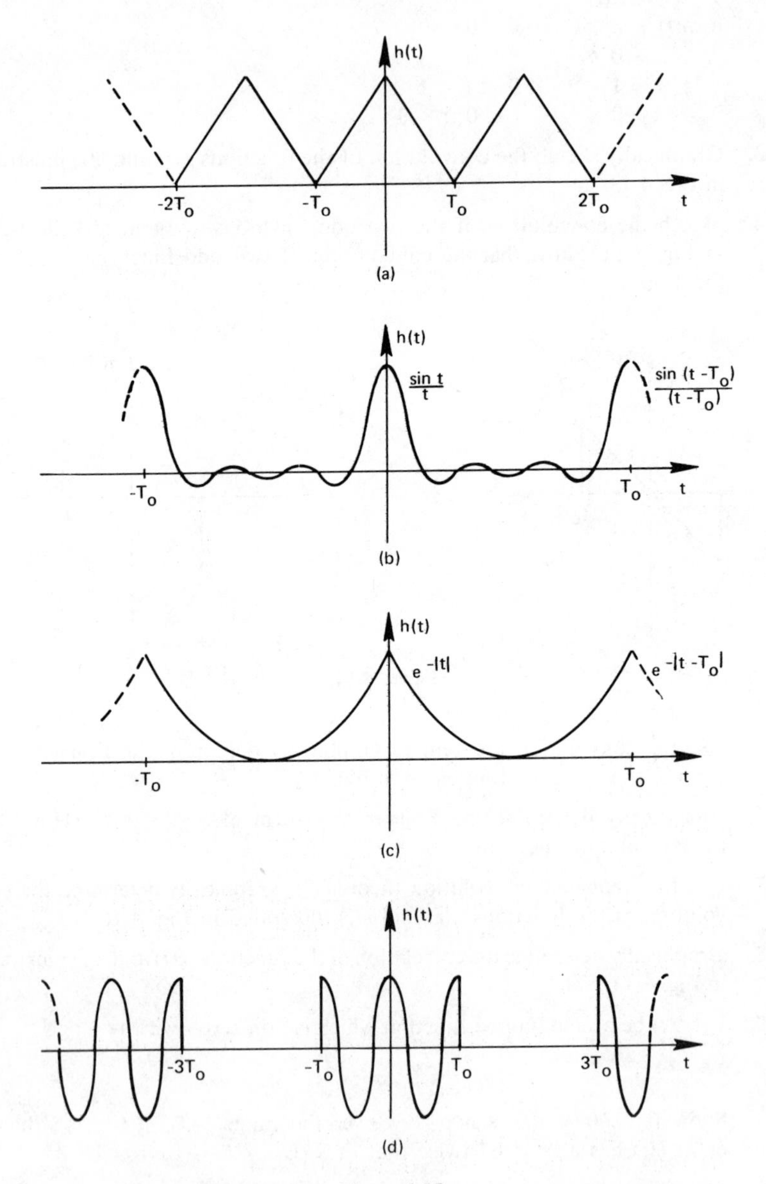
h(t)
-2T_o
-T_o
T_o
2T_o
t
(a)
h(t)
sin t / t
sin (t - T_o) / (t - T_o)
-T_o
T_o
t
(b)
h(t)
e^-|t|
e^-|t - T_o|
-T_o
T_o
t
(c)
h(t)
-3T_o
-T_o
T_o
3T_o
t
(d)

Figure 4-15.

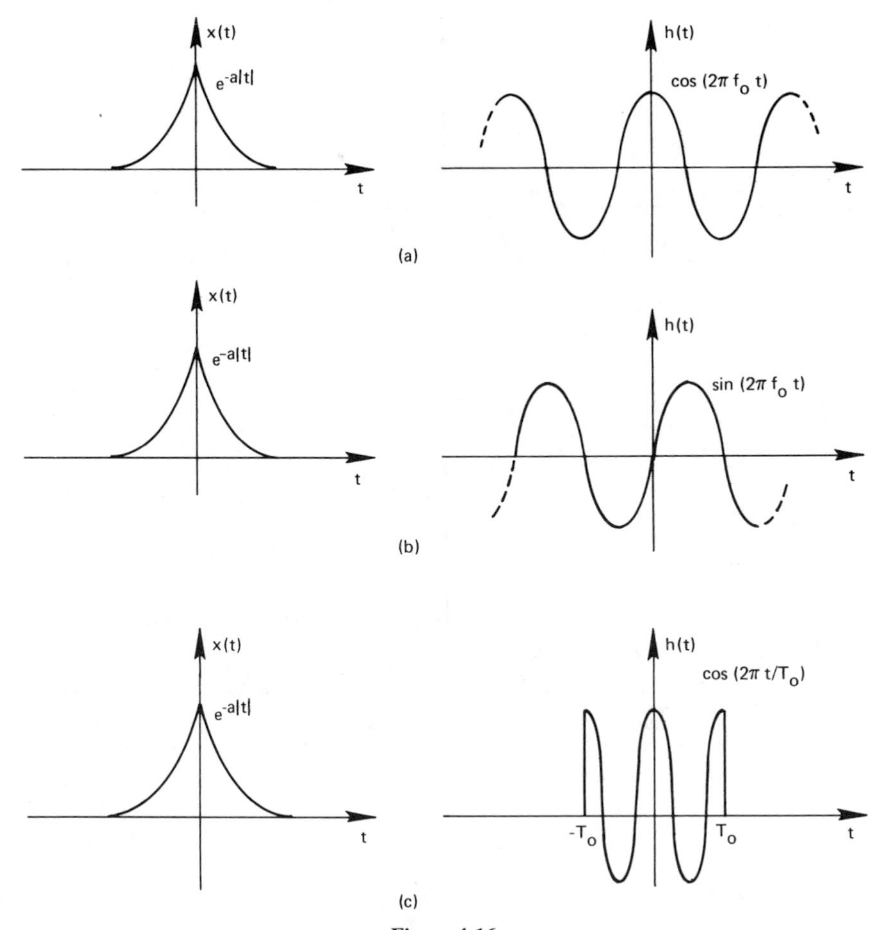

Figure 4-16.

4-12. By means of the frequency convolution theorem, graphically determine the Fourier transform of the half-wave rectified waveform shown in Fig. 4-17(a). Using this result incorporate the shifting theorem to determine the Fourier transform of the full-wave rectified waveform shown in Fig. 4-17(b).

4-13. Graphically find the Fourier transform of the following functions:

a. $h(t) = A \cos^2 (2\pi f_0 t)$
b. $h(t) = A \sin^2 (2\pi f_0 t)$
c. $h(t) = A \cos^2 (2\pi f_0 t) + A \cos^2 (\pi f_0 t)$

4-14. Find graphically the inverse Fourier transform of the following functions:

a. $\left[\frac{\sin (2\pi f)}{(2\pi f)}\right]^2$

b. $\frac{1}{(1 + j\,2\pi f)^2}$

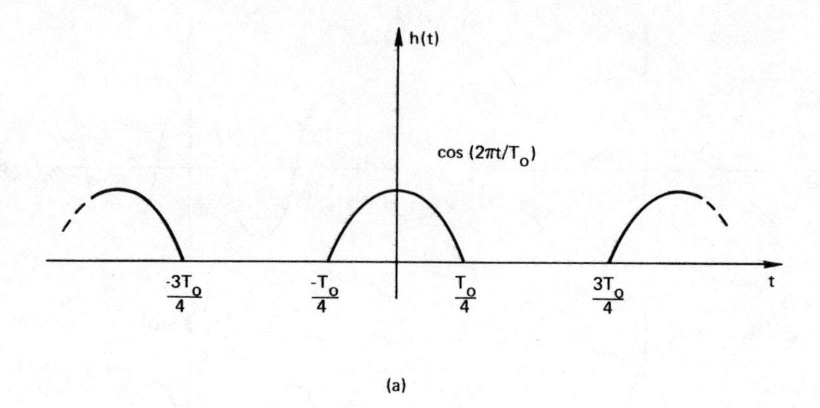

(a)

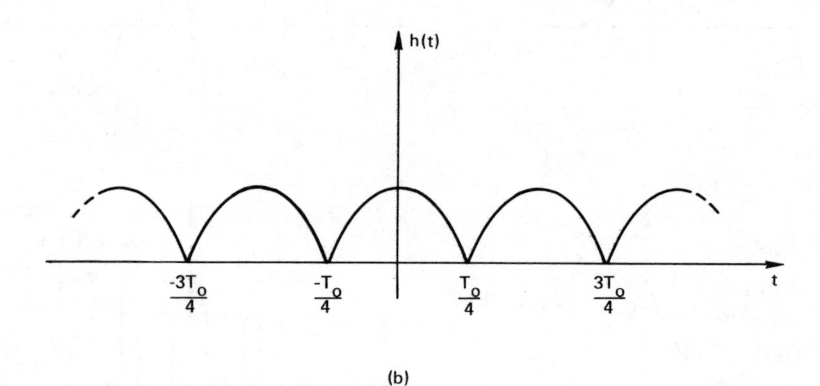

(b)

Figure 4-17.

c. $e^{-|2\pi f|}$

d. $1 - e^{-|f|}$

4-15. Develop a set of rules for determining the limits of integration for the correlation integral.

REFERENCES

1. Bracewell, R., *The Fourier Transform and Its Applications.* New York: McGraw-Hill, 1956.

2. Gupta, S., *Transform and State Variable Methods in Linear Systems.* New York: Wiley, 1966.

3. Healy, T. J., "Convolution Revisited," *IEEE Spectrum* (April 1969), Vol. 6, No. 4, pp. 87–93.

4. Papoulis, A., *The Fourier Integral and Its Applications.* New York: McGraw-Hill, 1962.

5

FOURIER SERIES AND SAMPLED WAVEFORMS

In the technical literature, Fourier series is normally developed independently of the Fourier integral. However, with the introduction of distribution theory, Fourier series can be theoretically derived as a special case of the Fourier integral. This approach is significant in that it is fundamental in considering the discrete Fourier transform as a special case of the Fourier integral. Also fundamental to an understanding of the discrete Fourier transform is the Fourier transform of sampled waveforms. In this chapter we relate both of these relationships to the Fourier transform and thereby provide the framework for the development of the discrete Fourier transform in Chapter 6.

5-1 FOURIER SERIES

A periodic function $y(t)$ with period T_0 expressed as a Fourier series is given by the expression

$$y(t) = \frac{a_0}{2} + \sum_{n=1}^{\infty} [a_n \cos(2\pi n f_0 t) + b_n \sin(2\pi n f_0 t)] \tag{5-1}$$

where f_0 is the fundamental frequency equal to $1/T_0$. The magnitude of the sinusoids or coefficients are given by the integrals

$$a_n = \frac{2}{T_0} \int_{-T_0/2}^{T_0/2} y(t) \cos(2\pi n f_0 t)\, dt \qquad n = 0, 1, 2, 3, \ldots \tag{5-2}$$

$$b_n = \frac{2}{T_0} \int_{-T_0/2}^{T_0/2} y(t) \sin(2\pi n f_0 t)\, dt \qquad n = 1, 2, 3, \ldots \tag{5-3}$$

By applying the identities

$$\cos(2\pi n f_0 t) = \frac{1}{2}(e^{j2\pi n f_0 t} + e^{-j2\pi n f_0 t}) \tag{5-4}$$

and

$$\sin(2\pi n f_0 t) = \frac{1}{2j}(e^{j2\pi n f_0 t} - e^{-j2\pi n f_0 t}) \tag{5-5}$$

expression (5-1) may be written as

$$y(t) = \frac{a_0}{2} + \frac{1}{2}\sum_{n=1}^{\infty}(a_n - jb_n)\, e^{j2\pi n f_0 t} + \frac{1}{2}\sum_{n=1}^{\infty}(a_n + jb_n)\, e^{-j2\pi n f_0 t} \tag{5-6}$$

To simplify this expression, negative values of n are introduced in Eqs. (5-2) and (5-3).

$$\begin{aligned} a_{-n} &= \frac{2}{T_0}\int_{-T_0/2}^{T_0/2} y(t)\cos(-2\pi n f_0 t)\,dt \\ &= \frac{2}{T_0}\int_{-T_0/2}^{T_0/2} y(t)\cos(2\pi n f_0 t)\,dt \\ &= a_n \qquad n = 1, 2, 3, \ldots \end{aligned} \tag{5-7}$$

$$\begin{aligned} b_{-n} &= \frac{2}{T_0}\int_{-T_0/2}^{T_0/2} y(t)\sin(-2\pi n f_0 t)\,dt \\ &= -\frac{2}{T_0}\int_{-T_0/2}^{T_0/2} y(t)\sin(2\pi n f_0 t)\,dt \\ &= -b_n \qquad n = 1, 2, 3, \ldots \end{aligned} \tag{5-8}$$

Hence we can write

$$\sum_{n=1}^{\infty} a_n\, e^{-j2\pi n f_0 t} = \sum_{n=-1}^{-\infty} a_n\, e^{j2\pi n f_0 t} \tag{5-9}$$

and

$$\sum_{n=1}^{\infty} jb_n\, e^{-j2\pi n f_0 t} = -\sum_{n=-1}^{-\infty} jb_n\, e^{j2\pi n f_0 t} \tag{5-10}$$

Substitution of (5-9) and (5-10) into Eq. (5-6) yields

$$\begin{aligned} y(t) &= \frac{a_0}{2} + \frac{1}{2}\sum_{n=-\infty}^{\infty}(a_n - jb_n)\, e^{j2\pi n f_0 t} \\ &= \sum_{n=-\infty}^{\infty} \alpha_n\, e^{j2\pi n f_0 t} \end{aligned} \tag{5-11}$$

Equation (5-11) is the Fourier series expressed in exponential form; coefficients α_n are, in general, complex. Since

$$\alpha_n = \frac{1}{2}(a_n - jb_n) \qquad n = 0, \pm 1, \pm 2, \ldots$$

the combination of Eqs. (5-2), (5-3), (5-7), and (5-8) yields

$$\alpha_n = \frac{1}{T_0}\int_{-T_0/2}^{T_0/2} y(t)e^{-j2\pi nf_0 t}\,dt \qquad n = 0, \pm 1, \pm 2, \ldots \tag{5-12}$$

The expression of the Fourier series in exponential form (5-11) and the complex coefficients in the form (5-12) is normally the preferred approach in analysis.

EXAMPLE 5-1

Determine the Fourier series of the periodic function illustrated in Fig. 5-1.

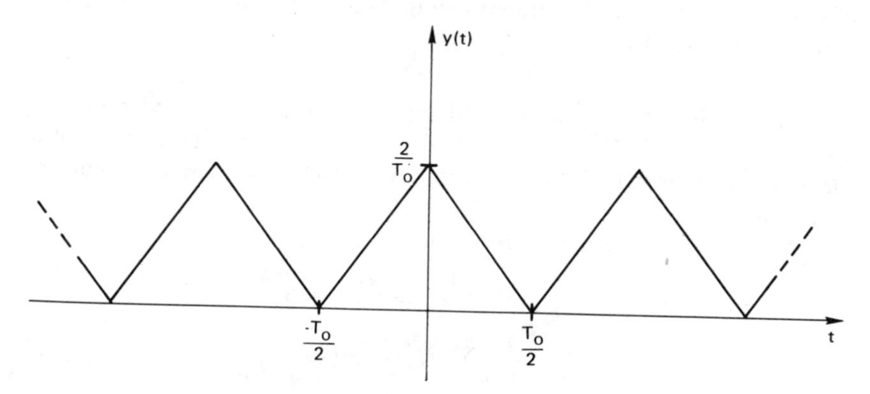

Figure 5-1. Periodic triangular waveform.

From (5-12), since $y(t)$ is an even function then

$$\alpha_n = \begin{cases} \dfrac{1}{T_0}\displaystyle\int_{-T_0/2}^{T_0/2} y(t)\cos(2\pi nf_0 t)\,dt \\ \dfrac{1}{T_0}\displaystyle\int_{-T_0/2}^{0}\left(\frac{2}{T_0}+\frac{4}{T_0^2}t\right)\cos(2\pi nf_0 t)\,dt + \frac{1}{T_0}\int_{0}^{T_0/2}\left(\frac{2}{T_0}-\frac{4}{T_0^2}t\right)\cos(2\pi nf_0 t)\,dt \\ \qquad\qquad n = 0, 1, 3, 5, \ldots \\ 0 \qquad n = 2, 4, 6, \ldots \end{cases}$$

$$\alpha_n = \begin{cases} \dfrac{4}{\pi^2 T_0}\dfrac{1}{n^2} & n = 1, 3, 5, \ldots \\ \dfrac{1}{T_0} & n = 0 \end{cases} \tag{5-13}$$

Hence

$$y(t) = \frac{1}{T_0} + \frac{8}{\pi^2 T_0}\left[\cos(2\pi f_0 t) + \frac{1}{3^2}\cos(6\pi f_0 t) + \frac{1}{5^2}\cos(10\pi f_0 t) + \cdots\right] \tag{5-14}$$

where $f_0 = 1/T_0$.

5-2 FOURIER SERIES AS A SPECIAL CASE OF THE FOURIER INTEGRAL

Consider the periodic triangular function illustrated in Fig. 5-2(e). From Ex. 5-1 we know that the Fourier series of this waveform is an infinite set of sinusoids. We will now show that an identical relationship can be obtained from the Fourier integral.

To accomplish the derivation we utilize the convolution theorem (4-11). Note that the periodic triangular waveform (period T_0) is simply the convolution of the single triangle shown in Fig. 5-2(a), and the infinite sequence of equidistant impulses illustrated in Fig. 5-2(b). Periodic function $y(t)$ can then be expressed by

$$y(t) = h(t) * x(t) \tag{5-15}$$

Fourier transforms of both $h(t)$ and $x(t)$ have been determined previously and are illustrated in Figs. 5-2(c) and (d), respectively. From the convolution theorem, the desired Fourier transform is the product of these two frequency functions

$$\begin{aligned} Y(f) &= H(f)X(f) \\ &= H(f)\frac{1}{T_0}\sum_{n=-\infty}^{\infty}\delta\left(f-\frac{n}{T_0}\right) \\ &= \frac{1}{T_0}\sum_{n=-\infty}^{\infty}H\left(\frac{n}{T_0}\right)\delta\left(f-\frac{n}{T_0}\right) \end{aligned} \tag{5-16}$$

Equations (2-40) and (4-16) were used to develop (5-16).

The Fourier transform of the periodic function is then an infinite set of sinusoids (i.e., an infinite sequence of equidistant impulses) with amplitudes of $H(n/T_0)$. Recall that the Fourier series of a periodic function is an infinite sum of sinusoids with amplitudes given by α_n, (5-12). But note that since the limit of integration of (5-12) is from $-T_0/2$ to $T_0/2$ and since

$$h(t) = y(t) \qquad -\frac{T_0}{2} < t < \frac{T_0}{2} \tag{5-17}$$

the function $y(t)$ can be replaced by $h(t)$ and (5-12) rewritten in the form

$$\begin{aligned} \alpha_n &= \frac{1}{T_0}\int_{-T_0/2}^{T_0/2} h(t)\, e^{-j2\pi n f_0 t}\, dt \\ &= \frac{1}{T_0} H(nf_0) = \frac{1}{T_0} H\left(\frac{n}{T_0}\right) \end{aligned} \tag{5-18}$$

Thus the coefficients as derived by means of the Fourier integral and those of the conventional Fourier series are the same for a periodic function. Also, a comparison of Figs. 5-2(c) and (f) reveals that except for a factor $1/T_0$, the coefficients α_n of the Fourier series expansion of $y(t)$ equal the values of the Fourier transform $H(f)$ evaluated at n/T_0.

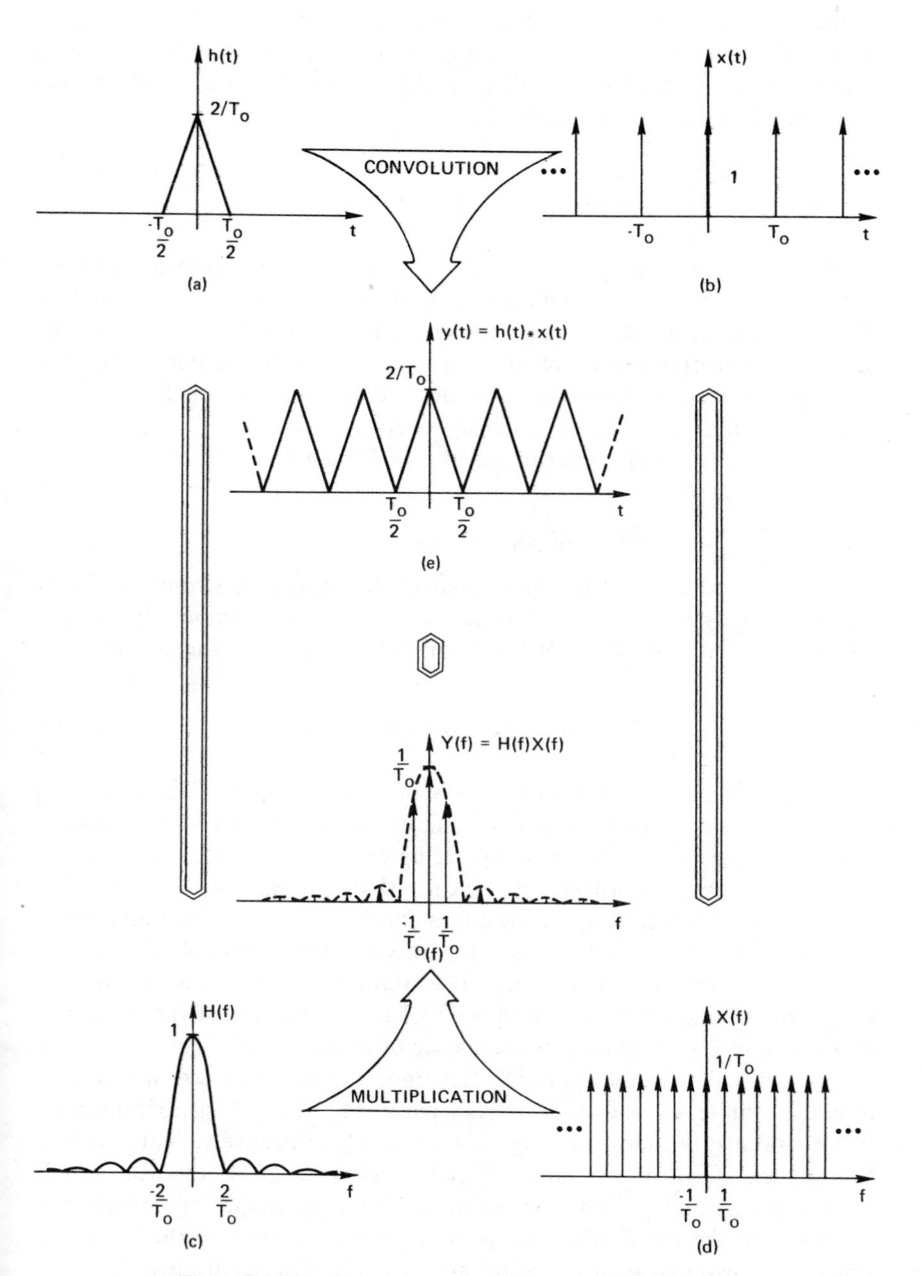

Figure 5-2. Graphical convolution theorem development of the Fourier transform of a periodic triangular waveform.

In summary, we point out again that the key to the preceding development is the incorporation of distribution theory into Fourier integral theory. As will be demonstrated in the discussions to follow, this unifying concept is basic to a thorough understanding of the discrete Fourier transform and hence the fast Fourier transform.

5-3 WAVEFORM SAMPLING

In the preceding chapters we have developed a Fourier transform theory which considers both continuous and impulse functions of time. Based on these developments, it is straightforward to extend the theory to include *sampled* waveforms which are of particular interest in this book. We have developed sufficient tools to investigate in detail the theoretical as well as the visual interpretations of *sampled* waveforms.

If the function $h(t)$ is continuous at $t = T$, then a sample of $h(t)$ at time equal to T is expressed as

$$\hat{h}(t) = h(t)\delta(t - T) = h(T)\delta(t - T) \tag{5-19}$$

where the product must be interpreted in the sense of distribution theory [Eq. (A-12)]. The impulse which occurs at time T has the amplitude equal to the function at time T. If $h(t)$ is continuous at $t = nT$ for $n = 0, \pm 1, \pm 2, \ldots,$

$$\hat{h}(t) = \sum_{n=-\infty}^{\infty} h(nT)\, \delta(t - nT) \tag{5-20}$$

is termed the sampled waveform $h(t)$ with sample interval T. Sampled $h(t)$ is then an infinite sequence of equidistant impulses, each of whose amplitude is given by the value of $h(t)$ corresponding to the time of occurrence of the impulse. Figure 5-3 illustrates graphically the sampling concept. Since Eq. (5-20) is the product of the continuous function $h(t)$ and the sequence of impulses, we can employ the frequency convolution theorem (4-17) to derive the Fourier transform of the sampled waveform. As illustrated in Fig. 5-3, the sampled function [Fig. 5-3(e)] is equal to the product of the waveform $h(t)$ shown in Fig. 5-3(a) and the sequence of impulses $\Delta(t)$ illustrated in Fig. 5-3(b). We call $\Delta(t)$ the sampling function; the notation $\Delta(t)$ will always imply an infinite sequence of impulses separated by T. The Fourier transforms of $h(t)$ and $\Delta(t)$ are shown in Fig. 5-3(c) and (d), respectively. Note that the Fourier transform of the sampling function $\Delta(t)$ is $\Delta(f)$; this function is termed the frequency sampling function. From the frequency convolution theorem, the desired Fourier transform is the convolution of the frequency functions illustrated in Figs. 5-3(c) and (d). The Fourier transform of the sampled waveform is then a periodic function where one period is equal,

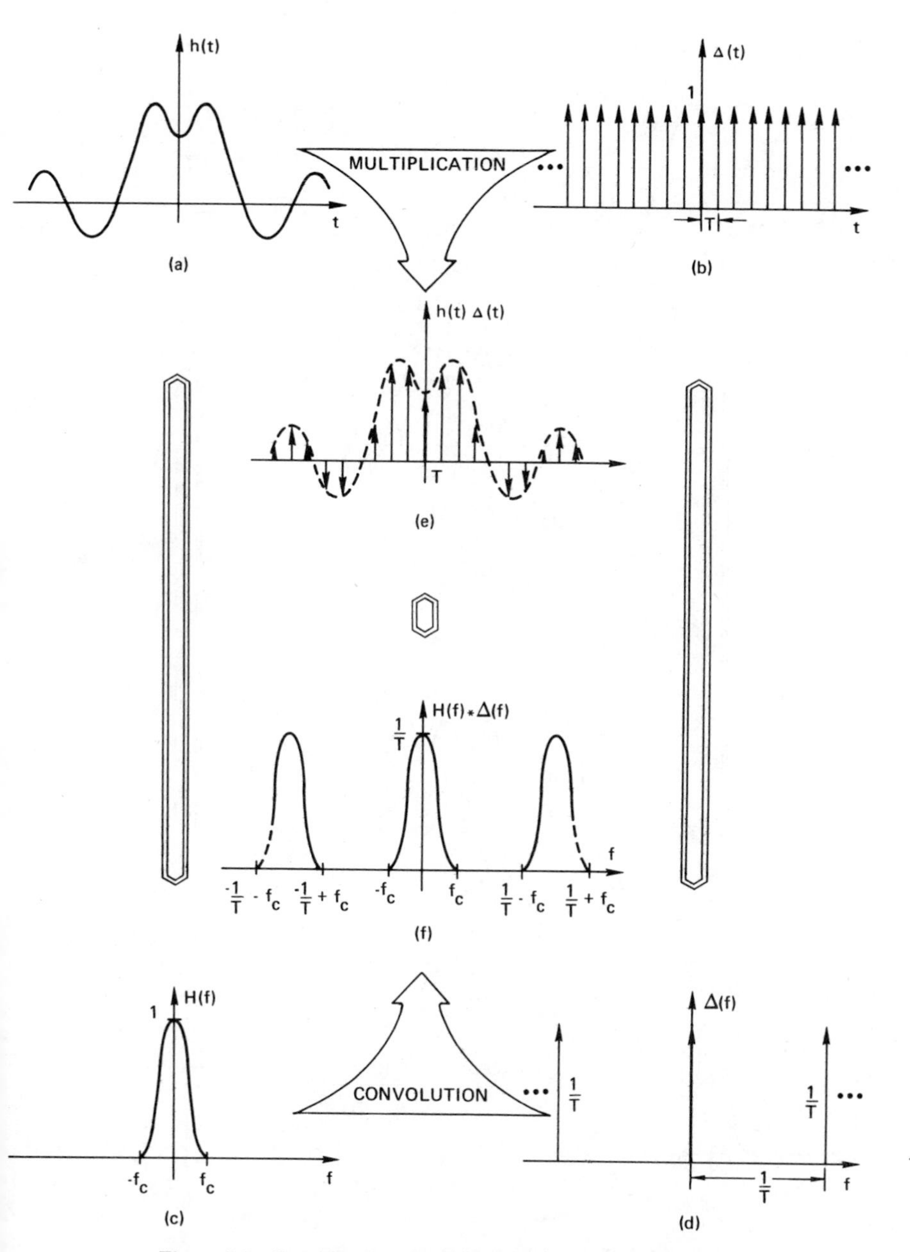

Figure 5-3. Graphical frequency convolution theorem development of the Fourier transform of a sampled waveform.

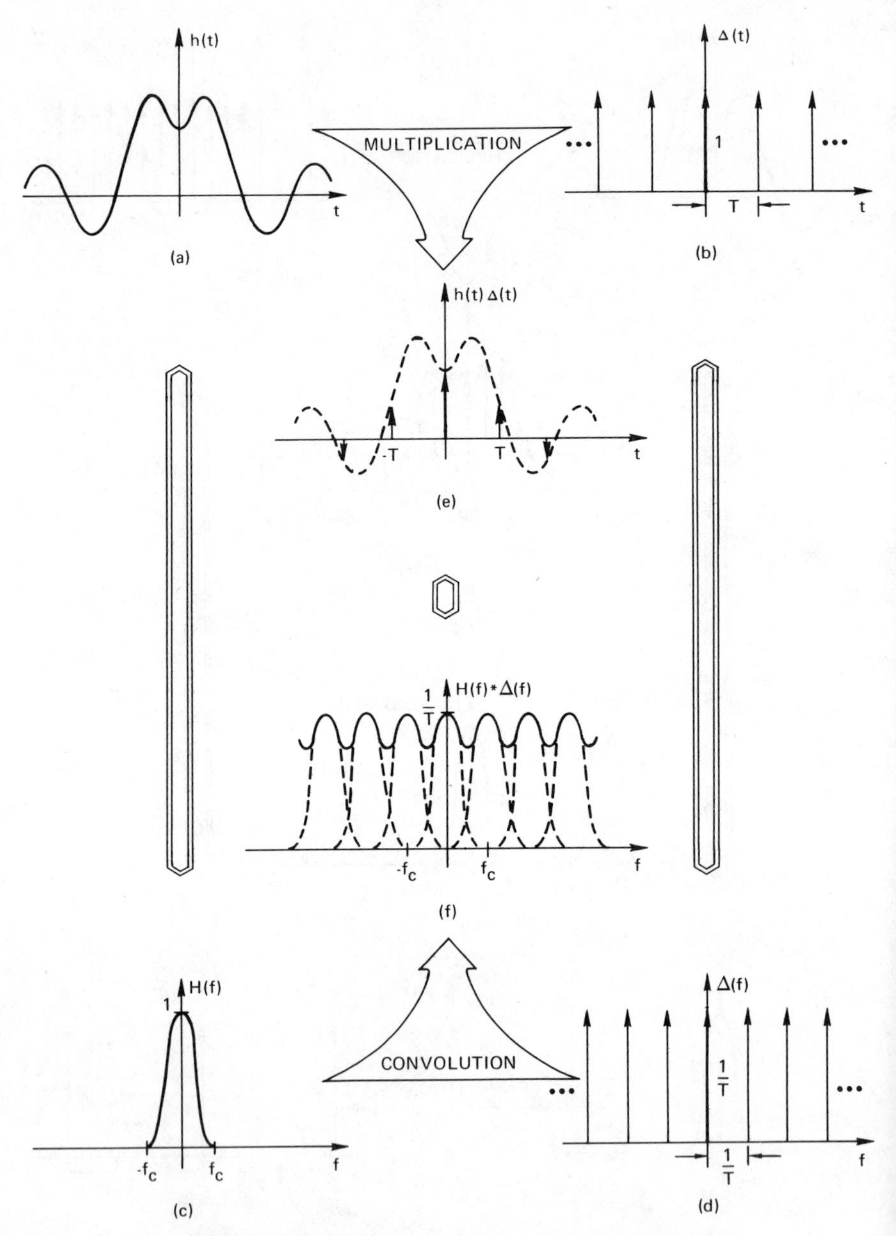

Figure 5-4. Aliased Fourier transform of a waveform sampled at an insufficient rate.

within a constant, to the Fourier transform of the continuous function $h(t)$. This last statement is valid only if the sampling interval T is sufficiently small.

If T is chosen too large, the results illustrated in Fig. 5-4 are obtained. Note that as the sample interval T is increased [Figs. 5-3(b) and 5-4(b)], the equidistant impulses of $\Delta(f)$ become more closely spaced [Figs. 5-3(d) and 5-4(d)]. Because of the decreased spacing of the frequency impulses, their convolution with the frequency function $H(f)$ [Fig. 5-4(c)] results in the overlapping waveform illustrated in Fig. 5-4(f). This distortion of the desired Fourier transform of a sampled function is known as *aliasing*. As described, aliasing occurs because the time function was not sampled at a sufficiently high rate, i.e., the sample interval T is too large. It is then natural to pose the question, "How does one guarantee himself that the Fourier transform of a sampled function is not aliased?" An examination of Figs. 5-4(c) and (d) points up the fact that convolution overlap will occur until the separation of the impulses of $\Delta(f)$ is increased to $1/T = 2f_c$, where f_c is the highest frequency component of the Fourier transform of the continuous function $h(t)$. That is, if the sample interval T is chosen equal to one-half the reciprocal of the highest frequency component, aliasing will not occur. This is an extremely important concept in many fields of scientific application; the reason lies in the fact that we need only retain samples of the continuous waveform to determine a replica of the continuous Fourier transform. Furthermore, if a waveform is sampled such that aliasing does not occur, these samples can be appropriately combined to reconstruct identically the continuous waveform. This is merely a statement of the sampling theorem which we will now investigate.

5-4 SAMPLING THEOREM

The sampling theorem states that if the Fourier transform of a function $h(t)$ is zero for all frequencies greater than a certain frequency f_c, then the continuous function $h(t)$ can be uniquely determined from a knowledge of its sampled values,

$$\hat{h}(t) = h(nT) \sum_{n=-\infty}^{\infty} \delta(t - nT) \tag{5-21}$$

where $T = \dfrac{1}{2fc}$.

In particular, $h(t)$ is given by

$$h(t) = T \sum_{n=-\infty}^{\infty} h(nT) \frac{\sin 2\pi f_c(t - nT)}{\pi(t - nT)} \tag{5-22}$$

Constraints of the theorem are illustrated graphically in Fig. 5-5. First, it is necessary that the Fourier transform of $h(t)$ be zero for frequencies greater than f_c. As shown in Fig. 5-5(c), the example frequency function is *band-limited* at the frequency f_c; the term *band-limited* is a shortened way of saying

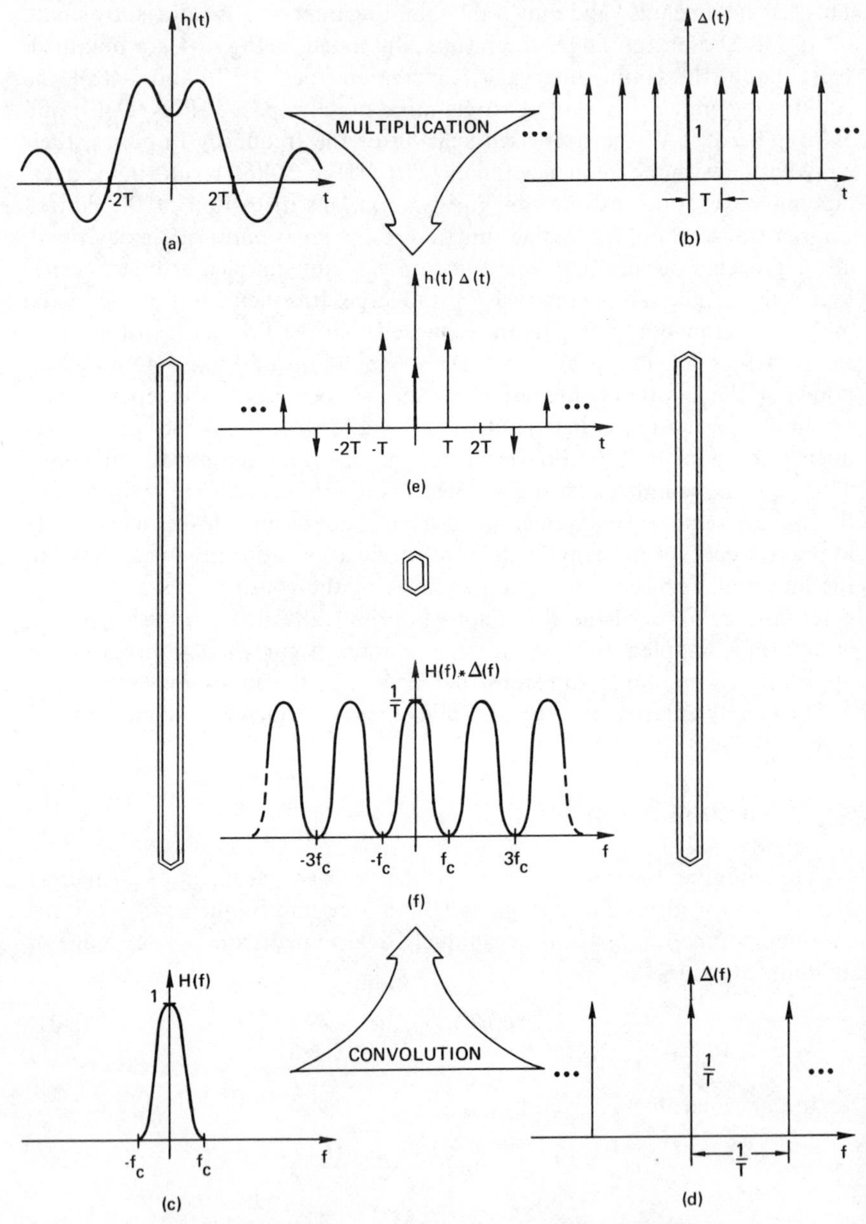

Figure 5-5. Fourier transform of a waveform sampled at the Nyquist sampling rate.

that the Fourier transform is zero for $|f| > f_c$. The second constraint is that the sample spacing be chosen as $T = \frac{1}{2fc}$; that is, the impulse functions of Fig. 5-5(d) are required to be separated by $1/T = 2f_c$. This spacing insures that when $\Delta(f)$ and $H(f)$ are convolved there will be no aliasing. Alternately, the functions $H(f)$ and $H(f) * \Delta(f)$ as illustrated in Figs. 5-5(*c*) and (*f*), respectively, are equal in the interval $|f| < f_c$, within the scaling constant T. If $T < \frac{1}{2fc}$, then aliasing will result; if $T > \frac{1}{2fc}$, the theorem still holds. The requirement that $\mathrm{T} = \frac{1}{2fc}$ is simply the maximum spacing between samples for which the theorem holds. Frequency $1/T = 2f_c$ is known as the *Nyquist sampling rate*. Given that these two constraints are true, the theorem states that $h(t)$ [Fig. 5-5(a)] can be reconstructed from a knowledge of the impulses illustrated in Fig. 5-5(e).

To construct a proof of the sampling theorem, recall from the discussion on constraints of the theorem that the Fourier transform of the sampled function is identical, within the constant T, to the Fourier transform of the unsampled function, in the frequency range $-f_c \leq f \leq f_c$. From Fig. 5-5(*f*), the Fourier transform of the sampled time function is given by $H(f) * \Delta(f)$. Hence, as illustrated in Figs. 5-6(a), (b), and (e), the multiplication of a rectangular frequency function of amplitude T with the Fourier transform of the sampled waveform is the Fourier transform $H(f)$;

$$H(f) = [H(f) * \Delta(f)]Q(f) \tag{5-23}$$

The inverse Fourier transform of $H(f)$ is the original waveform $h(t)$ as shown in Fig. 5-6(*f*). But from the convolution theorem, $h(t)$ is equal to the convolution of the inverse Fourier transforms of $H(f) * \Delta(f)$ and of the rectangular frequency function. Hence $h(t)$ is given by the convolution of $h(t)$ $\Delta(t)$ [Fig. 5-6(c)] and $q(t)$ [Fig. 5-6(d)];

$$\begin{aligned} h(t) &= [h(t)\,\Delta(t)] * q(t) \\ &= \sum_{n=-\infty}^{\infty} [h(nT)\,\delta(t - nT)] * q(t) \\ &= \sum_{n=-\infty}^{\infty} h(nT)\,q(t - nT) \\ &= T \sum_{n=-\infty}^{\infty} h(nT) \frac{\sin[2\pi f_c(t - nT)]}{\pi(t - nT)} \end{aligned} \tag{5-24}$$

Function $q(t)$ is given by the Fourier transform pair (2-27). Equation (5-24) is the desired expression for reconstructing $h(t)$ from a knowledge of only the samples of $h(t)$.

We should note carefully that it is possible to reconstruct a sampled waveform perfectly only if the waveform is band-limited. In practice, this condition rarely exists. The solution is to sample at such a rate that aliasing is negligible;

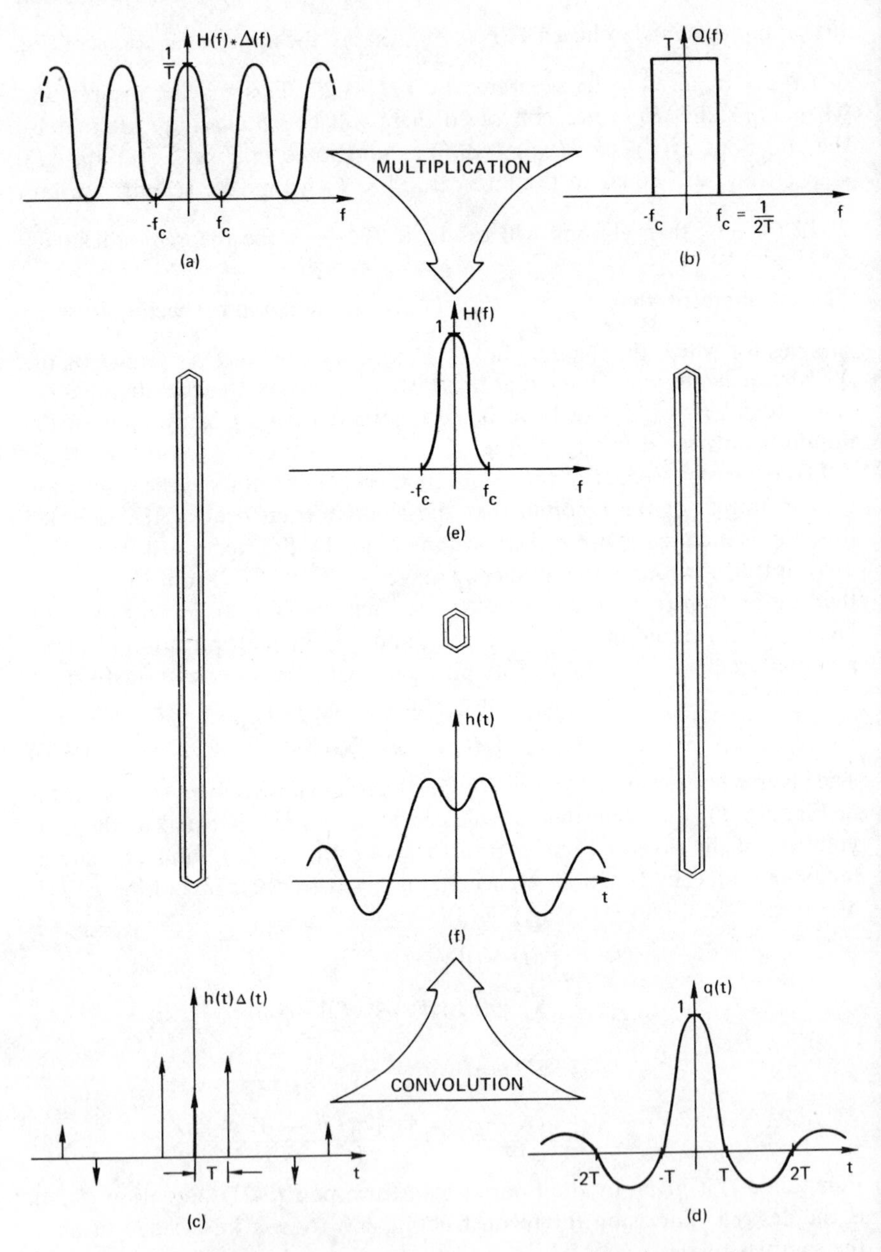

Figure 5-6. Graphical derivation of the sampling theorem.

it may be necessary to filter the signal prior to quantization to insure that there exists, to the extent possible, a band-limited function.

5-5 FREQUENCY SAMPLING THEOREM

Analogous to time domain sampling there exists a sampling theorem in the frequency domain. If a function $h(t)$ is time-limited, that is

$$h(t) = 0 \qquad |t| > T_c \tag{5-25}$$

then its Fourier transform $H(f)$ can be uniquely determined from equidistant samples of $H(f)$. In particular, $H(f)$ is given by

$$H(f) = \frac{1}{2T_c} \sum_{n=-\infty}^{\infty} H\left(\frac{n}{2T_c}\right) \frac{\sin\left[2\pi T_c(f - n/2T_c)\right]}{\pi(f - n/2T_c)} \tag{5-26}$$

The proof is similar to the proof of the time domain sampling theorem.

PROBLEMS

5-1. Find the Fourier series of the periodic waveforms illustrated in Fig. 5-7.

5-2. Determine the Fourier transform of the waveforms illustrated in Fig. 5-7. Compare these results with those of Problem 5-1.

5-3. By using graphical arguments similar to those of Fig. 5-4, determine the Nyquist sampling rate for the time functions whose Fourier transform magnitude functions are illustrated in Fig. 5-8. Are there cases where aliasing can be used to an advantage?

5-4. Graphically justify the bandpass sampling theorem which states that

$$\text{Critical sampling frequency} = \frac{2f_h}{\text{largest integer not exceeding } \dfrac{f_h}{(f_h - f_l)}}$$

where f_h and f_l are the upper and lower cutoff frequencies of the bandpass spectrum.

5-5. Assume that the function $h(t) = \cos(2\pi t)$ has been sampled at $t = n/4$; $n = 0, \pm 1, \pm 2, \ldots$. Sketch $h(t)$ and indicate the sampled values. Graphically and analytically determine Eq. (5-24) for $h(t = 7/8)$ where the summation is only over $n = 2, 3, 4$, and 5.

5-6. A frequency function (say a filter frequency response) has been determined experimentally in the laboratory and is given by a graphical curve. If it is desired to sample this function for computer storage purposes, what is the minimum frequency sampling interval if the frequency function is to later be totally reconstructed? State all assumptions.

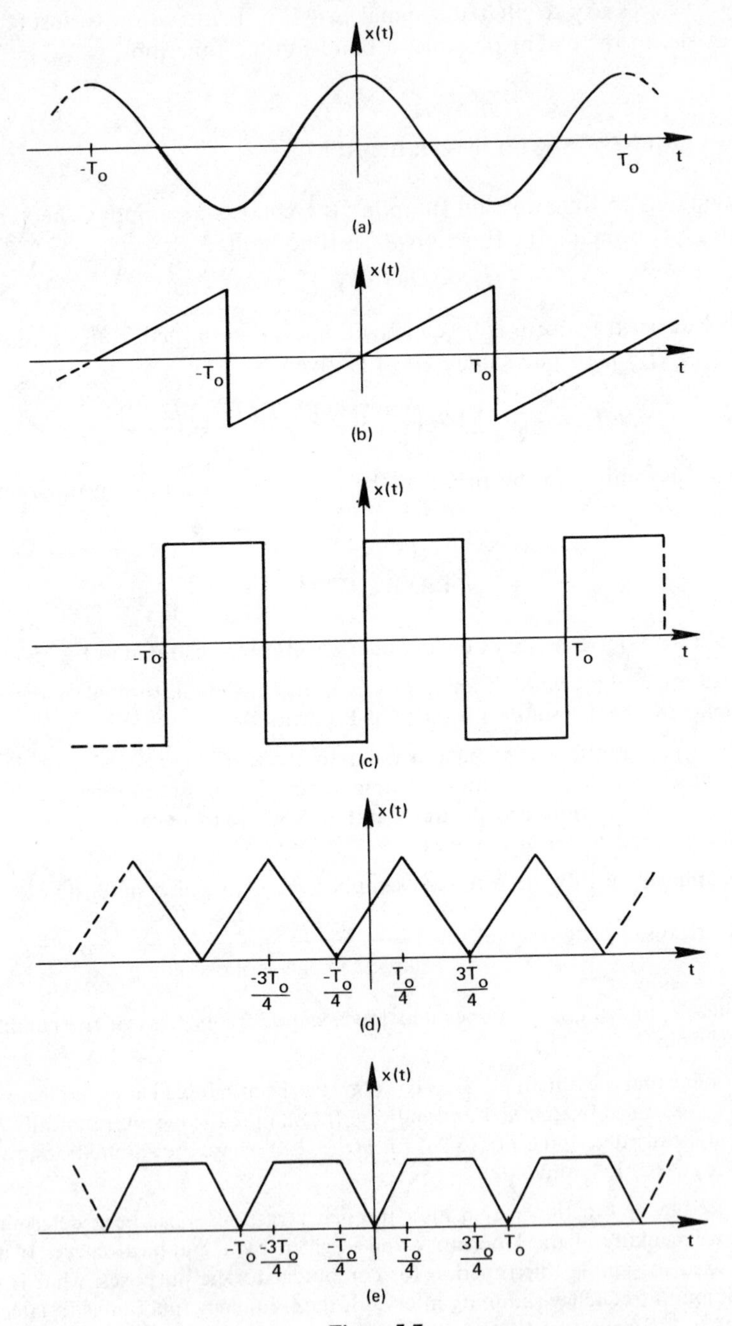
x(t)
-To
To
t
(a)
x(t)
-To
To
t
(b)
x(t)
-To
To
t
(c)
x(t)
-3To/4
-To/4
To/4
3To/4
t
(d)
x(t)
-To
-3To/4
-To/4
To/4
3To/4
To
t
(e)

Figure 5-7.

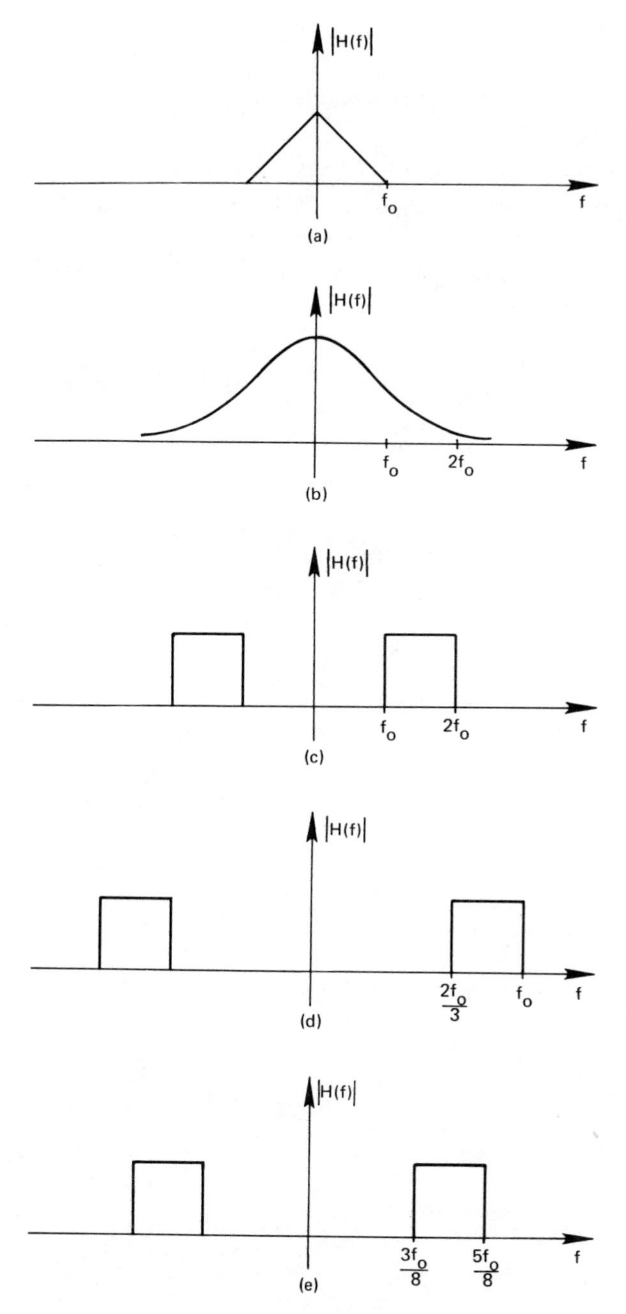

Figure 5-8.

REFERENCES

1. BRACEWELL, R., *The Fourier Transform and Its Applications.* New York: McGraw-Hill, 1965.

2. LEE, Y. W., *Statistical Theory of Communication.* New York: Wiley, 1964.

3. PANTER, P. F., *Modulation, Noise, and Spectral Analysis.* New York: McGraw-Hill, 1965.

4. PAPOULIS, A., *The Fourier Integral and Its Applications.* New York: McGraw-Hill, 1962.

6

THE DISCRETE FOURIER TRANSFORM

Normally a discussion of the discrete Fourier transform is based on an initial definition of the finite length discrete transform; from this assumed axiom those properties of the transform implied by this definition are derived. This approach is unrewarding in that at its conclusion there is always the unanswered question, "How does the discrete Fourier transform relate to the continuous Fourier transform?" To answer this question we find it preferable to derive the discrete Fourier transform as a special case of continuous Fourier transform theory.

In this chapter, we develop a special case of the continuous Fourier transform which is amenable to machine computation. The approach will be to develop the discrete Fourier transform from a graphical derivation based on continuous Fourier transform theory. These graphical arguments are then substantiated by a theoretical development. Both approaches emphasize the modifications of continuous Fourier transform theory which are necessary to define a computer-oriented transform pair.

6-1 A GRAPHICAL DEVELOPMENT

Consider the example function $h(t)$ and its Fourier transform $H(f)$ illustrated in Fig. 6-1(a). It is desired to modify this Fourier transform pair in such a manner that the pair is amenable to digital computer computation. This modified pair, termed the discrete Fourier transform is to approximate as closely as possibly the continuous Fourier transform.

To determine the Fourier transform of $h(t)$ by means of digital analysis

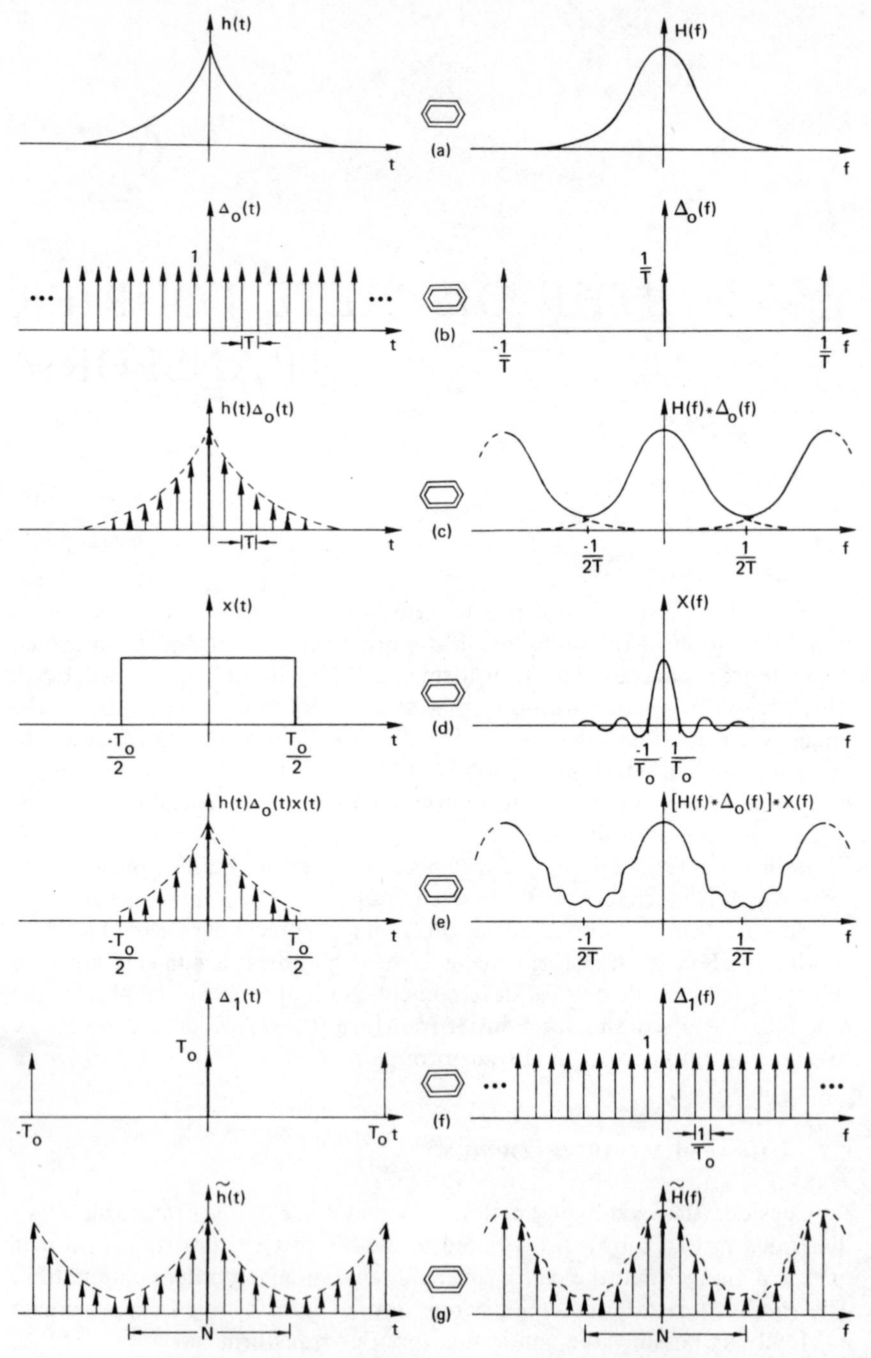

Figure 6-1. Graphical development of the discrete Fourier transform.

techniques, it is necessary to sample $h(t)$ as described in Chapter 5. Sampling is accomplished by multiplying $h(t)$ by the sampling function illustrated in Fig. 6-1(b). The sample interval is T. Sampled function $\hat{h}(t)$ and its Fourier transform are illustrated in Fig. 6-1(c). This Fourier transform pair represents the first modification to the original pair which is necessary in defining a discrete transform pair. Note that to this point the modified transform pair differs from the original transform pair only by the aliasing effect which result from sampling. As discussed in Sec. 5-3, if the waveform $h(t)$ is sampled at a frequency of at least twice the largest frequency component of $h(t)$, there is no loss of information as a result of sampling. If the function $h(t)$ is not band-limited; i.e., $H(f) \neq 0$ for some $|f| > f_c$, then sampling will introduce aliasing as illustrated in Fig. 6-1(c). To reduce this error we have only one recourse, and that is to sample faster; that is, choose T smaller.

The Fourier transform pair in Fig. 6-1(c) is not suitable for machine computation because an infinity of samples of $h(t)$ is considered; it is necessary to truncate the sampled function $h(t)$ so that only a finite number of points, say N, are considered. The rectangular or truncation function and its Fourier transform are illustrated in Fig. 6-1(d). The product of the infinite sequence of impulse functions representing $h(t)$ and the truncation function yields the finite length time function illustrated in Fig. 6-1(e). Truncation introduces the second modification of the original Fourier transform pair; this effect is to convolve the aliased frequency transform of Fig. 6-1(c) with the Fourier transform of the truncation function [Fig. 6-1(d)]. As shown in Fig. 6-1(e), the frequency transform now has a *ripple* to it; this effect has been accentuated in the illustration for emphasis. To reduce this effect, recall the inverse relation that exists between the *width* of a time function and its Fourier transform (Sec. 3-3). Hence, if the truncation (rectangular) function is increased in length, then the $\sin f/f$ function will approach an impulse; the more closely the $\sin f/f$ function approximates an impulse, the less ripple or error will be introduced by the convolution which results from truncation. Therefore, it is desirable to choose the length of the truncation function as long as possible. We will investigate in detail in Sec. 6-4 the effect of truncation.

The modified transform pair of Fig. 6-1(e) is still not an acceptable discrete Fourier transform pair because the frequency transform is a continuous function. For machine computation, only sample values of the frequency function can be computed; it is necessary to modify the frequency transform by the frequency sampling function illustrated in Fig. 6-1(f). The frequency sampling interval is $1/T_0$.

The discrete Fourier transform pair of Fig. 6-1(g) is acceptable for the purposes of digital machine computation since both the time and frequency domains are represented by discrete values. As illustrated in Fig. 6-1(g), the original time function is approximated by N samples; the original Fourier transform $H(f)$ is also approximated by N samples. These N samples define the discrete Fourier transform pair and approximate the original Fourier transform pair. Note that sampling in the time domain resulted in a periodic function of frequency; sampling in the frequency domain resulted in a periodic function of time. Hence, the discrete Fourier transform requires that both the original time and frequency functions be modified such that they become periodic functions. N time samples and N frequency values represent one period of the time and frequency domain waveforms, respectively. Since the N values of time and frequency are related by the continuous Fourier transform, then a discrete relationship can be derived.

6-2 THEORETICAL DEVELOPMENT

The preceding graphical development illustrates the point that if a continuous Fourier transform pair is suitably modified, then the modified pair is acceptable for computation on a digital computer. Thus, to develop this discrete Fourier transform pair, it is only necessary to derive the mathematical relationships which result from each of the required modifications: time domain sampling, truncation, and frequency domain sampling.

Consider the Fourier transform pair illustrated in Fig. 6-2(a). To discretize this transform pair it is first necessary to sample the waveform $h(t)$; the sampled waveform can be written as $h(t)\ \Delta_0(t)$ where $\Delta_0(t)$ is the time domain sampling function illustrated in Fig. 6-2(b). The sampling interval is T. From Eq. (5-20) the sampled function can be written as

$$\begin{aligned} h(t)\Delta_0(t) &= h(t) \sum_{k=-\infty}^{\infty} \delta(t-kT) \\ &= \sum_{k=-\infty}^{\infty} h(kT)\,\delta(t-kT) \qquad (6\text{-}1) \end{aligned}$$

The result of this multiplication is illustrated in Fig. 6-2(c). Note the aliasing effect which results from the choice of T.

Next, the sampled function is truncated by multiplication with the rectangular function $x(t)$ illustrated in Fig. 6-2(d):

$$\begin{aligned} x(t) &= 1 \qquad -\frac{T}{2} < t < T_0 - \frac{T}{2} \\ &= 0 \qquad \text{otherwise} \qquad (6\text{-}2) \end{aligned}$$

where T_0 is the duration of the truncation function. An obvious question at this point is, "Why is the rectangular function $x(t)$ not centered at zero or $T_0/2$?" Centering of $x(t)$ at zero is avoided to alleviate notation problems. The reason for not centering the rectangular function at $T_0/2$ will become obvious later in the development.

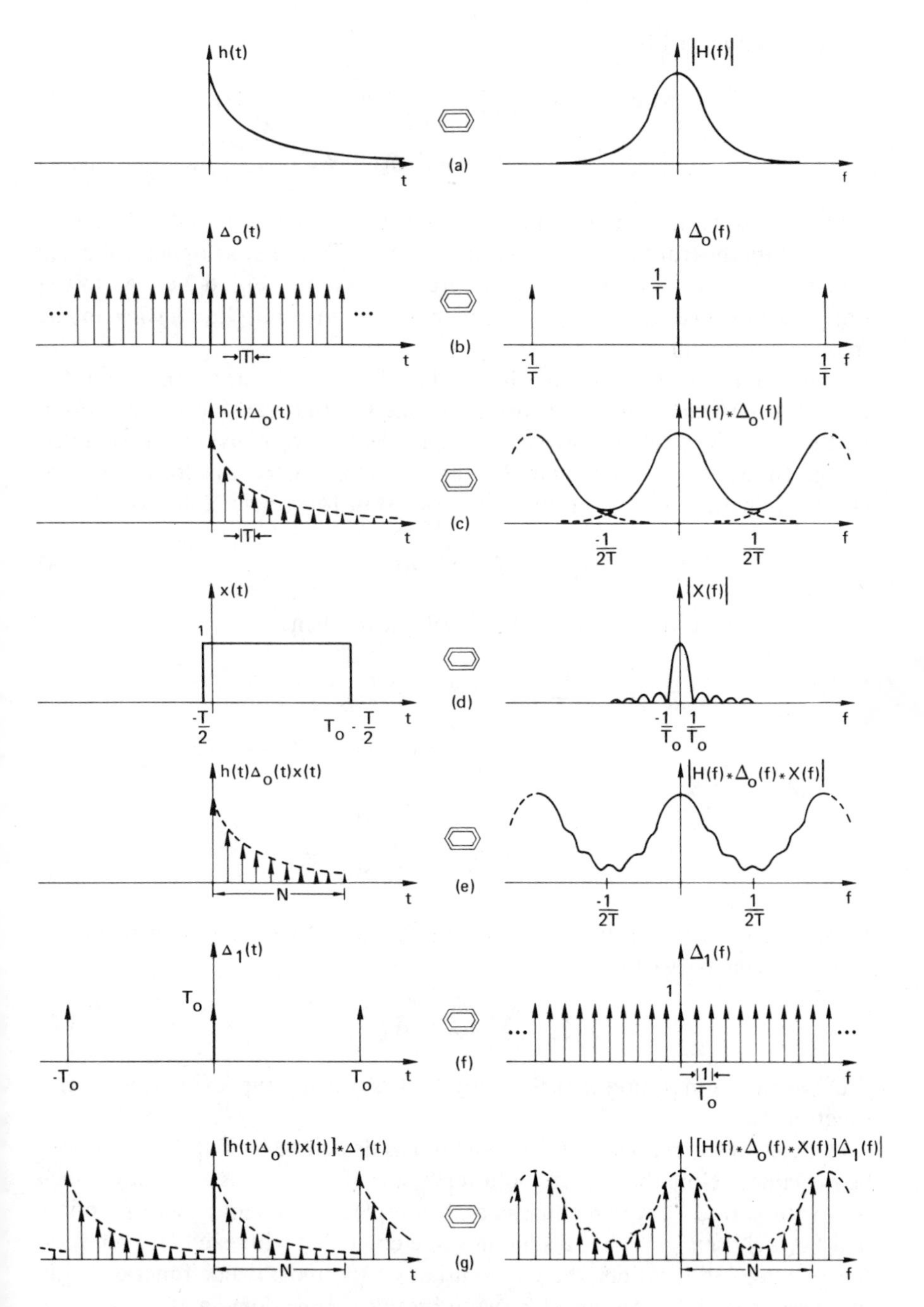

Figure 6-2. Graphical derivation of the discrete Fourier transform pair.

Truncation yields

$$h(t)\Delta_0(t)x(t) = \left[\sum_{k=-\infty}^{\infty} h(kT)\,\delta(t-kT)\right]x(t)$$
$$= \sum_{k=0}^{N-1} h(kT)\,\delta(t-kT) \tag{6-3}$$

where it has been assumed that there are N equidistant impulse functions lying within the truncation interval; that is, $N = T_0/T$. The sampled truncated waveform and its Fourier transform are illustrated in Fig. 6-2(e). As in the previous example, truncation in the time domain results in *rippling* in the frequency domain.

The final step in modifying the original Fourier transform pair to a discrete Fourier transform pair is to sample the Fourier transform of Eq. (6-3). In the time domain this product is equivalent to convolving the sampled truncated waveform (6-3) and the time function $\Delta_1(t)$, illustrated in Fig. 6-2(f). Function $\Delta_1(t)$ is given by Fourier transform pair (2-40) as

$$\Delta_1(t) = T_0 \sum_{r=-\infty}^{\infty} \delta(t - rT_0) \tag{6-4}$$

The desired relationship is $[h(t)\Delta_0(t)x(t)] * \Delta_1(t)$; hence

$$[h(t)\Delta_0(t)x(t)] * \Delta_1(t) = \left[\sum_{k=0}^{N-1} h(kT)\,\delta(t-kT)\right] * \left[T_0 \sum_{r=-\infty}^{\infty} \delta(t-rT_0)\right]$$
$$= \cdots + T_0 \sum_{k=0}^{N-1} h(kT)\,\delta(t+T_0-kT)$$
$$+ T_0 \sum_{k=0}^{N-1} h(kT)\,\delta(t-kT)$$
$$+ T_0 \sum_{k=0}^{N-1} h(kT)\,\delta(t-T_0-kT) + \cdots \tag{6-5}$$

Note that (6-5) is periodic with period T_0; in compact notation form the equation can be written as

$$\tilde{h}(t) = T_0 \sum_{r=-\infty}^{\infty} \left[\sum_{k=0}^{N-1} h(kT)\,\delta(t-kT-rT_0)\right] \tag{6-6}$$

We choose the notation $\tilde{h}(t)$ to imply that $\tilde{h}(t)$ is an approximation to the function $h(t)$.

Choice of the rectangular function $x(t)$ as described by Eq. (6-2) can now be explained. Note that the convolution result of Eq. (6-6) is a periodic function with period T_0 which consists of N samples. If the rectangular function had been chosen such that a sample value coincided with each end point of the rectangular function, the convolution of the rectangular function with impulses spaced at intervals of T_0 would result in time domain aliasing. That is, the Nth point of one period would coincide with (and add to) the first point of the next period. To insure that time domain aliasing does not occur, it is necessary to choose the truncation interval as illustrated in Fig. 6-2(d). (The truncation function may also be chosen as illustrated in Fig. 6-1(d), but note that the end points of the truncation function lie at the mid-point of two adjacent sample values to avoid time domain aliasing.)

To develop the Fourier transform of Eq. (6-6), recall from the discussion on Fourier series, Sec. 5-1, that the Fourier transform of a periodic function $h(t)$ is a sequence of equidistant impulses

$$\tilde{H}\left(\frac{n}{T_0}\right) = \sum_{n=-\infty}^{\infty} \alpha_n \, \delta(f - nf_0) \qquad f_0 = \frac{1}{T_0} \tag{6-7}$$

where

$$\alpha_n = \frac{1}{T_0} \int_{-T/2}^{T_0 - T/2} \tilde{h}(t) \, e^{-j2\pi nt/T_0} \, dt \qquad n = 0, \pm 1, \pm 2, \ldots \tag{6-8}$$

Substitution of (6-6) in (6-8) yields

$$\alpha_n = \frac{1}{T_0} \int_{-T/2}^{T_0 - T/2} T_0 \sum_{r=-\infty}^{\infty} \sum_{k=0}^{N-1} h(kT) \, \delta(t - kT - rT_0) \, e^{-j2\pi nt/T_0} \, dt$$

Integration is only over one period, hence

$$\begin{aligned} \alpha_n &= \int_{-T/2}^{T_0 - T/2} \sum_{k=0}^{N-1} h(kT) \, \delta(t - kT) \, e^{-j2\pi nt/T_0} \, dt \\ &= \sum_{k=0}^{N-1} h(kT) \int_{-T/2}^{T_0 - T/2} e^{-j2\pi nt/T_0} \, \delta(t - kT) \, dt \\ &= \sum_{k=0}^{N-1} h(kT) \, e^{-j2\pi knT/T_0} \end{aligned} \tag{6-9}$$

Since $T_0 = NT$, Eq. (6-9) can be rewritten as

$$\alpha_n = \sum_{k=0}^{N-1} h(kT) \, e^{-j2\pi kn/N} \qquad n = 0, \pm 1, \pm 2, \ldots \tag{6-10}$$

and the Fourier transform of Eq. (6-6) is

$$\tilde{H}\left(\frac{n}{NT}\right) = \sum_{n=-\infty}^{\infty} \sum_{k=0}^{N-1} h(kT) \, e^{-j2\pi kn/N} \tag{6-11}$$

From a cursory evaluation of (6-11), it is not obvious that the Fourier transform $\tilde{H}(n/NT)$ is periodic as illustrated in Fig. 6-2(g). However, there are only N distinct complex values computable from Eq. (6-11). To establish this fact let $n = r$ where r is an arbitrary integer; Eq. (6-11) becomes

$$\tilde{H}\left(\frac{r}{NT}\right) = \sum_{k=0}^{N-1} h(kT) \, e^{-j2\pi kr/N} \tag{6-12}$$

Now let $n = r + N$; note that

$$\begin{aligned} e^{-j2\pi k(r+N)/N} &= e^{-j2\pi kr/N} \, e^{-j2\pi k} \\ &= e^{-j2\pi kr/N} \end{aligned} \tag{6-13}$$

since $e^{-j2\pi k} = \cos(2\pi k) - j\sin(2\pi k) = 1$ for k integer valued. Thus for $n = r + N$

$$\tilde{H}\left(\frac{r+N}{NT}\right) = \sum_{k=0}^{N-1} h(kT)\, e^{-j2\pi k(r+N)/N}$$
$$= \sum_{k=0}^{N-1} h(kT)\, e^{-j2\pi kr/N}$$
$$= \tilde{H}\left(\frac{r}{NT}\right) \tag{6-14}$$

Therefore, there are only N distinct values for which Eq. (6-11) can be evaluated; $\tilde{H}(n/NT)$ is periodic with a period of N samples. Fourier transform (6-11) can be expressed equivalently as

$$\tilde{H}\left(\frac{n}{NT}\right) = \sum_{k=0}^{N-1} h(kT)\, e^{-j2\pi nk/N} \qquad n = 0, 1, \ldots, N-1 \tag{6-15}$$

Eq. (6-15) is the desired discrete Fourier transform; the expression relates N samples of time and N samples of frequency by means of the continuous Fourier transform. The discrete Fourier transform is then a special case of the continuous Fourier transform. If it is assumed that the N samples of the original function $h(t)$ are one period of a periodic waveform, the Fourier transform of this periodic function is given by the N samples as computed by Eq. (6-15). Notation $\tilde{H}(n/NT)$ is used to indicate that the discrete Fourier transform is an approximation to the continuous Fourier transform. Normally, Eq. (6-15) is written as

$$G\left(\frac{n}{NT}\right) = \sum_{k=0}^{N-1} g(kT)\, e^{-j2\pi nk/N} \qquad n = 0, 1, \ldots, N-1 \tag{6-16}$$

since the Fourier transform of the sampled periodic function $g(kT)$ is identically $G(n/NT)$.

6-3 DISCRETE INVERSE FOURIER TRANSFORM

The discrete inverse Fourier transform is given by

$$g(kT) = \frac{1}{N}\sum_{n=0}^{N-1} G\left(\frac{n}{NT}\right) e^{j2\pi nk/N} \qquad k = 0, 1, \ldots, N-1 \tag{6-17}$$

To prove that (6-17) and the transform relation (6-16) form a discrete Fourier transform pair, substitute (6-17) into Eq. (6-16).

$$G\left(\frac{n}{NT}\right) = \sum_{k=0}^{N-1}\left[\frac{1}{N}\sum_{r=0}^{N-1} G\left(\frac{r}{NT}\right) e^{j2\pi rk/N}\right] e^{-j2\pi nk/N}$$
$$= \frac{1}{N}\sum_{r=0}^{N-1} G\left(\frac{r}{NT}\right)\left[\sum_{k=0}^{N-1} e^{j2\pi rk/N}\, e^{-j2\pi nk/N}\right]$$
$$= G\left(\frac{n}{NT}\right) \tag{6-18}$$

Identity (6-18) follows from the orthogonality relationship

$$\sum_{k=0}^{N-1} e^{j2\pi rk/N} e^{-j2\pi nk/N} = \begin{cases} N \text{ if } r = n \\ 0 \text{ otherwise} \end{cases} \qquad (6\text{-}19)$$

The discrete inversion formula (6-17) exhibits periodicity in the same manner as the discrete transform; the period is defined by N samples of $g(kT)$. This property results from the periodic nature of $e^{j2\pi nk/N}$. Hence, $g(kT)$ is actually defined on the complete set of integers $k = 0, \pm 1, \pm 2, \ldots$ and is constrained by the identity

$$g(kT) = g[(rN + k)T] \qquad r = 0, \pm 1, \pm 2, \ldots \qquad (6\text{-}20)$$

In summary, the discrete Fourier transform pair is given by

$$g(kT) = \frac{1}{N} \sum_{n=0}^{N-1} G\left(\frac{n}{NT}\right) e^{j2\pi nk/N} \Longleftrightarrow G\left(\frac{n}{NT}\right) = \sum_{k=0}^{N-1} g(kT)\, e^{-j2\pi nk/N} \qquad (6\text{-}21)$$

It is important to remember that the pair (6-21) requires both the time and frequency domain functions to be periodic;

$$G\left(\frac{n}{NT}\right) = G\left[\frac{(rN + n)}{NT}\right] \qquad r = 0, \pm 1, \pm 2, \ldots \qquad (6\text{-}22)$$

$$g(kT) = g[(rN + k)T] \qquad r = 0, \pm 1, \pm 2, \ldots \qquad (6\text{-}23)$$

6-4 RELATIONSHIP BETWEEN THE DISCRETE AND CONTINUOUS FOURIER TRANSFORM

The discrete Fourier transform is of interest primarily because it approximates the continuous Fourier transform. Validity of this approximation is strictly a function of the waveform being analyzed. In this section we use graphical analysis to indicate for general classes of functions the degree of equivalence between the discrete and continuous transform. As will be stressed, differences in the two transforms arise because of the discrete transform requirement for sampling and truncation.

Band-Limited Periodic Waveforms: Truncation Interval Equal to Period

Consider the function $h(t)$ and its Fourier transform illustrated in Fig. 6-3(a). We wish to sample $h(t)$, truncate the sampled function to N samples, and apply the discrete Fourier transform Eq. (6-16). Rather than applying this equation directly, we will develop its application graphically. Waveform $h(t)$ is sampled by multiplication with the sampling function illustrated in Fig. 6-3(b). Sampled waveform $h(kT)$ and its Fourier transform are illustrated in Fig. 6-3(c). Note that for this example there is no aliasing. Also

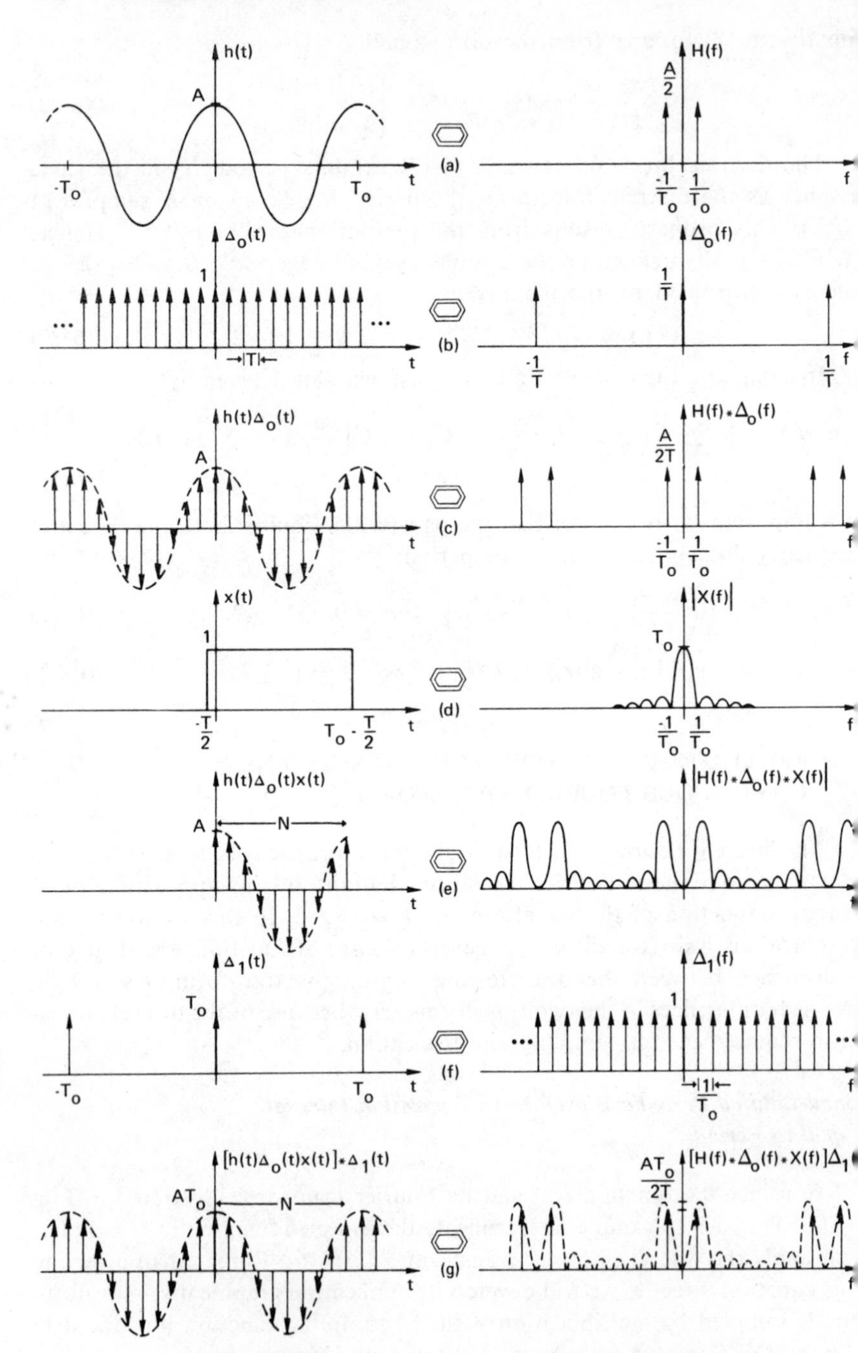

Figure 6-3. Discrete Fourier transform of a band-limited periodic waveform: truncation interval equal to period.

observe that as a result of time domain sampling, the frequency domain has been scaled by the factor $1/T$; the Fourier transform impulse now has an area of $A/2T$ rather than the original area of $A/2$. The sampled waveform is truncated by multiplication with the rectangular function illustrated in Fig. 6-3(d); Fig. 6-3(e) illustrates the sampled and truncated waveform. As shown, we chose the rectangular function so that the N sample values remaining after truncation equate to one period of the original waveform $h(t)$.

The Fourier transform of the finite length sampled waveform [Fig. 6-3(e)] is obtained by convolving the frequency domain impulse functions of Fig. 6-3(c) and the $\sin f/f$ frequency function of Fig. 6-3(d). Figure 6-3(e) illustrates the convolution results; an expanded view of this convolution is shown in Fig. 6-4(b). A $\sin f/f$ function (dashed line) is centered on each impulse of Fig. 6-4(a) and the resultant waveforms are additively combined (solid line) to form the convolution result.

With respect to the original transform $H(f)$, the convolved frequency function [Fig. 6-4(b)] is significantly distorted. However, when this function is sampled by the frequency sampling function illustrated in Fig. 6-3(f) the distortion is eliminated. This follows because the equidistant impulses of the frequency sampling function are separated by $1/T_0$; at these frequencies

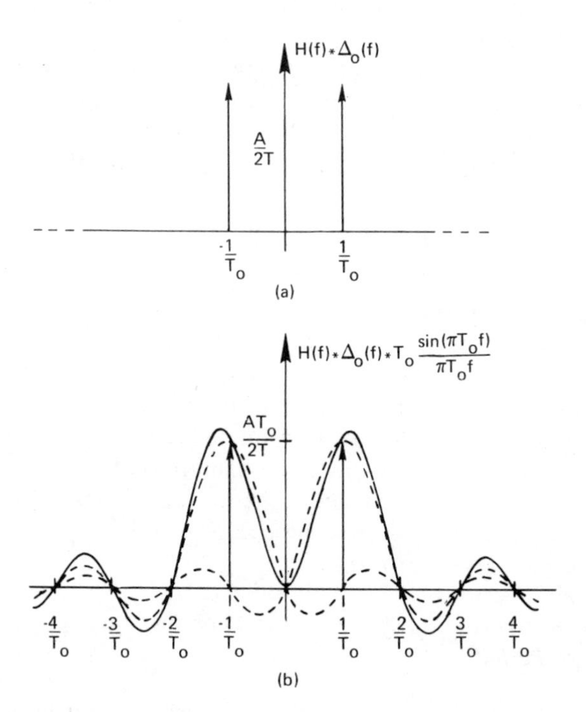

Figure 6-4. Expanded illustration of the convolution of Fig. 6-3(e).

the solid line of Fig. 6-4(b) is zero except at the frequency $\pm 1/T_0$. Frequency $\pm 1/T_0$ corresponds to the frequency domain impulses of the original frequency function $H(f)$. Because of time domain truncation, these impulses now have an area of $AT_0/2T$ rather than the original area of $A/2$. (Fig. 6-4(b) does not take into account that the Fourier transform of the truncation function $x(t)$ illustrated in Fig. 6-4(d) is actually a complex frequency function; however, had we considered a complex function, similar results would have been obtained.)

Multiplication of the frequency function of Fig. 6-3(e) and the frequency sampling function $\Delta_1(f)$ implies the convolution of the time functions shown in Figs. 6-3(e) and (f). Because the sampled truncated waveform [Fig. 6-3(e)] is exactly one period of the original waveform $h(t)$ and since the time domain impulse functions of Fig. 6-3(f) are separated by T_0, then their convolution yields a periodic function as illustrated in Fig. 6-3(g). This is simply the time domain equivalent to the previously discussed frequency sampling which yielded only a single impulse or frequency component. The time function of Fig. 6-3(g) has a maximum amplitude of AT_0, compared to the original maximum value of A as a result of frequency domain sampling.

Examination of Fig. 6-3(g) indicates that we have taken our original time function, sampled it, and then multiplied each sample by T_0. The Fourier transform of this function is related to the original frequency function by the factor $AT_0/2T$. Factor T_0 is common and can be eliminated. If we desire to compute the Fourier transform by means of the discrete Fourier transform, it is necessary to multiply the discrete time function by the factor T which yields the desired $A/2$ area for the frequency function; Eq. (6-16) thus becomes

$$H\left(\frac{n}{NT}\right) = T\sum_{k=0}^{N-1} h(kT)e^{-j2\pi nk/N} \tag{6-24}$$

We expect this result since the relationship (6-24) is simply the rectangular rule for integration of the continuous Fourier transform.

This example represents the only class of waveforms for which the discrete and continuous Fourier transforms are exactly the same within a scaling constant. Equivalence of the two transforms requires: (1) the time function $h(t)$ must be periodic, (2) $h(t)$ must be band-limited, (3) the sampling rate must be at least two times the largest frequency component of $h(t)$, and (4) the truncation function $x(t)$ must be non-zero over exactly one period (or integer multiple period) of $h(t)$.

Band-Limited Periodic Waveforms: Truncation Interval Not Equal to Period

If a periodic, band-limited function is sampled and truncated to consist of other than an integer multiple of the period, the resulting discrete and continuous Fourier transform will differ considerably. To examine this

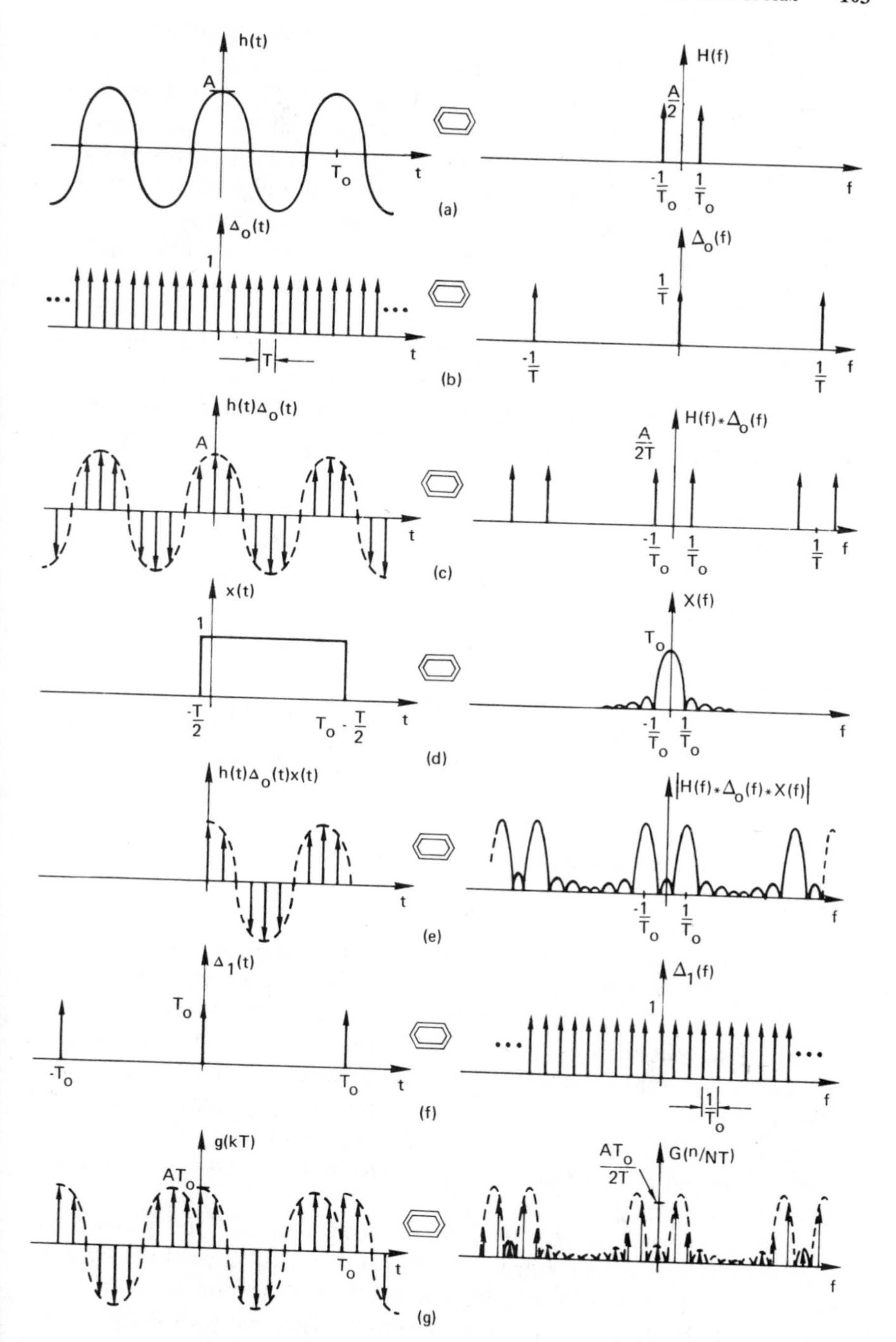

Figure 6-5. Discrete Fourier transform of a band-limited periodic waveform: truncation interval not equal to period.

effect, consider the illustrations of Fig. 6-5. This example differs from the preceding only in the frequency of the sinusoidal waveform $h(t)$. As before, function $h(t)$ is sampled [Fig. 6-5(c)] and truncated [Fig. 6-5(e)]. Note that the sampled, truncated function is not an integer multiple of the period of $h(t)$; therefore, when the time functions of Figs. 6-5(e) and (f) are convolved, the periodic waveform of Fig. 6-5(g) results. Although this function is periodic, it is not a replica of the original periodic function $h(t)$. We would not expect the Fourier transform of the time waveforms of Figs. 6-5(a) and (g) to be equivalent. It is of value to examine these same relationships in the frequency domain.

Fourier transform of the sampled truncated waveform of Fig. 6-5(e) is obtained by convolving the frequency domain impulse functions of Fig. 6-5(c) and the sin f/f function illustrated in Fig. 6-5(d). This convolution is graphically illustrated in an expanded view in Fig. 6-6. Sampling of the

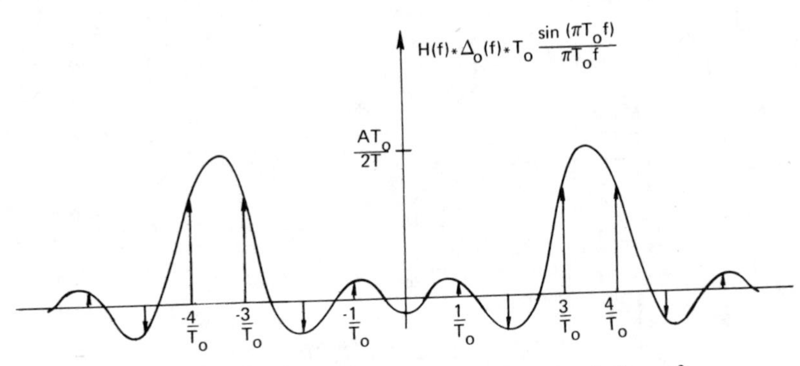

Figure 6-6. Expanded illustration of the convolution of Fig. 6-5(e).

resulting convolution at frequency intervals of $1/T_0$ yields the impulses as illustrated in Fig. 6-6 and, equivalently, Fig. 6-5(g). These sample values represent the Fourier transform of the periodic time waveform of Fig 6-5(g). Note that there is an impulse at zero frequency. This componen represents the average value of the truncated waveform; since the truncate waveform is not an even number of cycles, the average value is not expecte to be zero. The remaining frequency domain impulses occur because th zeros of the sin f/f function are not coincident with each sample value a was the case in the previous example.

This discrepancy between the continuous and discrete Fourier transform is probably the one most often encountered and least understood by user of the discrete Fourier transform. The effect of truncation at other than multiple of the period is to create a periodic function with sharp *discon tinuities* as illustrated in Fig. 6-5(g). Intuitively, we expect the introduction o these sharp changes in the time domain to result in additional frequency components in the frequency domain. Viewed in the frequency domain, time domain truncation is equivalent to the convolution of a $\sin f/f$ function with the single impulse representing the original frequency function $H(f)$. Consequently, the frequency function is no longer a single impulse but rather a continuous function of frequency with a local maximum centered at the original impulse and a series of other peaks which are termed sidelobes. These sidelobes are responsible for the additional frequency components which occur after frequency domain sampling. This effect is termed *leakage* and is inherent in the discrete Fourier transform because of the required time domain truncation. Techniques for reducing leakage will be explored in Sec. 9-5.

Finite Duration Waveforms

The preceding two examples have explored the relationship between the discrete and continuous Fourier transforms for band-limited periodic functions. Another class of functions of interest is that which is of finite duration such as the function $h(t)$ illustrated in Fig. 6-7. If $h(t)$ is time-limited, its Fourier transform cannot be band-limited; sampling must result in aliasing. It is necessary to choose the sample interval T such that aliasing is reduced to an acceptable range. As illustrated in Fig. 6-7(c), the sample interval T was chosen too large and as a result there is significant aliasing.

If the finite-length waveform is sampled and if N is chosen equal to the number of samples of the time-limited waveform, then it is not necessary to truncate in the time domain. Truncation is omitted and the Fourier transform of the time sampled function [Fig. 6-7(c)] is multiplied by $\Delta_1(f)$, the frequency domain sampling function. The time domain equivalent to this product is the convolution of the time functions shown in Figs. 6-7(c) and (d). The resulting waveform is periodic where a period is defined by the N samples of the original function, and thus is a replica of the original function. The Fourier transform of this periodic function is the sampled function illustrated in Fig. 6-7(e).

For this class of functions, if N is chosen equal to the number of samples of the finite-length function, then the only error is that introduced by aliasing. Errors introduced by aliasing are reduced by choosing the sample interval T sufficiently small. For this case the discrete Fourier transform sample values will agree (within a constant) reasonably well with samples of the continuous Fourier transform. Unfortunately, there exist few applications of discrete Fourier transform for this class of functions.

General Periodic Waveforms

Figure 6-7 can also be used to illustrate the relationship between the discrete and continuous Fourier transform for periodic functions which are

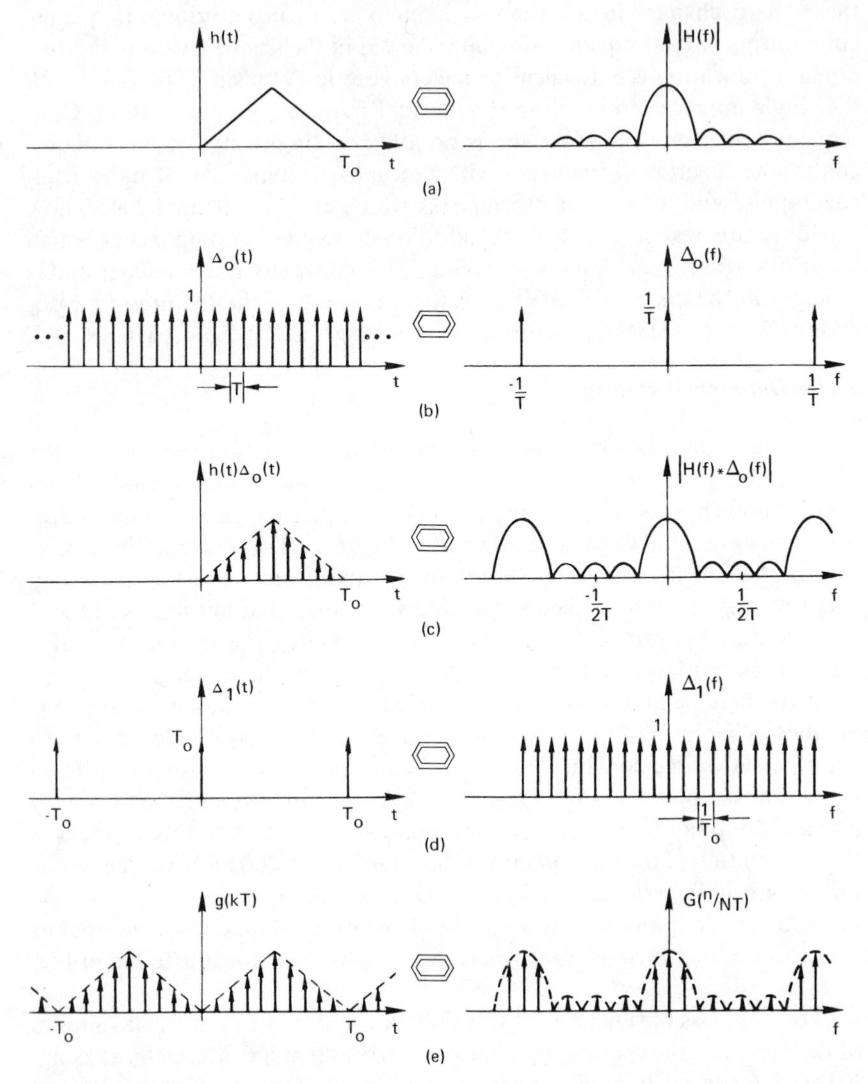

Figure 6-7. Discrete Fourier transform of a time-limited waveform.

not band-limited. Assume that $h(t)$, as illustrated in Fig. 6-7(a), is only one period of a periodic waveform. If this periodic waveform is sampled and truncated at exactly the period, then the resulting waveform will be identical to the time waveform of Fig. 6-7(c). Instead of the continuous frequency function as illustrated in Fig. 6-7(c), the frequency transform will be an

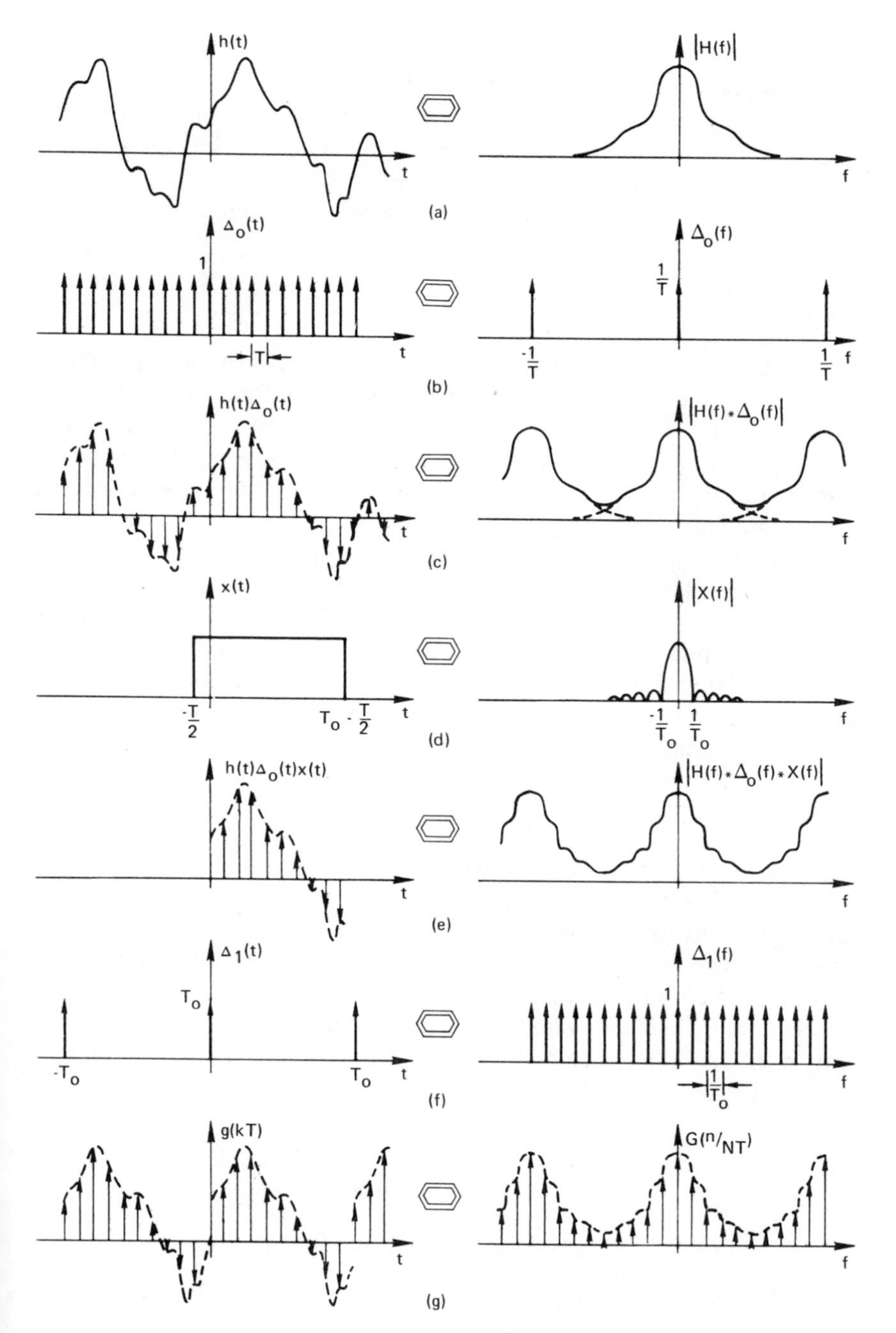

Figure 6-8. Discrete Fourier transform of a general waveform.

infinite series of equidistant impulses separated by $1/T_0$ whose areas ar given exactly by the continuous frequency function. Since the frequenc sampling function $\Delta_1(f)$, as illustrated in Fig. 6-7(d), is an infinite series c equidistant impulses separated by $1/T_0$ then the result is identical to those c Fig. 6-7(e). As before, the only error source is that of aliasing if the trunca tion function is chosen exactly equal to an integer multiple of the period. I the time domain truncation is not equal to a period, then results as describe previously are to be expected.

General Waveforms

The most important class of functions are those which are neither time limited nor band-limited. An example of this class of functions is illustrate in Fig. 6-8(a). Sampling results in the aliased frequency function illustrate in Fig. 6-8(c). Time domain truncation introduces rippling in the frequenc domain of Fig. 6-8(e). Frequency sampling results in the Fourier transforr pair illustrated in Fig. 6-8(g). The time domain function of this pair is periodic function where the period is defined by the N points of the origina function after sampling and truncation. The frequency domain function c the pair is also a periodic function where a period is defined by N points whos values differ from the original frequency function by the errors introduced i aliasing and time domain truncation. The aliasing error can be reduced to a acceptable level by decreasing the sample interval T. Procedures for reducin time domain truncation errors will be addressed in Sec. 9-5.

Summary

We have shown that if care is exercised, then there exist many application where the discrete Fourier transform can be employed to derive result essentially equivalent to the continuous Fourier transform. The most impor tant concept to keep in mind is that the discrete Fourier transform implie periodicity in both the time and frequency domain. If one will always remem ber that the N sample values of the time domain function represent on sample of a periodic function, then application of the discrete Fourie transform should result in few surprises.

PROBLEMS

6-1. Repeat the graphical development of Fig. 6-1 for the following functions:

a. $h(t) = |t| e^{-a|t|}$

b. $h(t) = 1 - |t| \qquad |t| \leq 1$

$\quad\;\; = 0 \qquad\qquad |t| > 1$

c. $h(t) = \cos(t)$

6-2. Retrace the development of the discrete Fourier transform [Eqs. (6-1) through (6-16)]. Write out in detail all steps of the derivation.

6-3. Repeat the graphical derivation of Fig. 6-3 for $h(t) = \sin(2\pi f_0 t)$. Show the effect of setting the truncation interval unequal to the period. What is the result of setting the truncation interval equal to two periods?

6-4. Consider Fig. 6-7. Assume that $h(t)\,\Delta_0(t)$ is represented by N non-zero samples. What is the effect of truncating $h(t)\,\Delta_0(t)$ so that only $3N/4$ non-zero samples are considered? What is the effect of truncating $h(t)\,\Delta_0(t)$ so that the N non-zero samples and $N/4$ zero samples are considered?

6-5. Repeat the graphical derivation of Fig. 6-7 for $h(t) = \sum_{n=-\infty}^{\infty} e^{-a|t-nT_0|}$. What are the error sources?

6-6. To establish the concept of rippling, perform the following graphical convolutions:

a. An impulse with $\dfrac{\sin t}{t}$

b. A narrow pulse with $\dfrac{\sin t}{t}$

c. A wide pulse with $\dfrac{\sin t}{t}$

d. A single triangle waveform with $\dfrac{\sin t}{t}$

6-7. Write out several terms of Eq. (6-19) to establish the orthogonality relationship.

6-8. The truncation interval is often termed the "record length." In terms of the record length, write an equation defining the "resolution" or frequency spacing of the frequency domain samples of the discrete Fourier transform.

6-9. Comment on the following: The discrete Fourier transform is analogous to a bank of band-pass filters.

REFERENCES

1. Cooley, J. W., P. A. W. Lewis, and P. D. Welch, "The Finite Fourier Transform," *IEEE Transactions on Audio and Electroacoustics* (June 1969), Vol. AU-17, No. 2, pp. 77–85.

2. Bergland, G. D., "A Guided Tour of the Fast Fourier Transform," *IEEE Spectrum* (July 1969), Vol. 6, No. 7, pp. 41–52.

3. Swick, D. A., "Discrete Finite Fourier Transforms—a Tutorial Approach." Washington, D. C., Naval Research Labs, NRL Dept. 6557, June 1967.

7

DISCRETE CONVOLUTION AND CORRELATION

Possibly the most important discrete Fourier transform properties are those of convolution and correlation. This follows from the fact that the importance of the fast Fourier transform is primarily a result of its efficiency in computing discrete convolution or correlation. In this chapter we examine analytically and graphically the discrete convolution and correlation equations. The relationship between discrete and continuous convolution is also explored in detail.

7-1 DISCRETE CONVOLUTION

Discrete convolution is defined by the summation

$$y(kT) = \sum_{i=0}^{N-1} x(iT)h[(k-i)T] \tag{7-1}$$

where both $x(kT)$ and $h(kT)$ are periodic functions with period N,

$$\begin{aligned} x(kT) &= x[(k+rN)T] \qquad r = 0, \pm 1, \pm 2, \ldots \\ h(kT) &= h[(k+rN)T] \qquad r = 0, \pm 1, \pm 2, \ldots \end{aligned} \tag{7-2}$$

For convenience of notation, discrete convolution is normally written as

$$y(kT) = x(kT) * h(kT) \tag{7-3}$$

To examine the discrete convolution equation, consider the illustrations of Fig. 7-1. Both functions $x(kT)$ and $h(kT)$ are periodic with period $N = 4$. From Eq. (7-1) functions $x(iT)$ and $h[(k-i)T]$ are required. Function $h(-iT)$ is the image of $h(iT)$ about the ordinate axis as illustrated in Fig.

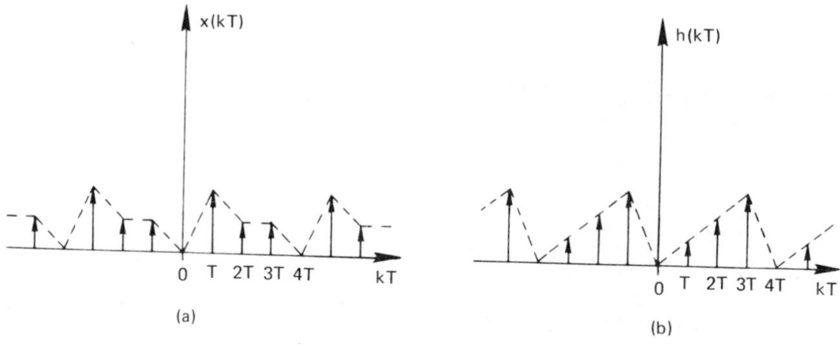

Figure 7-1. Example sampled waveforms to be convolved discretely.

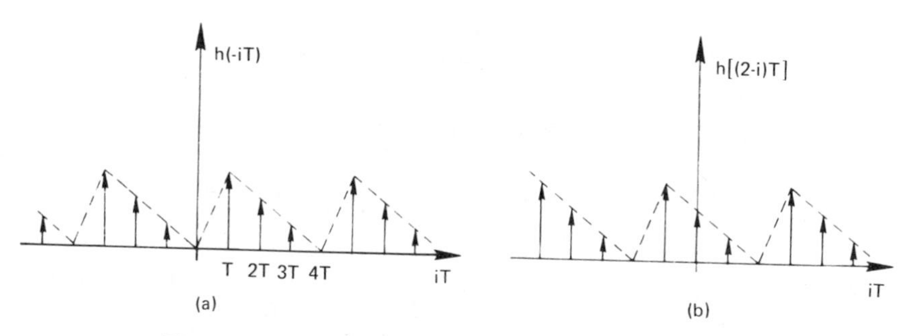

Figure 7-2. Graphical description of discrete convolution shifting operation.

7-2(a); function $h[(k - i)T]$ is simply the function $h(-iT)$ shifted by the amount kT. Figure 7-2(b) illustrates $h[(k - i)T]$ for the shift $2T$. Equation (7-1) is evaluated for each shift kT by performing the required multiplications and additions.

7-2 GRAPHICAL DISCRETE CONVOLUTION

The discrete convolution process is illustrated graphically in Fig. 7-3. Sample values of $x(kT)$ and $h(kT)$ are denoted by *dots* and *crosses*, respectively. Figure 7-3(a) illustrates the desired computation for $k = 0$. The value of each *dot* is multiplied by the value of the *cross* which occurs at the same abscissa value; these products are summed over the $N = 4$ discrete values indicated. Computation of Eq. (7-1) is graphically evaluated for $k = 1$ in Fig. 7-3(b); multiplication and summation is over the N points indicated. Figures 7-3(c) and (d) illustrate the convolution computation for $k = 2$ and $k = 3$, respectively. Note that for $k = 4$ [Fig. 7-3(e)], the terms multiplied

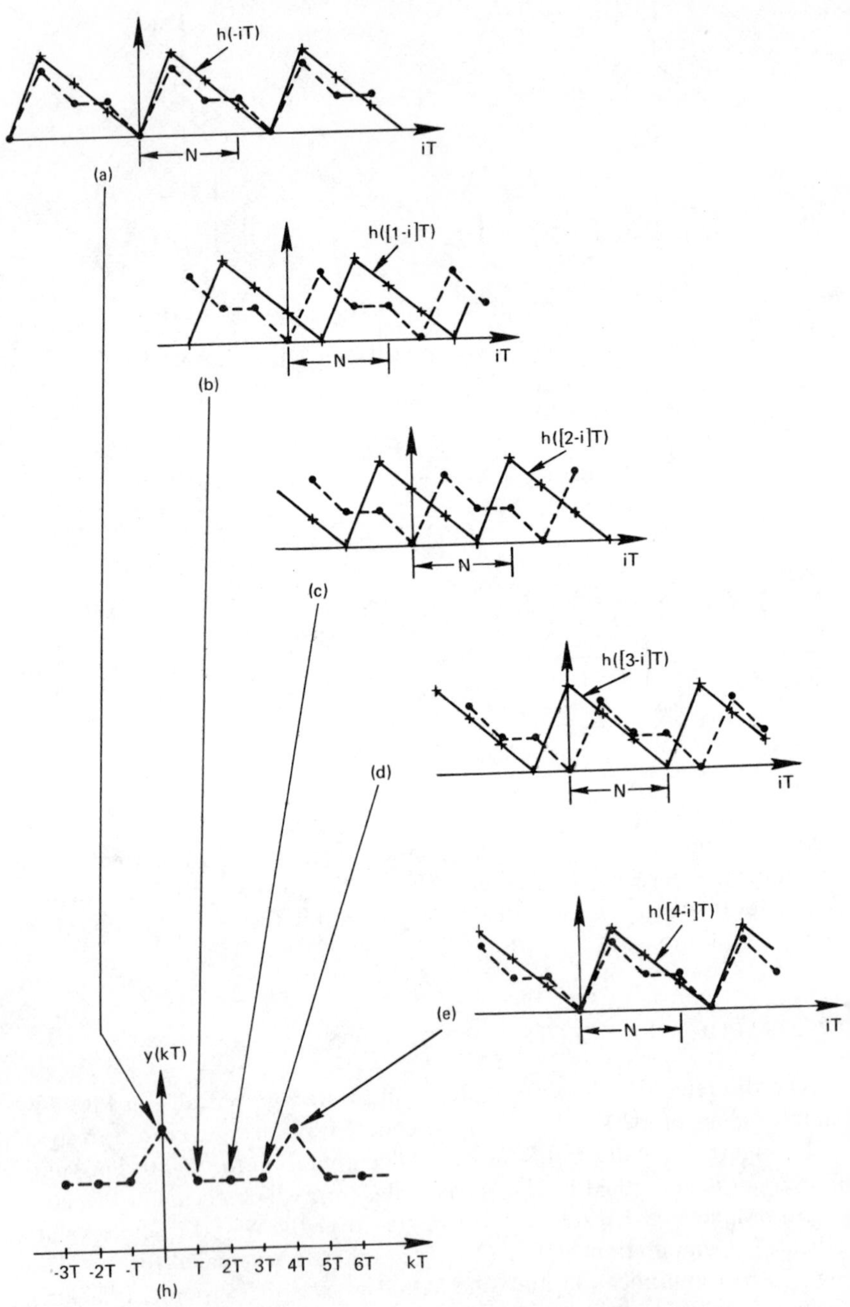

Figure 7-3. Graphical illustration of discrete convolution.

and summed are identical to those of Fig. 7-3(a). This is expected because both $x(kT)$ and $h(kT)$ are periodic with a period of four terms. Therefore

$$y(kT) = y[(k + rN)T] \qquad r = 0, \pm 1, \pm 2, \ldots \tag{7-4}$$

Steps for graphically computing the discrete convolution differ from those of continuous convolution only in that integration is replaced by summation. For discrete convolution these steps are: (1) folding, (2) displacement or shifting, (3) multiplication, and (4) summation. As in the convolution of continuous functions, either the sequence $x(kT)$ or $h(kT)$ can be selected for displacement. Equation (7-1) can be written equivalently as

$$y(kT) = \sum_{i=0}^{N-1} x[(k - i)T]h(iT) \tag{7-5}$$

7-3 RELATIONSHIP BETWEEN DISCRETE AND CONTINUOUS CONVOLUTION

If we only consider periodic functions represented by equally spaced impulse functions, discrete convolution relates identically to its continuous equivalent. This follows since we show in Appendix A (Eq. A-14) that continuous convolution is well defined for impulse functions.

The most important application of discrete convolution is not to sampled periodic functions but rather to approximate the continuous convolutions of general waveforms. For this reason, we will now explore in detail the relationship between discrete and continuous convolution.

Discrete Convolution of Finite Duration Waveforms

Consider the functions $x(t)$ and $h(t)$ as illustrated in Fig. 7-4(a). We wish to convolve these two functions both continuously and discretely, and to compare these results. Continuous convolution $y(t)$ of the two functions is also shown in Fig. 7-4(a). To evaluate the discrete convolution, we sample both $x(t)$ and $h(t)$ with sample interval T and we assume that both sample functions are periodic with period N. As illustrated in Fig. 7-4(b), the period has been chosen as $N = 9$ and both $x(kT)$ and $h(kT)$ are represented by $P = Q = 6$ samples; the remaining samples defining a period are set to zero. Figure 7-4(b) also illustrates the discrete convolution $y(kT)$ for the period $N = 9$; for this choice of N, the discrete convolution is a very poor approximation of the continuous case because the periodicity constraint results in an overlap of the desired periodic output. That is, we did not choose the period sufficiently large so that the convolution result of one period would not *interfere* or *overlap* the convolution result of the succeeding period. It is obvious that if we wish the discrete convolution to approximate continuous

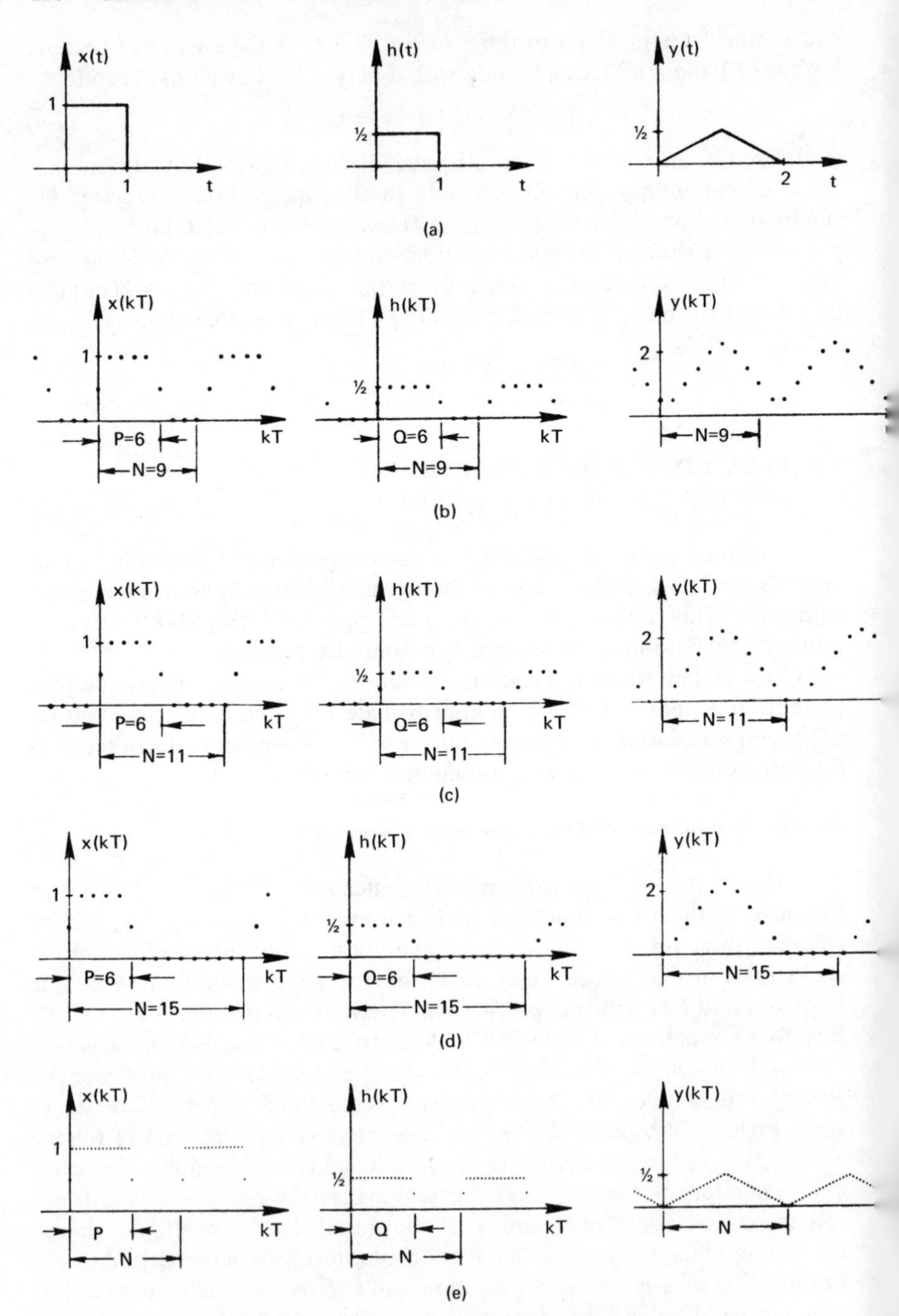

Figure 7-4. Relationship between discrete and continuous convolution: finite duration waveforms.

convolution then it is necessary that the period be chosen so that there is no overlap.

Choose the period according to the relationship

$$N = P + Q - 1 \tag{7-6}$$

This situation is illustrated in Fig. 7-4(c) where $N = P + Q - 1 = 11$. Note that for this choice of N there is no overlap in the resulting convolution. Equation (7-6) is based on the fact that the convolution of a function represented by P samples and a function represented by Q samples is a function described by $P + Q - 1$ samples.

There is no advantage in choosing $N > P + Q - 1$; as shown in Fig. 7-4(d), for $N = 15$ the non-zero values of the discrete convolution are identical to those of Fig. 7-4(c). As long as N is chosen according to Eq. (7-6), discrete convolution results in a periodic function where each period approximates the continuous convolution results.

Figure 7-4(c) illustrates the fact that discrete convolution results are scaled differently than that of continuous convolution. This scaling constant is T; modifying the discrete convolution Eq. (7-1) we obtain

$$y(kT) = T \sum_{i=0}^{N-1} x(iT)h[(k - i)T] \tag{7-7}$$

Relation (7-7) is simply the continuous convolution integral for time-limited functions evaluated by rectangular integration. Thus, for finite-length time functions, discrete convolution approximates continuous convolution within the error introduced by rectangular integration. As illustrated in Fig. 7-4(e), if the sample interval T is made sufficiently small, then the error introduced by the discrete convolution Eq. (7-7) is negligible.

Discrete Convolution of an Infinite and a Finite Duration Waveform

The previous example considered the case for which both $x(kT)$ and $h(kT)$ were of finite duration. Another case of interest is that where only one of the time functions to be convolved is finite. To explore the relationship of the discrete and continuous convolution for this case consider the illustrations of Fig. 7-5. As illustrated in Fig. 7-5(a), function $h(t)$ is assumed to be of finite duration and $x(t)$ infinite in duration; convolution of these two functions is shown in Fig. 7-5(b). Since the discrete convolution requires that both the sampled functions $x(kT)$ and $h(kT)$ be periodic, we obtain the illustrations of Fig. 7-5(c); period N has been chosen [Figs. 7-5(a) and (c)]. For $x(kT)$ infinite in duration, the imposed periodicity introduces what is known as *end effect*.

Compare the discrete convolution of Fig. 7-5(d) and the continuous convolution [Fig. 7-5(b)]. As illustrated, the two results agree reasonably well, with the exception of the first $Q - 1$ samples of the discrete convolution. To establish this fact more clearly consider the illustrations of Fig. 7-6. We

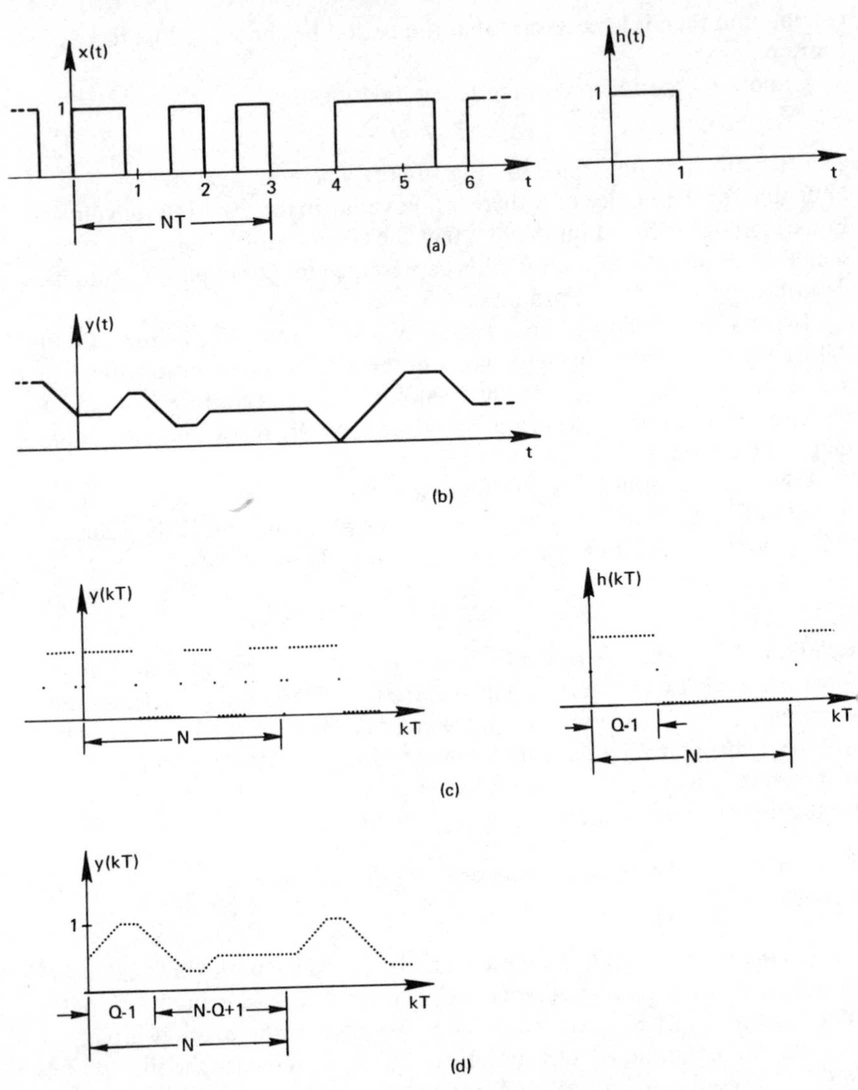

Figure 7-5. Relationship between discrete and continuous convolution: finite and infinite duration waveforms.

show only one period of $x(iT)$ and $h[(5 - i)T]$. To compute the discret convolution (7-1) for this shift, we multiply those samples of $x(iT)$ an $h[(5 - i)T]$ which occur at the same time [Fig. 7-6(a)] and add. The con volution result is a function of $x(iT)$ at both ends of the period. Such a con dition obviously has no meaningful interpretation in terms of the desire continuous convolution. Similar results are obtained for each shift valu

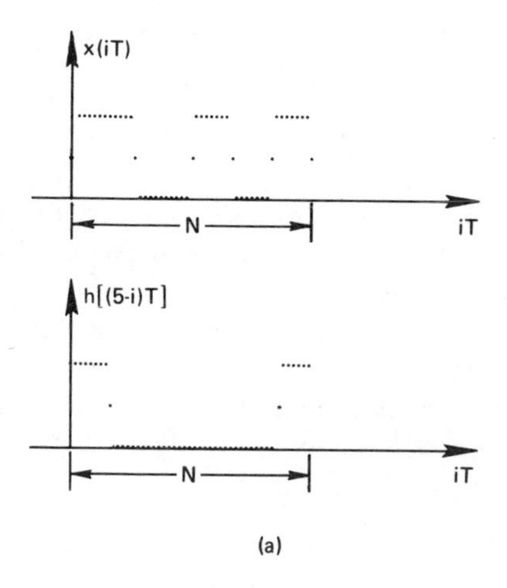

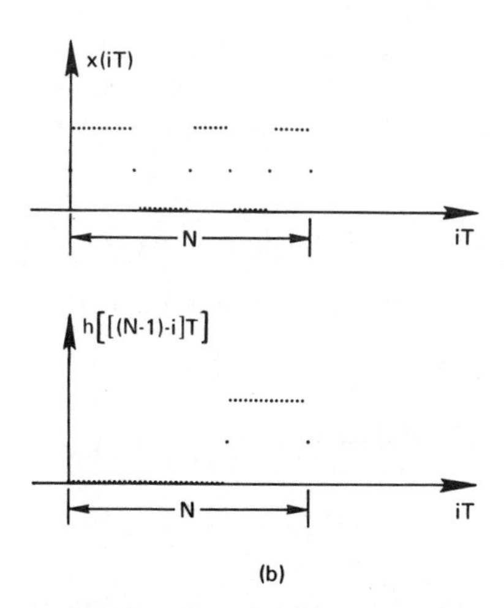

Figure 7-6. Illustration of the *end effect*.

until the Q points of $h(iT)$ are shifted by $Q - 1$; that is, the *end effect* exists until shift $k = Q - 1$.

Note that the *end effect* does not occur at the right end of the N sample values; functions $h(iT)$ for the shift $k = N - 1$ (therefore maximum shift) and $x(iT)$ are illustrated in Fig. 7-6(b). Multiplication of those values of $x(iT)$ and $h[(N - 1 - i)T]$ which occur at the same time and subsequent addition yields the desired convolution; the result is only a function of the correct values of $x(iT)$.

If the sample interval T is chosen sufficiently small, then the discrete convolution for this example class of functions will closely approximate the continuous convolution except for the *end effect.*

Summary

We have emphasized the point that discrete convolution is defined only for periodic functions. However, as illustrated graphically, the implications of this requirement are negligible if at least one of the functions to be convolved is of finite duration. For this case, discrete convolution is approximately equivalent to continuous convolution where the differences in the two methods are due to rectangular integration and to the *end effect.*

In general, it is impossible to discretely convolve two functions of infinite duration.

7-4 DISCRETE CONVOLUTION THEOREM

Analogous to Fourier transform theory, one of the most important properties of the discrete Fourier transform is exhibited by the discrete Fourier transform of Eq. (7-1). Discrete Fourier transformation of the convolution Eq. (7-1) yields the discrete convolution theorem which is expressed as

$$\sum_{i=0}^{N-1} x(iT)h[(k-i)T] \quad \Longleftrightarrow \quad X\left(\frac{n}{NT}\right)H\left(\frac{n}{NT}\right) \tag{7-8}$$

To establish this result, substitute (6-17) into the left-hand side of (7-8)

$$\begin{aligned}\sum_{i=0}^{N-1} x(iT)h[(k-i)T] &= \sum_{i=0}^{N-1} \frac{1}{N} \sum_{n=0}^{N-1} X\left(\frac{n}{NT}\right) e^{j2\pi ni/N} \frac{1}{N} \sum_{m=0}^{N-1} H\left(\frac{m}{NT}\right) e^{j2\pi m(k-i)/N} \\ &= \frac{1}{N} \sum_{n=0}^{N-1} \sum_{m=0}^{N-1} X\left(\frac{n}{NT}\right) H\left(\frac{m}{NT}\right) e^{j2\pi mk/N} \\ &\quad \times \frac{1}{N}\left[\sum_{i=0}^{N-1} e^{j2\pi in/N} e^{-j2\pi im/N}\right]\end{aligned} \tag{7-9}$$

The bracketed term of (7-9) is simply the orthogonality relationship (6-19) and is equal to N if $m = n$; therefore,

$$\sum_{i=0}^{N-1} x(iT)h[(k-i)T] = \frac{1}{N} \sum_{n=0}^{N-1} X\left(\frac{n}{NT}\right) H\left(\frac{n}{NT}\right) e^{j2\pi nk/N} \tag{7-10}$$

Thus, the discrete Fourier transform of the convolution of two periodic sampled functions with period N is equal to the product of the discrete Fourier transform of the periodic functions.

Recall that the discrete convolution Eq. (7-1) was defined in such a manner that those functions being convolved were assumed to be periodic. The underlying reason for this assumption is to enable the discrete convolution theorem (7-8) to hold. Since the discrete Fourier transform is only defined for periodic functions of time, then for (7-8) to hold, it is required that the time functions be periodic. Reiterating, discrete convolution is simply a special case of continuous convolution; discrete convolution requires both functions to be sampled and to be periodic.

The convolution waveforms illustrated in Sec. 7-3 could have been computed equivalently by means of the convolution theorem. If we compute the discrete Fourier transform of each of the periodic sequences $x(kT)$ and $h(kT)$, multiply the resulting transforms, and then compute the inverse discrete Fourier transform of this product, we will obtain identical results to those illustrated. As will be discussed in Chapter 13, it is normally faster computationally to use the discrete Fourier transform to compute the discrete convolution if the fast Fourier transform algorithm is employed.

7-5 DISCRETE CORRELATION

Discrete correlation is defined as

$$z(kT) = \sum_{i=0}^{N-1} x(iT)h[(k+i)T] \tag{7-11}$$

where $x(kT)$, $h(kT)$, and $z(kT)$ are periodic functions.

$$\begin{aligned} z(kT) &= z[(k+rN)T] \qquad r = 0, \pm 1, \pm 2, \ldots \\ x(kT) &= x[(k+rN)T] \qquad r = 0, \pm 1, \pm 2, \ldots \\ h(kT) &= h[(k+rN)T] \qquad r = 0, \pm 1, \pm 2, \ldots \end{aligned} \tag{7-12}$$

As in the continuous case, discrete correlation differs from convolution in that there is no folding operation. Hence, the remaining rules for displacement, multiplication and summation are performed exactly as for the case of discrete convolution.

The discrete correlation theorem transform pair is given by

$$\sum_{i=0}^{N-1} x(iT)h[(k+i)T] \quad \Longleftrightarrow \quad X^*\left(\frac{n}{NT}\right)H\left(\frac{n}{NT}\right) \tag{7-13}$$

The proof of (7-13) is delayed until Sec. 8-11 to take advantage of additional properties developed in that section.

7-6 GRAPHICAL DISCRETE CORRELATION

To illustrate the process of discrete correlation or *lagged products*, as it sometimes is referred, consider Fig. 7-7. The discrete functions to be correlated are shown in Fig. 7-7(a). According to the rules for correlation, we shift, multiply, and sum as illustrated in Figs. 7-7(b), (c), and (d), respectively. Compare with the results of Ex. 4-7.

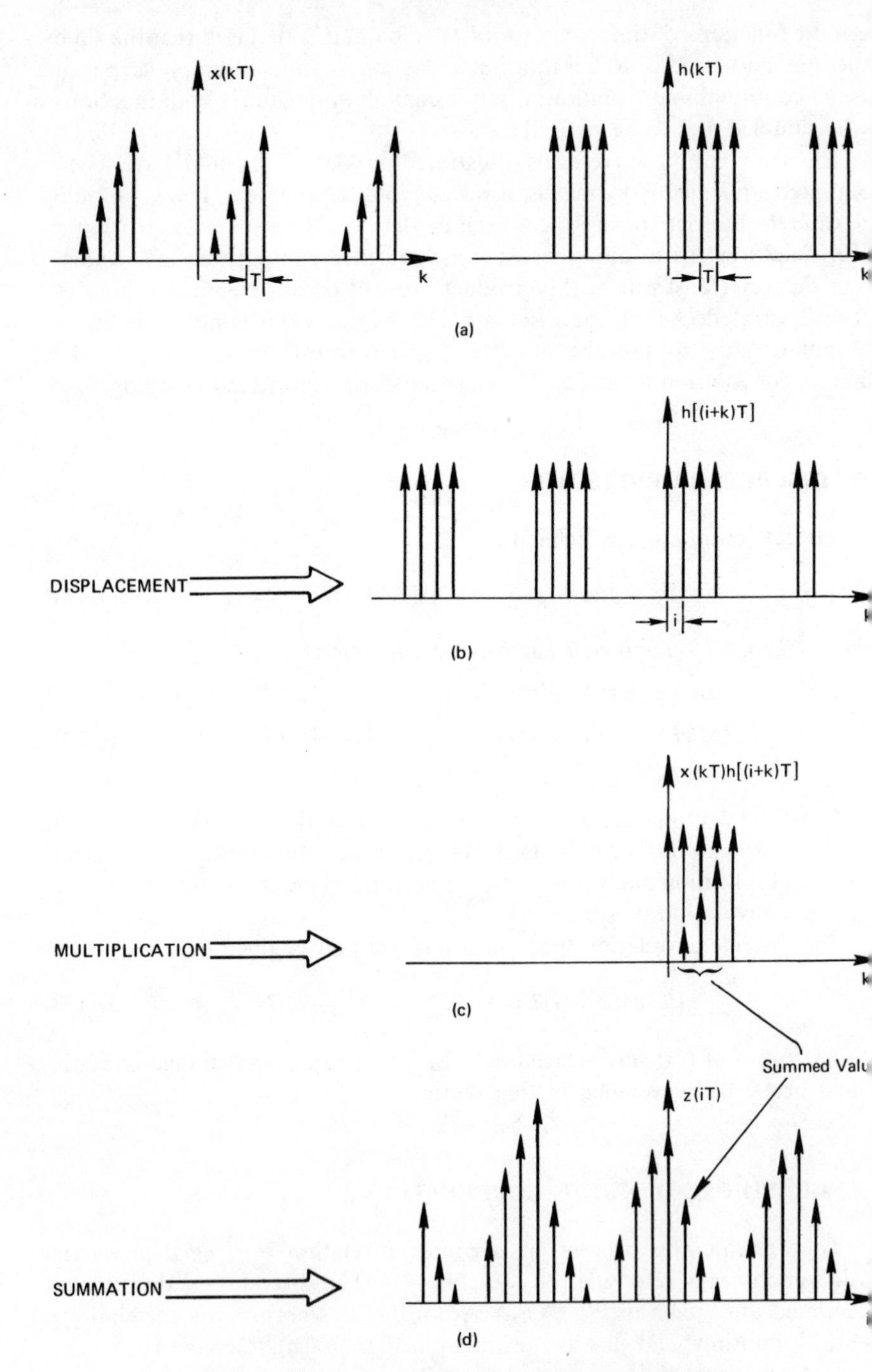

Figure 7-7. Graphical illustration of discrete correlation.

PROBLEMS

7-1. Let

$$
\begin{aligned}
x(kT) &= e^{-kT} && k = 0, 1, 2, 3\\
&= 0 && k = 4, 5, \dots, N\\
&= x[(k + rN)T] && r = 0, \pm 1, \pm 2, \dots
\end{aligned}
$$

and

$$
\begin{aligned}
h(kT) &= 1 && k = 0, 1, 2\\
&= 0 && k = 3, 4, \dots, N\\
&= h[(k + rN)T] && r = 0, \pm 1, \pm 2, \dots
\end{aligned}
$$

With $T = 1$, graphically and analytically determine $x(kT) * h(kT)$. Choose N less than, equal to, and greater than Eq. (7-6).

7-2. Consider the continuous functions $x(t)$ and $h(t)$ illustrated in Fig. 4-13(a). Sample both functions with sample interval $T = T_0/4$ and assume both sample functions are periodic with period N. Choose N according to relationship (7-6). Determine $x(kT) * h(kT)$ both analytically and graphically. Investigate the results of an incorrect choice of N. Compare results with continuous convolution results.

7-3. Repeat Problem 7-2 for Figs. 4-13(b), (c), (d), (e), and (f).

7-4. Refer to Fig. 7-5. Let $x(t)$ be defined as illustrated in Fig. 7-5(a). Function $h(t)$ is given as:

a. $h(t) = \delta(t)$

b. $h(t) = \delta(t) + \delta\left(t - \frac{3}{2}\right)$

c.
$$
\begin{aligned}
h(t) &= 0 && t < 0\\
&= 1 && 0 < t < \tfrac{1}{2}\\
&= 0 && \tfrac{1}{2} < t < 1\\
&= 1 && 1 < t < \tfrac{3}{2}\\
&= 0 && t > \tfrac{3}{2}
\end{aligned}
$$

Following Fig. 7-5, graphically determine the discrete convolution in each case. Compare the discrete and continuous convolution in each case. Investigate the end effect in each case.

7-5. It is desired to discretely convolve a finite duration and an infinite duration waveform. Assume that a hardware device is to be used which is limited in capacity to N sample values of each function. Describe a procedure which allows one to perform successive N point discrete convolutions and combine the two to eliminate the end effect. Demonstrate your concept by repeating the illustrations of Fig. 7-5 for the case $NT = 1.5$. Successively apply the developed technique to determine the discrete convolution $y(kT)$ for $0 \leq kT \leq 3$.

7-6. Derive the discrete convolution theorem for the following:

a. $\sum_{i=0}^{N-1} h(iT)x[(k - i)T]$

b. $\sum_{i=0}^{N-1} h(iT)h[(k-i)T]$

7-7. Let $x(kT)$ and $h(kT)$ be defined by Problem 7-1. Determine the discrete correlation (7-11) both analytically and graphically. What are the contraints on N?

7-8. Repeat Problem 7-2 for discrete correlation.

7-9. Repeat Problem 7-3 for discrete correlation.

7-10. Repeat Problem 7-4 for discrete correlation.

REFERENCES

1. Cooley, J. W., P. A. W. Lewis, and P. D. Welch, "The Finite Fourier Transform," *IEEE Transactions on Audio and Electroacoustics* (June 1969), Vol. AU–17, No. 2, pp. 77–85.

2. Gold, B., and C. Rader, *Digital Processing of Signals.* New York: McGraw-Hill, 1969.

3. Gupta, S., *Transform and State Variable Methods in Linear Systems.* New York: Wiley, 1966.

4. Papoulis, A., *The Fourier Integral and Its Applications.* New York: McGraw-Hill, 1962.

8

DISCRETE FOURIER TRANSFORM PROPERTIES

Those properties established for the Fourier transform in Chapter 3 can be extended to the discrete Fourier transform. This follows from the fact that we have shown that the discrete Fourier transform is simply a special case of the Fourier transform. It is possible then to simply restate these properties with the appropriate notation as required by the discrete Fourier transform. We believe that familiarization with the manipulation of discrete functions is invaluable for future developments and for this reason the pertinent properties are developed from a discrete viewpoint.

We replace kT by k and n/NT by n for convenience of notation. This concise notation will be used throughout the remainder of this book.

8-1 LINEARITY

If $x(k)$ and $y(k)$ have discrete Fourier transform $X(n)$ and $Y(n)$, respectively, then

$$x(k) + y(k) \quad \Longleftrightarrow \quad X(n) + Y(n) \tag{8-1}$$

Discrete Fourier transform pair (8-1) follows directly from the discrete Fourier transform pair (6-21).

8-2 SYMMETRY

If $h(k)$ and $H(n)$ are a discrete Fourier transform pair then

$$\frac{1}{N}H(k) \quad \Longleftrightarrow \quad h(-n) \tag{8-2}$$

Discrete Fourier transform pair (8-2) is established by rewriting Eq. (6-17

$$h(-k) = \frac{1}{N} \sum_{k=0}^{N-1} H(n)\, e^{j2\pi n(-k)/N} \tag{8-3}$$

and by interchanging the parameters k and n

$$h(-n) = \frac{1}{N} \sum_{n=0}^{N-1} H(k)\, e^{-j2\pi nk/N} \tag{8-4}$$

8-3 TIME SHIFTING

If $h(k)$ is shifted by the integer i then

$$h(k-i) \quad \Longleftrightarrow \quad H(n)\, e^{-j2\pi ni/N} \tag{8-5}$$

To verify (8-5), substitute $r = k - i$ into the inverse discrete Fourier transform

$$h(r) = \frac{1}{N} \sum_{n=0}^{N-1} H(n)\, e^{j2\pi nr/N}$$

$$h(k-i) = \frac{1}{N} \sum_{n=0}^{N-1} H(n)\, e^{j2\pi n(k-i)/N}$$

$$= \frac{1}{N} \sum_{n=0}^{N-1} [H(n)\, e^{-j2\pi ni/N}]\, e^{j2\pi nk/N} \tag{8-6}$$

8-4 FREQUENCY SHIFTING

If $H(n)$ is shifted by the integer i, then its inverse discrete Fourier transform is multiplied by $e^{j2\pi ik/N}$

$$h(k)\, e^{j2\pi ik/N} \quad \Longleftrightarrow \quad H(n-i) \tag{8-7}$$

This discrete Fourier transform pair is established by substituting $r = n - i$ into the discrete Fourier transform

$$H(r) = \sum_{k=0}^{N-1} h(k)\, e^{-j2\pi rk/N}$$

$$H(n-i) = \sum_{k=0}^{N-1} h(k)\, e^{-j2\pi(n-i)k/N}$$

$$= \sum_{k=0}^{N-1} [h(k)\, e^{j2\pi ik/N}]\, e^{-j2\pi nk/N} \tag{8-8}$$

8-5 ALTERNATE INVERSION FORMULA

The discrete inversion formula (6-17) may also be written as

$$h(k) = \frac{1}{N} \left[\sum_{k=0}^{N-1} H^*(n)\, e^{-j2\pi nk/N} \right]^* \tag{8-9}$$

where* implies conjugation. To prove (8-9) we simply perform the indicated conjugation. Let $H(n) = R(n) + jI(n)$; hence $H^*(n) = R(n) - jI(n)$ and (8-9) becomes

$$\begin{aligned}
h(k) &= \frac{1}{N}\left[\sum_{n=0}^{N-1} [R(n) - jI(n)]\, e^{-j2\pi nk/N}\right]^* \\
&= \frac{1}{N}\left[\sum_{n=0}^{N-1} [R(n) - jI(n)]\left[\cos\frac{2\pi nk}{N} - j\sin\frac{2\pi nk}{N}\right]\right]^* \\
&= \frac{1}{N}\left[\sum_{n=0}^{N-1} R(n)\cos\frac{2\pi nk}{N} - I(n)\sin\frac{2\pi nk}{N}\right. \\
&\qquad \left. - j\sum_{n=0}^{N-1} R(n)\sin\frac{2\pi nk}{N} + I(n)\cos\frac{2\pi nk}{N}\right]^* \\
&= \frac{1}{N}\left[\sum_{n=0}^{N-1} R(n)\cos\frac{2\pi nk}{N} - I(n)\sin\frac{2\pi nk}{N}\right. \\
&\qquad \left. + j\sum_{n=0}^{N-1} R(n)\sin\frac{2\pi nk}{N} + I(n)\cos\frac{2\pi nk}{N}\right] \\
&= \frac{1}{N}\sum_{n=0}^{N-1} [R(n) + jI(n)]\left[\cos\frac{2\pi nk}{N} + j\sin\frac{2\pi nk}{N}\right] \\
&= \frac{1}{N}\sum_{n=0}^{N-1} H(n)\, e^{j2\pi nk/N}
\end{aligned} \tag{8-10}$$

The significance of the alternate inversion formula is that the discrete transform Eq. (8-10) can be used to compute both the Fourier transform and its inversion. If the computation is to be performed on a digital machine, one only needs to develop a single computer program.

8-6 EVEN FUNCTIONS

If $h_e(k)$ is an even function then $h_e(k) = h_e(-k)$ and the discrete Fourier transform of $h_e(k)$ is an even function and is real

$$h_e(k) \quad \Longleftrightarrow \quad R_e(n) = \sum_{n=0}^{N-1} h_e(k)\cos\frac{2\pi nk}{N} \tag{8-11}$$

To verify (8-11) we simply manipulate the defining relationships,

$$\begin{aligned}
H_e(n) &= \sum_{k=0}^{N-1} h_e(k)\, e^{-j2\pi nk/N} \\
&= \sum_{k=0}^{N-1} h_e(k)\cos\frac{2\pi nk}{N} + j\sum_{k=0}^{N-1} h_e(k)\sin\frac{2\pi nk}{N} \\
&= \sum_{k=0}^{N-1} h_e(k)\cos\frac{2\pi nk}{N} \\
&= R_e(n)
\end{aligned} \tag{8-12}$$

The imaginary summation is zero since the summation is over an even number of cycles of an odd function. Because $h_e(k)\cos(2\pi nk/N) = h_e(k)[\cos(2\pi(-n)k/N)]$, then $H_e(n) = H_e(-n)$ and the frequency function is even. The inversion formula is proven similarly. Hence, if $H(n)$ is given as a real and even function, then its inverse discrete Fourier transform is an even function.

8-7 ODD FUNCTIONS

If $h_0(k) = -h_0(-k)$ then $h_0(k)$ is an odd function and its discrete Fourier transform is an odd and imaginary function;

$$\begin{aligned} H_0(n) &= \sum_{k=0}^{N-1} h_0(k)\, e^{-j2\pi nk/N} \\ &= \sum_{k=0}^{N-1} h_0(k) \cos \frac{2\pi nk}{N} - j \sum_{k=0}^{N-1} h_0(k) \sin \frac{2\pi nk}{N} \\ &= -j \sum_{k=0}^{N-1} h_0(k) \sin \frac{2\pi nk}{N} \\ &= jI_0(n) \end{aligned} \tag{8-13}$$

The real summation is zero since summation is over an even number of cycles of an odd function. For $H(n)$ given as an odd and imaginary function, the proof that $h_0(k)$ is an odd function is established similarly; therefore

$$h_0(k) \quad ⬡ \quad jI_0(n) = -j \sum_{k=0}^{N-1} h_0(k) \sin \frac{2\pi nk}{N} \tag{8-14}$$

8-8 WAVEFORM DECOMPOSITION

To decompose an arbitrary function $h(k)$ into an even and an odd function we simply add and subtract the common function $h(-k)/2$.

$$\begin{aligned} h(k) &= \frac{h(k)}{2} + \frac{h(k)}{2} \\ &= \left[\frac{h(k)}{2} + \frac{h(-k)}{2}\right] + \left[\frac{h(k)}{2} - \frac{h(-k)}{2}\right] \\ &= h_e(k) + h_0(k) \end{aligned} \tag{8-15}$$

Terms in brackets satisfy the definition of an even and an odd function, respectively. Since $h(k)$ is periodic with period N then

$$h(-k) = h(N-k) \tag{8-16}$$

and

$$\begin{aligned} h_e(k) &= \frac{h(k)}{2} + \frac{h(N-k)}{2} \\ h_0(k) &= \frac{h(k)}{2} - \frac{h(N-k)}{2} \end{aligned} \tag{8-17}$$

For discrete periodic functions Eq. (8-17) is the desired relationship for decomposition. From (8-11) and (8-14) the discrete Fourier transform of (8-15) is

$$H(n) = R(n) + jI(n) = H_e(n) + H_0(n) \tag{8-18}$$

where

$$H_e(n) = R(n) \quad \text{and} \quad H_0(n) = jI(n) \tag{8-19}$$

8-9 COMPLEX TIME FUNCTIONS

If $h(k) = h_r(k) + jh_i(k)$ where $h_r(k)$ and $h_i(k)$ are respectively the real and imaginary part of the complex time function $h(k)$, then the discrete Fourier transform becomes

$$\begin{aligned} H(n) &= \sum_{k=0}^{N-1} [h_r(k) + jh_i(k)]\, e^{-j2\pi nk/N} \\ &= \sum_{k=0}^{N-1} h_r(k) \cos \frac{2\pi nk}{N} + h_i(k) \sin \frac{2\pi nk}{N} \\ &\quad - j\left[\sum_{k=0}^{N-1} h_r(k) \sin \frac{2\pi nk}{N} - h_i(k) \cos \frac{2\pi nk}{N}\right] \end{aligned} \tag{8-20}$$

The first expression of (8-20) is $R(n)$, the real part of the discrete transform, and the latter expression is $I(n)$, the imaginary part of the discrete transform. If $h(k)$ is real, then $h(k) = h_r(k)$, and from (8-20)

$$R_e(n) = \sum_{k=0}^{N-1} h_r(k) \cos \frac{2\pi nk}{N} \tag{8-21}$$

$$I_0(n) = -j \sum_{k=0}^{N-1} h_r(k) \sin \frac{2\pi nk}{N} \tag{8-22}$$

Note that $\cos(2\pi nk/N) = \cos(-2\pi nk/N)$; thus $R_e(n) = R_e(-n)$, and $R_e(n)$ is an even function. Similarly, $I_0(n) = -I_0(-n)$ and $I_0(n)$ is an odd function.

For $h(k)$ purely imaginary, $h(k) = jh_i(k)$ and from (8-20)

$$R_0(n) = \sum_{k=0}^{N-1} h_i(k) \sin \frac{2\pi nk}{N} \tag{8-23}$$

$$I_e(n) = \sum_{k=0}^{N-1} h_i(k) \cos \frac{2\pi nk}{N} \tag{8-24}$$

For $h(k)$ imaginary, the real part of its transform is odd and the imaginary part of its transform is even.

8-10 FREQUENCY CONVOLUTION THEOREM

Consider the frequency convolution

$$Y(n) = \sum_{i=0}^{N-1} X(i)H(n-i) \tag{8-25}$$

We can establish the frequency convolution theorem by substitution into (8-25)

$$\sum_{i=0}^{N-1} X(i)H(n-i) = \sum_{i=0}^{N-1}\left[\sum_{m=0}^{N-1} x(m)e^{-j2\pi mi/N}\right]\left[\sum_{k=0}^{N-1} h(k)e^{-j2\pi k(n-i)/N}\right]$$
$$= \sum_{m=0}^{N-1}\sum_{k=0}^{N-1} x(m)h(k)e^{-j2\pi kn/N}\left[\sum_{i=0}^{N-1} e^{-j2\pi mi/N} e^{j2\pi ki/N}\right] \quad (8\text{-}26)$$

The bracketed term of (8-26) is the orthogonality relationship (6-19) and is equal to N if $m = k$; therefore

$$\sum_{i=0}^{N-1} X(i)H(n-i) = N \sum_{k=0}^{N-1} x(k)h(k)e^{-j2\pi nk/N} \quad (8\text{-}27)$$

and the discrete transform pair is established

$$x(k)h(k) \quad ⬡ \quad \frac{1}{N}\sum_{i=0}^{N-1} X(i)H(n-i) \quad (8\text{-}28)$$

8-11 DISCRETE CORRELATION THEOREM

The transform pair

$$\sum_{i=0}^{N-1} x(i)\,h(k+i) \quad ⬡ \quad X^*(n)\,H(n) \quad (8\text{-}29)$$

is termed the discrete correlation theorem. By means of the correlation theorem, correlation may be determined equivalently in the transform domain. To verify this relationship, substitute the discrete Fourier transform into the left-hand side of (8-29)

$$\sum_{i=0}^{N-1} x(i)\,h(k+i) = \sum_{i=0}^{N-1}\left[\frac{1}{N}\sum_{n=0}^{N-1} X(n)e^{j2\pi in/N}\right]\frac{1}{N}\sum_{m=0}^{N-1} H(m)e^{j2\pi m(k+i)/N}$$
$$= \sum_{i=0}^{N-1}\left[\frac{1}{N}\sum_{n=0}^{N-1} X^*(n)e^{-j2\pi in/N}\right]^*\left[\frac{1}{N}\sum_{m=0}^{N-1} H(m)e^{j2\pi m(k+i)/N}\right]$$
$$(8\text{-}30)$$

where the alternate inversion formula (8-9) has been utilized to introduce the conjugate of $X(n)$. Note that the second conjugation indicated in (8-9) can be omitted if only real functions are considered. For this case Eq. (8-30) can be rewritten as

$$\sum_{i=0}^{N-1} x(i)\,h(k+i) = \frac{1}{N}\sum_{n=0}^{N-1}\sum_{m=0}^{N-1} X^*(n)\,H(m)\,e^{j2\pi mk/N}\left[\frac{1}{N}\sum_{i=0}^{N-1} e^{-j2\pi in/N} e^{j2\pi im/N}\right]$$
$$(8\text{-}31)$$

From the orthogonality relationship (6-19), the bracketed term is equal to N if $n = m$. Hence (8-31) becomes

$$\sum_{i=0}^{N-1} x(i)\,h(k+i) = \frac{1}{N}\sum_{n=0}^{N-1} X^*(n)\,H(n)\,e^{j2\pi nk/N} \quad (8\text{-}32)$$

TABLE 8–1

Fourier transform	Property	Discrete fourier transform
$x(t) + y(t) \Longleftrightarrow X(f) + Y(f)$	Linearity (3–2) (8–1)	$x(k) + y(k) \Longleftrightarrow X(n) + Y(n)$
$H(t) \Longleftrightarrow h(-f)$	Symmetry (3–6) (8–2)	$\frac{1}{N} H(k) \Longleftrightarrow h(-n)$
$h(t - t_0) \Longleftrightarrow H(f)e^{-j2\pi f t_0}$	Time shifting (3–21) (8–5)	$h(k - i) \Longleftrightarrow H(n)e^{-j2\pi ni/N}$
$h(t)e^{j2\pi t f_0} \Longleftrightarrow H(f - f_0)$	Frequency shifting (3–23) (8–7)	$h(k)e^{j2\pi ki/N} \Longleftrightarrow H(n - i)$
$\left[\int_{-\infty}^{\infty} H^*(f)e^{-j2\pi ft}\,df\right]^*$	Alternate inversion formula (3–25) (8–9)	$\left[\frac{1}{N}\sum_{n=0}^{N-1} H^*(n)e^{-j2\pi kn/N}\right]^*$
$h_e(t) \Longleftrightarrow R_e(f)$	Even functions (3–27) (8–11)	$h_e(k) \Longleftrightarrow R_e(n)$
$h_0(t) \Longleftrightarrow jI_0(f)$	Odd functions (3–32) (8–14)	$h_0(k) \Longleftrightarrow jI_0(n)$
$h(t) = h_e(t) + h_0(t) = \left[\frac{h(t)}{2} + \frac{h(-t)}{2}\right] + \left[\frac{h(t)}{2} - \frac{h(-t)}{2}\right]$	Decomposition (3–33) (8–15)	$h(k) = h_e(k) + h_0(k) = \left[\frac{h(k)}{2} + \frac{h(N-k)}{2}\right] + \left[\frac{h(k)}{2} - \frac{h(N-k)}{2}\right]$
$y(t) = \int_{-\infty}^{\infty} x(\tau)h(t - \tau)\,d\tau = x(t) * h(t)$	Convolution (4–1) (7–1)	$y(k) = \sum_{i=0}^{N-1} x(i)h(k - i) = x(k) * h(k)$
$y(t) * h(t) \Longleftrightarrow Y(f)H(f)$	Time convolution theorem (4–11) (7–8)	$y(k) * h(k) \Longleftrightarrow Y(n)H(n)$
$y(t) = \int_{-\infty}^{\infty} x(\tau)h(t + \tau)\,d\tau$	Correlation (4–20) (7–11)	$y(k) = \sum_{i=0}^{N-1} x(i)h(k + i)$
$y(t)h(t) \Longleftrightarrow Y(f) * H(f)$	Frequency convolution theorem (4–17) (8–28)	$y(k)h(k) \Longleftrightarrow \frac{1}{N} Y(n) * H(n)$
$\int_{-\infty}^{\infty} h^2(t)\,dt = \int_{-\infty}^{\infty} \lvert H(f)\rvert^2\,df$	Parseval's theorem (4–19) (8–33)	$\sum_{k=0}^{N-1} h^2(k) = \frac{1}{N}\sum_{n=0}^{N-1} \lvert H(n)\rvert^2$

8-12 PARSEVAL'S THEOREM

For discrete functions, the relationship between power as computed in the time domain and as computed in the frequency domain is given by

$$\sum_{k=0}^{N-1} h^2(k) = \frac{1}{N} \sum_{n=0}^{N-1} |H(n)|^2 \tag{8-33}$$

To prove this relationship, let $y(k) = h(k)\,h(k)$. The discrete transform of $y(k)$ is given by the frequency convolution theorem (8-28) as

$$\sum_{k=0}^{N-1} h^2(k) e^{-j2\pi nk/N} = \frac{1}{N} \sum_{i=0}^{N-1} H(i)\,H(n-i) \tag{8-34}$$

If we set $n = 0$, then (8-34) becomes

$$\begin{aligned}\sum_{k=0}^{N-1} h^2(k) &= \frac{1}{N} \sum_{i=0}^{N-1} H(i)\,H(-i) \\ &= \frac{1}{N} \sum_{i=0}^{N-1} |H(i)|^2 \end{aligned} \tag{8-35}$$

The last equality follows from (8-21) and (8-22).

8-13 SUMMARY OF PROPERTIES

For future reference, the basic discrete Fourier transform properties are summarized in Table 8–1. The continuous Fourier transform properties are also tabled for purposes of comparison. Appropriate equation numbers are listed in order that one can easily locate the continuous or discrete development of each property.

PROBLEMS

Let $x(k)$ and $y(k)$ be periodic discrete functions:

$$x(k) = \begin{cases} \frac{1}{2} & k = 0, 4 \\ 1 & k = 1, 2, 3 \\ 0 & k = 5, 6, 7 \end{cases}$$

$$x(k + 8r) = x(k) \qquad r = 0, \pm 1, \pm 2, \ldots$$

$$y(k) = x(k)$$

$$y(k + 8r) = y(k) \qquad r = 0, \pm 1, \pm 2, \ldots$$

8-1. Compute $X(n)$ and $Y(n)$. Add these results to determine $[X(n) + Y(n)]$. Determine $z(k) = x(k) + y(k)$. Compute $Z(n)$. Discuss your results in terms of the linearity property.

8-2. Demonstrate the symmetry property (8-2) for $x(k)$.

8-3. Compute the discrete Fourier transform of $x(k - 3)$. Compare results with those obtained from the time-shifting relationship (8-5).

8-4. Compute the inverse discrete Fourier transform of $X(n - 1)$. Repeat this computation by applying the frequency shifting theorem (8-7) and compare results.

8-5. Compute the inverse discrete Fourier transform of $X(n)$ using the alternate inversion formula (8-9).

8-6. Compute the discrete Fourier transform of $x(k - 2)$. Investigate the even-odd relationship of $x(k - 2)$ and the real-imaginary relationship of its discrete transform. Is Eq. (8-11) applicable?

8-7. Let $z(k) = x(k) - y(k - 4)$. Compute the discrete Fourier transform of $z(k)$. Investigate these results in view of Sec. 8-7.

8-8. Let $z(k) = y(k) + y(k - 2) - x(k - 4)$. Decompose $z(k)$ into even and odd functions both analytically and graphically. Demonstrate Eq. (8-18) with $z(k)$.

8-9. Demonstrate the frequency convolution theorem using $x(k)$ and $y(k)$.

8-10. Demonstrate the discrete correlation theorem using $x(k)$ and $y(k)$.

8-11. Demonstrate Parseval's theorem using $x(k)$.

REFERENCES

1. COOLEY, J. W., P. A. W. LEWIS, and P. D. WELCH, "The Finite Fourier Transform," *IEEE Transactions on Audio and Electroacoustics* (June 1969) Vol. AU–17, No. 2, pp. 77–85.

2. COOLEY, J. W., P. A. W. LEWIS, and P. D. WELCH, "Application of the Fast Fourier Transform to Computation of Fourier Integrals, Fourier Series, and Convolution Integrals," *IEEE Transactions on Audio and Electroacoustics* (June 1967), Vol. AU–15, No. 2, pp. 79–83.

3. GENTLEMAN, W. M., and G. SANDE, "Fast Fourier transforms for fun and profit," *AFIPS Proc.*, **1966** *Fall Joint Computer Conf.*, Vol. 29, pp. 563–578, Washington, D. C.: Spartan, 1966.

4. PAPOULIS, A., *The Fourier Integral and Its Applications*. New York: McGraw-Hill 1962.

5. SCHOENBERG, I. J., "The Finite Fourier Series and Elementary Geometry," *Am. Math. Monthly* (June–July 1950), Vo. 57, No. 6.

9

APPLYING THE DISCRETE FOURIER TRANSFORM

In Chapter 6 we developed the relationship between the discrete and continuous Fourier transforms. In this chapter we explore the mechanics of applying the discrete Fourier transform to the computation of Fourier transforms and Fourier series. As we will see, the primary concern is one of correctly interpreting these results.

9-1 FOURIER TRANSFORMS

To illustrate the application of the discrete Fourier transform to the computation of Fourier transforms, consider Fig. 9-1. We show in Fig. 9-1(a) the function e^{-t}. We wish to compute by means of the discrete Fourier transform an approximation to the Fourier transform of this function.

The first step in applying the discrete transform is to choose the number of samples N and the sample interval T. For $N = 32$ and $T = 0.25$ we show the samples of e^{-t} in Fig. 9-1(a). Note that we have defined the sample value at $t = 0$ to be consistent with Eq. (2-43) which states that the value of the function at a discontinuity must be defined to be the mid-value if the inverse Fourier transform is to hold.

We next compute the discrete Fourier transform

$$H\left(\frac{n}{NT}\right) = T \sum_{k=0}^{N-1} [e^{-kT}] e^{-j2\pi nk/N} \qquad n = 0, 1, \ldots, N-1 \qquad \text{(9-1)}$$

Note the scale factor T which is introduced to produce equivalence between the continuous and discrete transforms. These results are shown in Figs. 9-1(b) and (c). In Fig. 9-1(b) we show the real part of Fourier transform as

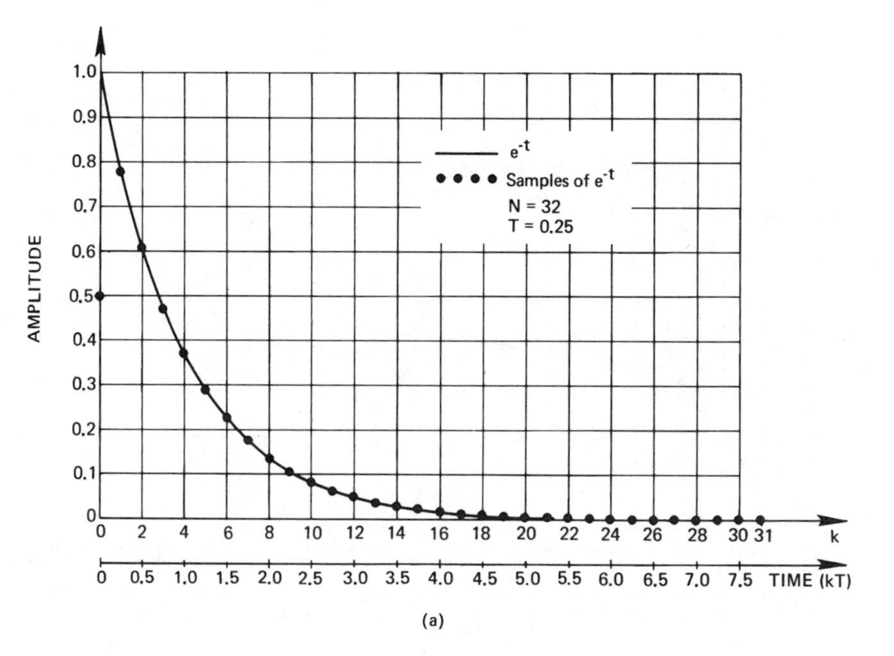

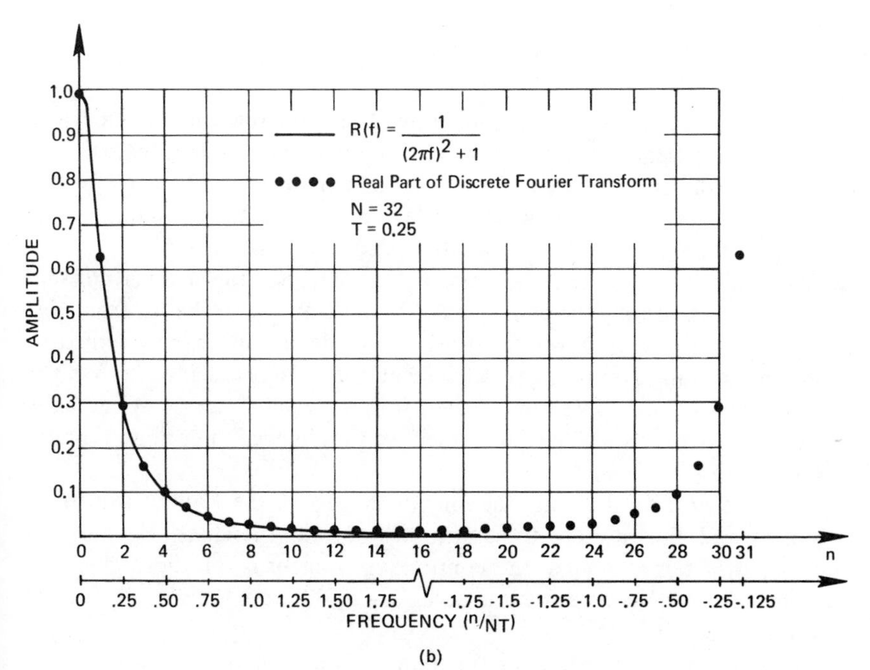

Figure 9-1. Example of Fourier transform computation via the discrete Fourier transform.

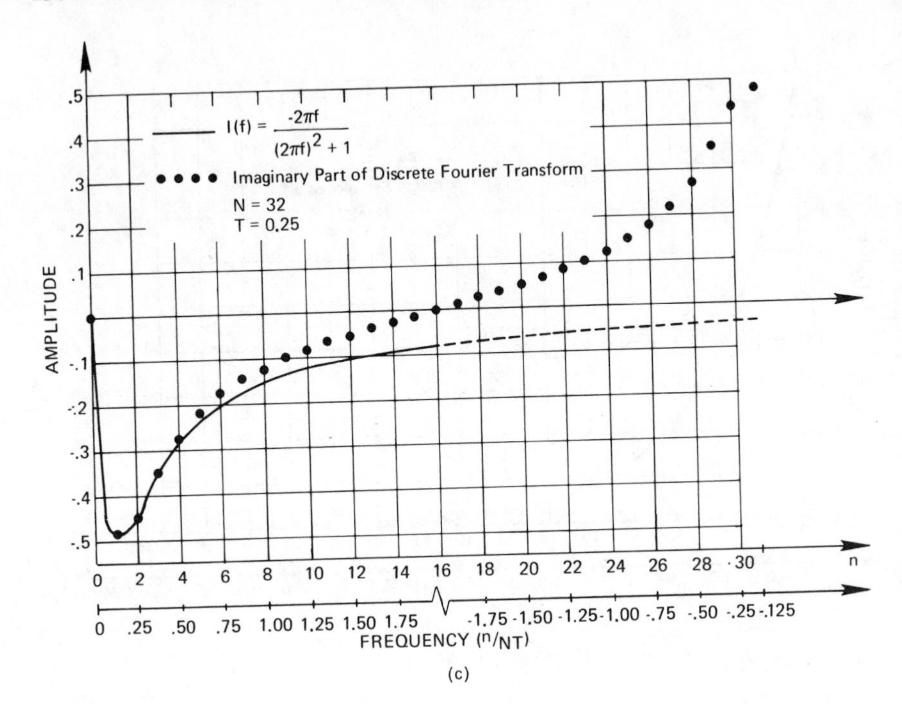

(c)

Figure 9-1. (continued).

determined in Ex. 2-1 and as computed by (9-1). Note that the discrete transform is symmetrical about $n = N/2$. This follows from the fact that the real part of the transform is even [Eq. (8-11)] and that the results for $n > N/2$ are simply negative frequency results. This latter point is emphasized by plotting a true frequency scale beneath the scale for parameter n.

We could have graphed the data of Fig. 9-1(b) in the manner conventionally used to display the continuous Fourier transform; that is, from $-f_0$ to $+f_0$. However, the conventional method of displaying results of the discrete Fourier transform is to graph the results of Eq. (9-1) as a function of the parameter n. As long as we remember that those results for $n > N/2$ actually relate to negative frequency results, then we should encounter no interpretation problems.

In Fig. 9-1(c) we illustrate the imaginary part of the Fourier transform (Ex. 2-1) and the discrete transform. As shown, the discrete transform approximates rather poorly the continuous transform for the higher frequencies. To reduce this error it is necessary to decrease the sample interval T and increase N.

We note that the imaginary function is odd with respect to $n = N/2$. This follows from Eq. (8-14). Repeating, those results for $n > N/2$ are to be interpreted as negative frequency results.

In summary, applying the discrete Fourier transform to the computation of the Fourier transform only requires that we exercise care in the choice of T and N and interpret the results correctly.

9-2 INVERSE FOURIER TRANSFORM APPROXIMATION

Assume that we are given the continuous real and imaginary frequency functions considered in the previous discussion and that we wish to determine the corresponding time function by means of the inverse discrete Fourier transform

$$h(kT) = \Delta f \sum_{n=0}^{N-1} [R(n\Delta f) + jI(n\Delta f)]e^{j2\pi nk/N} \qquad k = 0, 1, \ldots, N-1 \tag{9-2}$$

where Δf is the sample interval in frequency. Assume $N = 32$ and $\Delta f = 1/8$.

Since we know that $R(f)$, the real part of the complex frequency function, must be an even function then we *fold* $R(f)$ about the frequency $f = 2.0$ which corresponds to the sample point $n = N/2$. As shown in Fig. 9-2(a), we simply sample the frequency function up to the point $n = N/2$ and then *fold* these values about $n = N/2$ to obtain the remaining samples.

In Fig. 9-2(b) we illustrate the method for determining the N samples of the imaginary part of the frequency function. Because the imaginary frequency function is odd, we must not only *fold* about the sample value $N/2$ but also *flip* the results. To preserve symmetry, we set the sample at $n = N/2$ to zero.

Computation of (9-2) with the sampled function illustrated in Figs. 9-2(a) and (b) yields the inverse discrete Fourier transform. The result is a complex function whose imaginary part is approximately zero and whose real part is as shown in Fig. 9-2(c). We note that at $k = 0$ the result is approximately equal to the correct mid-value and reasonable agreement is obtained for all but the results for k large. Improvement can be obtained by reducing Δf and increasing N.

The key to using the discrete inverse Fourier transform for obtaining an approximation to continuous results is to specify the sampled frequency functions correctly. Figures 9-3(a) and (b) illustrate this correct method. One should observe the scale factor Δf which was required to give a correct approximation to continuous inverse Fourier transform results.

Equivalent results could have been obtained by using the alternate inversion formula (8-9). To use this relationship, we first conjugate the complex frequency function; that is, the imaginary sampled function illustrated in Fig. 9-2(b) is multiplied by -1. Since the resulting time function is

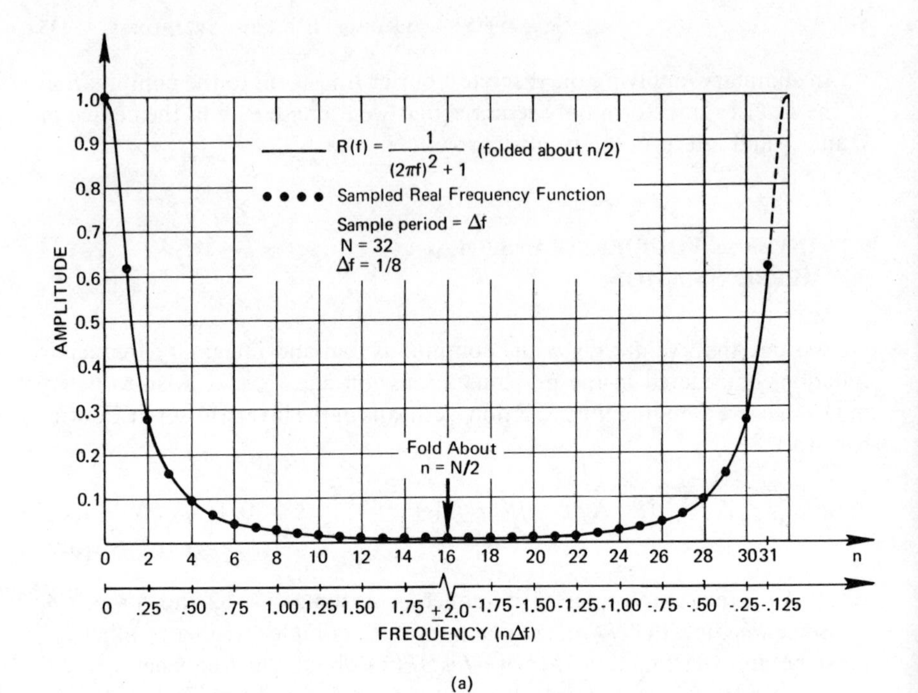

(a)

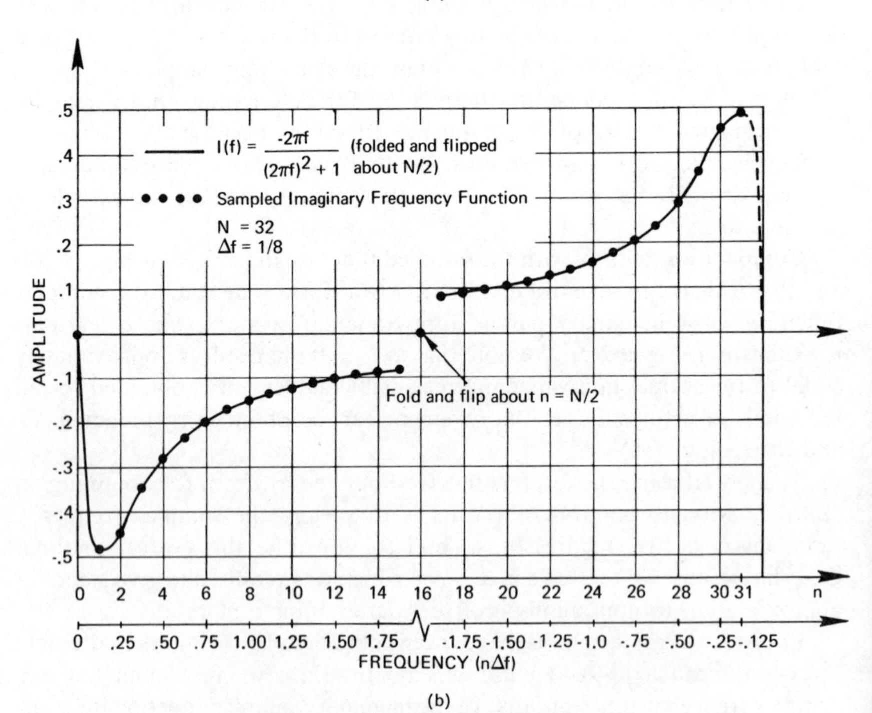

(b)

Figure 9-2. Example of inverse Fourier transform conputation via the discrete Fourier transform.

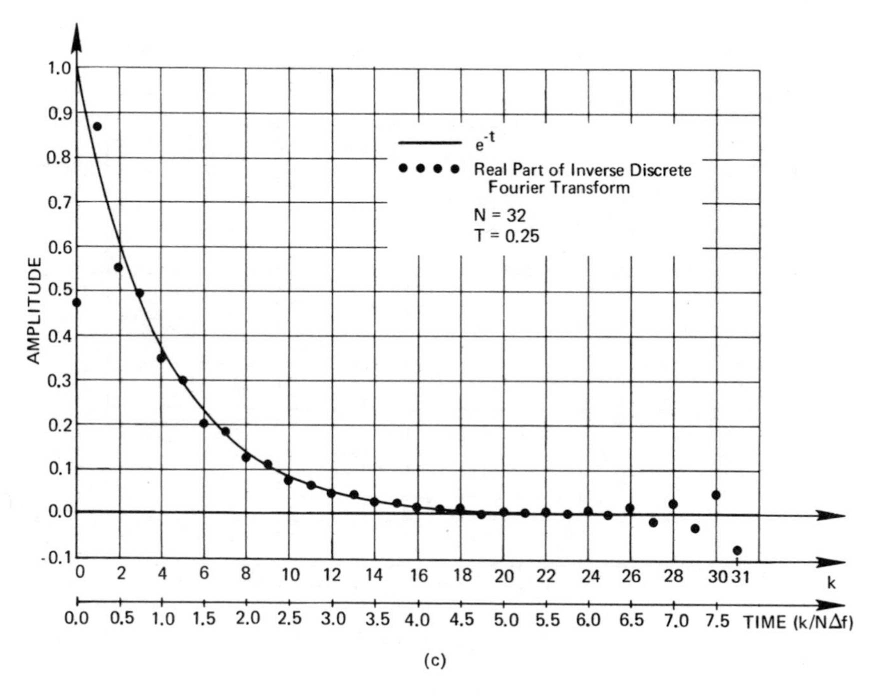

(c)

Figure 9-2. (continued).

real, the final conjugation operation illustrated in Eq. (8-9) can be omitted. Hence we compute

$$h(kT) = \Delta f \sum_{n=0}^{N-1} [R(n\Delta f) + j(-1)\, I(n\Delta f)]\, e^{-j2\pi nk/N} \tag{9-3}$$

which yields the time function illustrated in Fig. 9-2(c).

9-3 FOURIER SERIES HARMONIC ANALYSIS

Application of the discrete Fourier transform to the Fourier harmonic analysis of a waveform [Eq. (5-12)] requires that we compute

$$H\left(\frac{n}{NT}\right) = \frac{T}{(NT)} \sum_{k=0}^{N-1} h(kT)\, e^{-j2\pi nk/N} \tag{9-4}$$

where the divisor (NT) is the time duration or period of the lowest frequency harmonic to be determined. Recall from Chapter 6 that for (9-4) to yield valid results, the N sample values of $h(kT)$ must represent exactly one complete period of the periodic function $h(t)$.

Consider the square wave function illustrated in Fig. 9-3(a). As shown, the function has a period of 8 sec. Thus, if $N = 32$ then T must be chosen equal to 0.25 to insure that the 32 samples exactly equal one period.

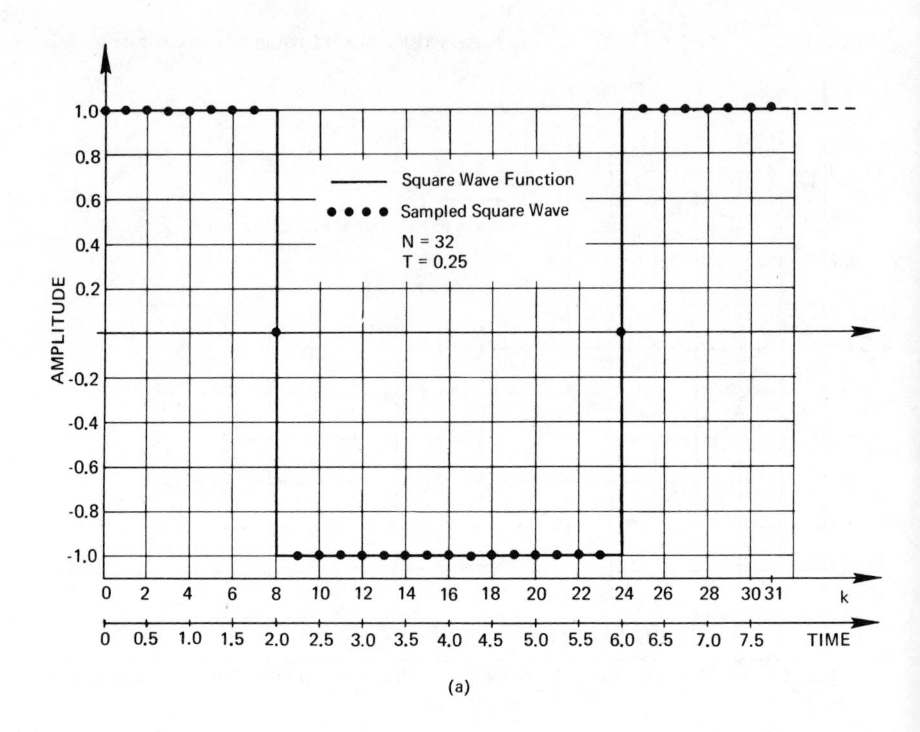

(a)

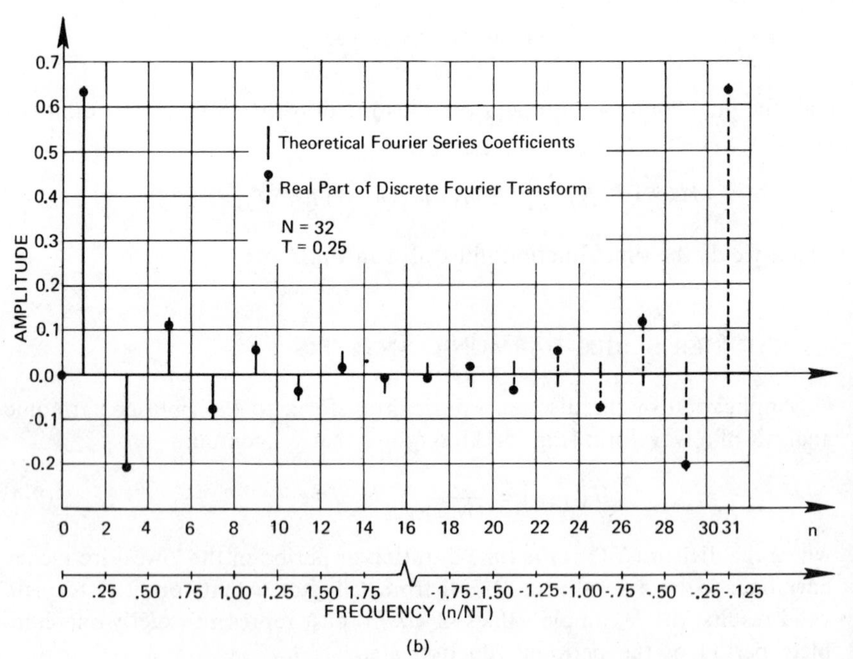

(b)

Figure 9-3. Example of Fourier series harmonic analysis via the discrete Fourier transform.

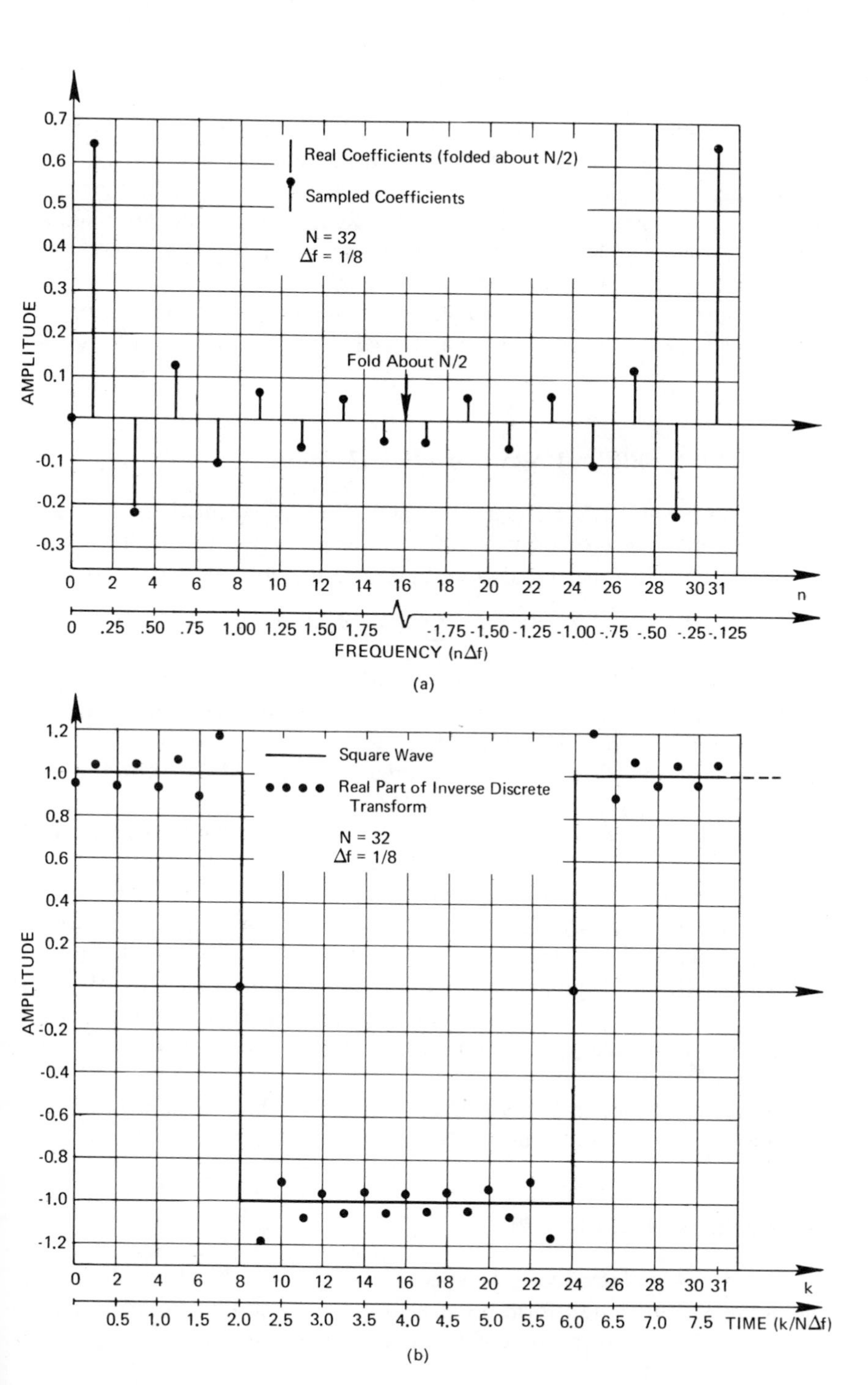

Figure 9-4. Example of Fourier series harmonic synthesis via the discrete Fourier transform.

Substitution of these sample values into (9-4) yields the results illustrated in Fig. 9-3(b). Vertical solid lines represent the magnitude of the harmonic coefficients as obtained theoretically from Eq. (5-12). As expected, the results are symmetrical about the point $n = N/2$. Reasonable results are obtained for the lower order harmonics. Accuracy can be improved for the higher harmonics by decreasing T and increasing N.

Note that we have introduced significant aliasing evidenced by the fact that the true coefficient values have appreciable magnitude at sample number $n = N/2$.

9-4 FOURIER SERIES HARMONIC SYNTHESIS

Harmonic synthesis refers to the procedure of calculating a periodic waveform given the coefficients of the Fourier series [Eq. (5-11)]. To accomplish this analysis using the discrete Fourier transform, we simply compute

$$h(kT) = \Delta f \sum_{n=0}^{N-1} H(n\Delta f)\, e^{j2\pi nk/N} \tag{9-5}$$

where Δf must be chosen as an integer multiple of the fundamental harmonic.

To apply (9-5) we must sample the real and imaginary coefficients consistent with the procedures discussed previously. If we consider the previous example, then only the real coefficients must be sampled. As shown in Fig. 9-4(a), these samples are folded about the point $N/2$. Note that we have in fact truncated the Fourier series because the sample values nearing $N/2$ still have appreciable magnitude.

Computation of (9-5) with the sample values shown in Fig. 9-4(a) yields the synthesized waveform illustrated in Fig. 9-4(b). As shown, the results tend to oscillate about the correct value. These oscillations are due to the well-known Gibbs phenomenon† which states that truncation in one domain leads to oscillations in the other domain. To decrease the magnitude of these oscillations, it is necessary to consider more hamonic coefficients; that is, increase N.

Results illustrated in Fig. 9-4(b) could also have been obtained by using the alternate inversion formula (8-9).

9-5 LEAKAGE REDUCTION

In Sec. 6-4 we introduced the effect termed leakage which is inherent in the discrete Fourier transform because of the required time domain truncation. Recall that the truncation of a periodic function at other than a multiple

†A. Papoulis, *The Fourier Integral and Its Applications* (New York: McGraw-Hill, 1962), p. 30.

of the period results in a sharp discontinuity in the time domain, or equivalently results in side-lobes in the frequency domain. These side-lobes are responsible for the additional frequency components which are termed leakage. In this section we will investigate the techniques for computing a discrete Fourier transform with minimum leakage.

For review, let us reconsider the developments illustrated in Fig. 6-3. Recall that time domain truncation of the sampled waveform [Fig. 6-3(d)] results in a frequency domain convolution with a $\sin(f)/f$ function. This convolution introduces additional components in the frequency domain because of the side-lobe characteristics of the $\sin(f)/f$ function. If the truncation interval is chosen equal to a multiple of the period, the frequency domain sampling function [Fig. 6-3(f)] is coincident with the zeros of the $\sin(f)/f$ function. As a result, the side-lobe characteristics of the $\sin(f)/f$ function do not alter the discrete Fourier transform results [Fig. 6-4(b)].

To illustrate this point we have computed the discrete Fourier transform of the cosine function illustrated in Fig. 9-5(a). For sample interval $T = 1.0$ and the number of samples $N = 32$, we also show in Fig. 9-5(a) samples of the cosine waveform. Note that the thirty-two samples define exactly four periods of the periodic waveform. In Fig. 9-5(b), we illustrate the magnitude of the discrete Fourier transform of these samples as computed by Eq. (9-4). The results are zero except at the desired frequency.

If the time truncation interval is not chosen equal to a multiple of the period, the side-lobe characteristics of the $\sin(f)/f$ frequency function result in a considerable difference in discrete and continuous Fourier transform results (Figs. 6-5 and 6-6). To illustrate this effect, consider the cosine waveform illustrated in Fig. 9-6(a). For $T = 1.0$ and $N = 32$, we also show the sampled waveform in Fig. 9-6(a). Note that the thirty-two points do not define a multiple of the period and as a result a sharp discontinuity has been introduced.

In Fig. 9-6(b) we show the magnitude of the discrete Fourier transform of the samples of Fig. 9-6(a). There exist non-zero frequency components at all discrete frequencies of the discrete transform. As stated previously, the additional frequency components are termed *leakage* and are a result of the side-lobe characteristics of the $\sin(f)/f$ function. To reduce leakage it is necessary to employ a time domain truncation function which has side-lobe characteristics which are of smaller magnitude than those of the $\sin(f)/f$ function. The smaller the side-lobes, the less leakage will affect the results of the discrete Fourier transform. Fortunately, there exist truncation functions which exhibit exactly the desired characteristics.

One particularly good truncation function is the Hanning [1] function illustrated in Fig. 9-7(a) and given by

$$x(t) = \frac{1}{2} - \frac{1}{2}\cos\frac{2\pi t}{T_c} \qquad 0 \leq t \leq T_c \tag{9-6}$$

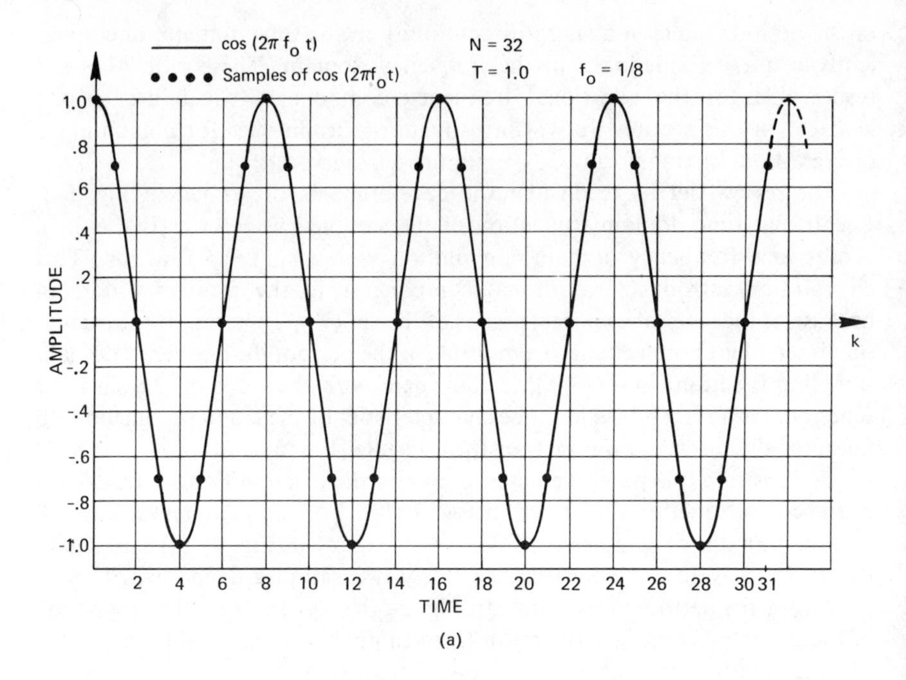

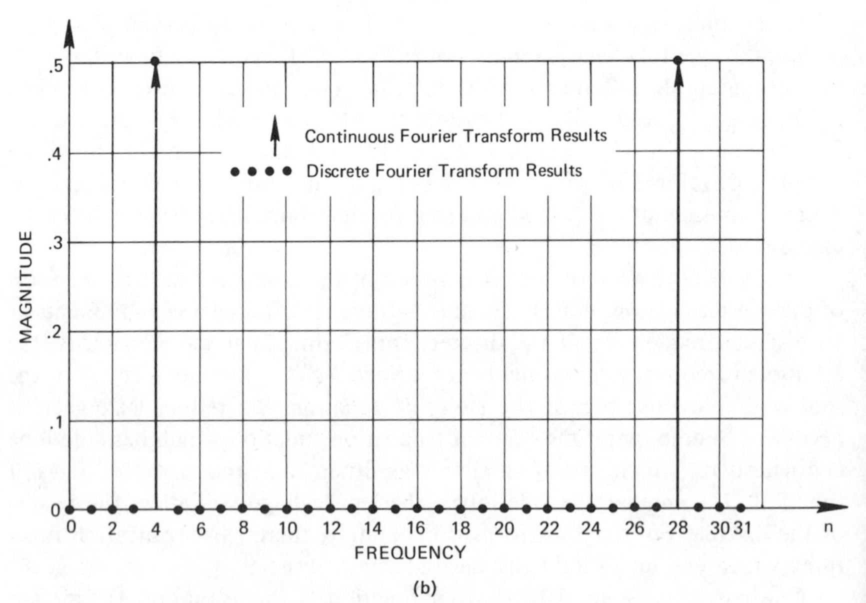

Figure 9-5. Fourier transform of a cosine waveform: truncation interval equal to a multitude of the period.

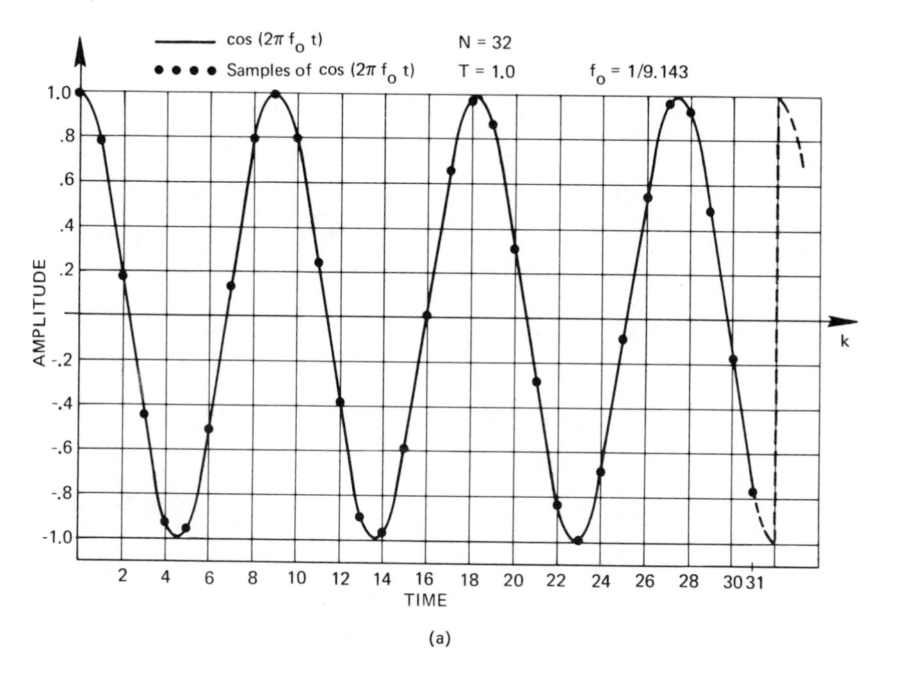

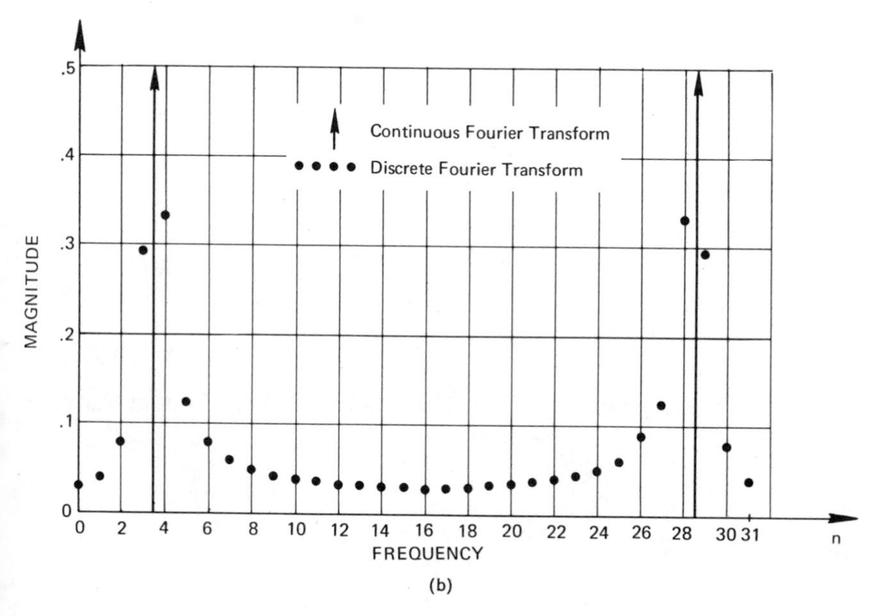

Figure 9-6. Fourier transform of a cosine waveform: truncation interval not equal to a multiple of the period.

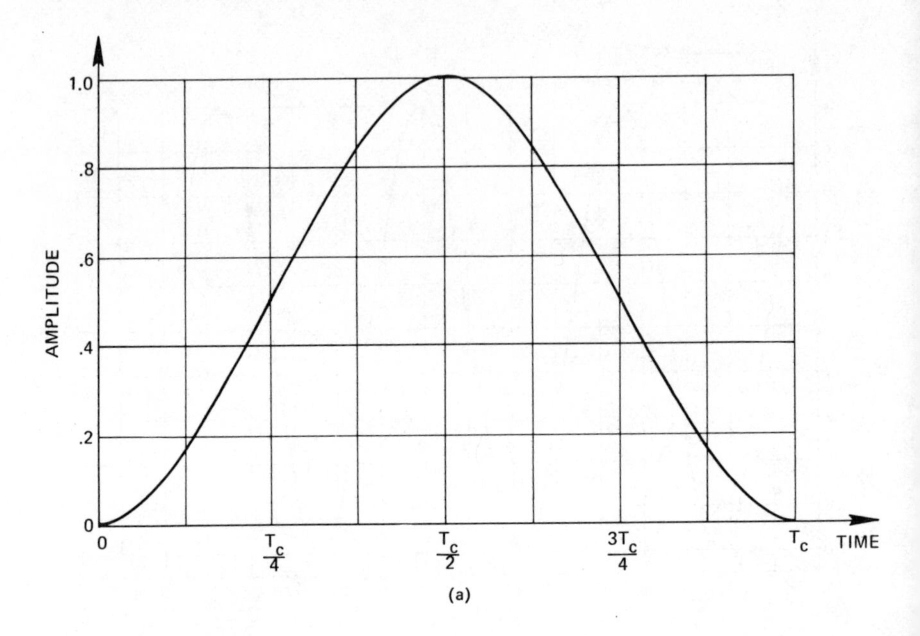

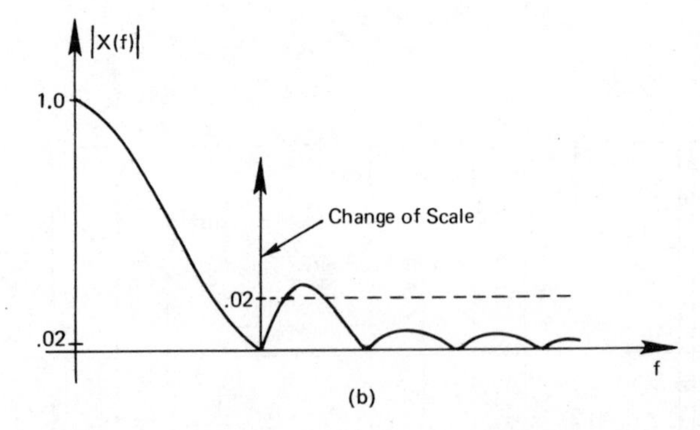

Figure 9-7. Hanning function Fourier transform pair.

where T_c is the truncation interval. The magnitude of the Fourier transform of the Hanning function is given by

$$|X(f)| = \frac{1}{2}Q(f) + \frac{1}{4}\left[Q\left(f + \frac{1}{T_c}\right) + Q\left(f - \frac{1}{T_c}\right)\right] \tag{9-7}$$

where

$$Q(f) = \frac{\sin(\pi T_c f)}{\pi f} \tag{9-8}$$

As shown in Fig. 9-7(b), this frequency function has very small side-lobes. Other truncation functions have similar properties [1]; however, we choose the Hanning function for its simplicity.

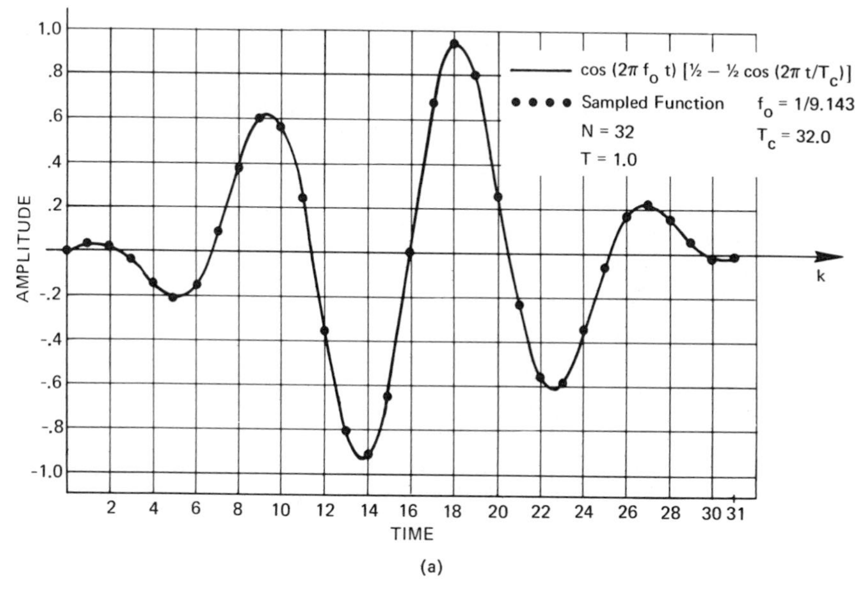

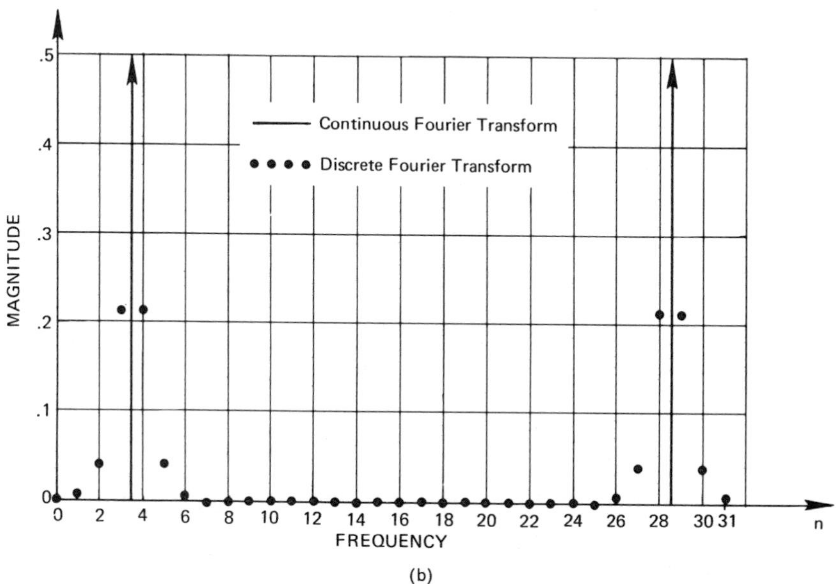

Figure 9-8. Example of applying the Hanning function to reduce leakage in the computation of the discrete Fourier transform.

Because of the low side-lobe characteristics of the Hanning function, we expect that its utilization will significantly reduce the leakage which results from time domain truncation. In Fig. 9-8(a) we show the cosine function of Fig. 9-6(a) multiplied by the Hanning truncation function illustrated in Fig. 9-7(a). Note that the effect of the Hanning function is to reduce the discontinuity which results from the rectangular truncation function.

Figure 9-8(b) illustrates the magnitude of the discrete Fourier transform of the samples of Fig. 9-8(a). As expected, the leakage illustrated in Fig. 9-6(b) has been significantly reduced because of the low side-lobe characteristics of the Hanning truncation function. The non-zero frequency components are considerably *broadened* or *smeared* with respect to the desired impulse function. Recall that this is to be expected since the effect of time domain truncation is to convolve the frequency impulse function with the Fourier transform of the truncation function. In general, the more one reduces the leakage, the broader or more smeared the results of the discrete Fourier transform appear. The Hanning function is an acceptable compromise.

PROBLEMS

9-1. Consider $h(t) = e^{-t}$. Sample $h(t)$ with $T = 1$ and $N = 4$. Set $h(0) = 0.5$. Compute Eq. (9-1) for $n = 0, 1, 2, 3$ and sketch the results. What is the correct interpretation of the results for $n = 2$ and 3? Note the real and imaginary relationships of the discrete frequency function. Do the results of Eq. (9-1) approximate closely continuous Fourier transform results? If not, why not?

9-2. Let $H(f) = R(f) + jI(f)$ where

$$R(f) = \frac{1}{(2\pi f)^2 + 1}$$

$$I(f) = \frac{-2\pi f}{(2\pi f)^2 + 1}$$

For $N = 4$ and $\Delta f = 1/4$, sketch the sampled functions $R(f)$ and $I(f)$ analogously to Figs. 9-2(a) and (b). Compute Eq. (9-2) and sketch the result.

9-3. Given the sampled functions $R(n\Delta f)$ and $I(n\Delta f)$ of Problem 9-2, compute the inverse discrete Fourier transform using the alternate inversion formula. Compare with the inverse transform results of Problem 9-2.

9-4. Consider the function $x(t)$ illustrated in Fig. 5-7(b). Sample $x(t)$ with $N = 6$. What is the sample interval T if we wish to apply the discrete Fourier transform to perform harmonic analysis of the waveform? Compute Eq. (9-4) and sketch the results. Compare with the Fourier series results of Chapter 5. Explain differences in the two results.

PROJECTS

The following projects will require access to a computer.

9-5. Develop a computer program which will compute the discrete Fourier transform of complex time domain waveforms. Use the alternate inversion formula to use this program to also compute the inverse discrete Fourier transform. Refer to this computer program as Program DFT (Discrete Fourier Transform).

9-6. Use Program DFT to validate each of the five example problems of Chapter 9.

9-7. Let $h(t) = e^{-t}$. Sample $h(t)$ with $T = 0.25$. Compute the discrete Fourier transform of $h(kT)$ for $N = 8$, 16, 32, and 64. Compare these results and explain the differences. Repeat for $T = 0.1$ and $T = 1.0$ and discuss the results.

9-8. Let $h(t) = \cos(2\pi t)$. Sample $h(t)$ with $T = \pi/8$. Compute the discrete Fourier transform with $N = 16$. Compare these results with those of Fig. 6-3(g). Repeat for $N = 24$. Compare these results to those of Fig. 6-5(g).

9-9. Consider $h(t)$ illustrated in Fig. 6-7(a). Let $T_0 = 1.0$. Sample $h(t)$ with $T = 0.1$ and $N = 10$. Compute the discrete Fourier transform. Repeat for $T = 0.2$ and $N = 5$, and for $T = 0.01$ and $N = 100$. Compare and explain these results.

9-10. Let $h(t) = te^{-t}$, $t > 0$. Compute the discrete Fourier transform. Give rationale for choice of T and N.

9-11. Let

$$\begin{aligned} h(t) &= 0 && t < 0 \\ &= \frac{1}{2} && t = 0 \\ &= 1 && 0 < t < 1 \\ &= \frac{1}{2} && t = 1 \\ &= 0 && t > 1 \\ x(t) &= h(t) \end{aligned}$$

Use the discrete convolution theorem to compute an approximation to $h(t) * x(t)$.

9-12. Consider $h(t)$ and $x(t)$ as defined in Problem 9-11. Compute the discrete approximation to the correlation of $h(t)$ and $x(t)$. Use the correlation theorem.

9-13. Validate the results of Fig. 7-4 using the discrete convolution theorem.

9-14. Apply the discrete convolution theorem to demonstrate the results of Fig. 7-5.

REFERENCES

1. Blackman, R. B., J. W. Tukey, *The Measurement of Power Spectra from the Point of View of Communications Engineering*. New York: Dover, 1959.

10

THE FAST FOURIER TRANSFORM (FFT)

Interpretation of fast Fourier transform results does not require a well-grounded education in the algorithm itself but rather a thorough understanding of the discrete Fourier transform. This follows from the fact that the FFT is simply an algorithm (i.e., a particular method of performing a series of computations) that can compute the discrete Fourier transform much more rapidly than other available algorithms. For this reason our discussion of the FFT addresses only the computational aspect of the algorithm.

A simple matrix factoring example is used to intuitively justify the FFT algorithm. The factored matrices are alternatively represented by signal flow graphs. From these graphs we construct the logic of an FFT computer program.

10-1 MATRIX FORMULATION

Consider the discrete Fourier transform (6-16)

$$X(n) = \sum_{k=0}^{N-1} x_0(k)\, e^{-j2\pi nk/N} \qquad n = 0, 1, \ldots, N-1 \tag{10-1}$$

where we have replaced kT by k and n/NT by n for convenience of notation. We note that (10-1) describes the computation of N equations. For example, if $N = 4$ and if we let

$$W = e^{-j2\pi/N} \tag{10-2}$$

then (10-1) can be written as

$$\begin{aligned}X(0) &= x_0(0)W^0 + x_0(1)W^0 + x_0(2)W^0 + x_0(3)W^0\\X(1) &= x_0(0)W^0 + x_0(1)W^1 + x_0(2)W^2 + x_0(3)W^3\\X(2) &= x_0(0)W^0 + x_0(1)W^2 + x_0(2)W^4 + x_0(3)W^6\\X(3) &= x_0(0)W^0 + x_0(1)W^3 + x_0(2)W^6 + x_0(3)W^9\end{aligned} \tag{10-3}$$

Equations (10-3) can be more easily represented in matrix form

$$\begin{bmatrix}X(0)\\X(1)\\X(2)\\X(3)\end{bmatrix} = \begin{bmatrix}W^0 & W^0 & W^0 & W^0\\W^0 & W^1 & W^2 & W^3\\W^0 & W^2 & W^4 & W^6\\W^0 & W^3 & W^6 & W^9\end{bmatrix}\begin{bmatrix}x_0(0)\\x_0(1)\\x_0(2)\\x_0(3)\end{bmatrix} \tag{10-4}$$

or more compactly as

$$\mathbf{X}(\boldsymbol{n}) = \mathbf{W}^{nk}\mathbf{x}_0(\boldsymbol{k}) \tag{10-5}$$

We will denote a matrix by boldface type.

Examination of (10-4) reveals that since $\mathbf{W}$ and possibly $\mathbf{x}_0(\boldsymbol{k})$ are complex, then N^2 complex multiplications and $(N)(N-1)$ complex additions are necessary to perform the required matrix computation. The FFT owes its success to the fact that the algorithm reduces the number of multiplications and additions required in the computation of (10-4). We will now discuss, on an intuitive level, how this reduction is accomplished. A proof of the FFT algorithm will be delayed until Chapter 11.

10-2 INTUITIVE DEVELOPMENT

To illustrate the FFT algorithm, it is convenient to choose the number of sample points of $x_0(k)$ according to the relation $N = 2^\gamma$, where γ is an integer. Later developments will remove this restriction. Recall that Eq. (10-4) results from the choice of $N = 4 = 2^\gamma = 2^2$; therefore, we can apply the FFT to the computation of (10-4).

The first step in developing the FFT algorithm for this example is to rewrite (10-4) as

$$\begin{bmatrix}X(0)\\X(1)\\X(2)\\X(3)\end{bmatrix} = \begin{bmatrix}1 & 1 & 1 & 1\\1 & W^1 & W^2 & W^3\\1 & W^2 & W^0 & W^2\\1 & W^3 & W^2 & W^1\end{bmatrix}\begin{bmatrix}x_0(0)\\x_0(1)\\x_0(2)\\x_0(3)\end{bmatrix} \tag{10-6}$$

Matrix Eq. (10-6) was derived from (10-4) by using the relationship $W^{nk} = W^{nk \bmod (N)}$. Recall that $[nk \bmod (N)]$ is the remainder upon division of nk

by N; hence if $N = 4$, $n = 2$, and $k = 3$ then

$$W^6 = W^2 \tag{10-7}$$

since

$$\begin{aligned} W^{nk} = W^6 &= \exp\left[\left(\frac{-j2\pi}{4}\right)(6)\right] = \exp[-j3\pi] \\ &= \exp[-j\pi] = \exp\left[\left(\frac{-j2\pi}{4}\right)(2)\right] = W^2 = W^{nk \bmod N} \end{aligned} \tag{10-8}$$

The second step in the development is to factor the square matrix in (10-6) as follows:

$$\begin{bmatrix} X(0) \\ X(2) \\ X(1) \\ X(3) \end{bmatrix} = \begin{bmatrix} 1 & W^0 & 0 & 0 \\ 1 & W^2 & 0 & 0 \\ 0 & 0 & 1 & W^1 \\ 0 & 0 & 1 & W^3 \end{bmatrix} \begin{bmatrix} 1 & 0 & W^0 & 0 \\ 0 & 1 & 0 & W^0 \\ 1 & 0 & W^2 & 0 \\ 0 & 1 & 0 & W^2 \end{bmatrix} \begin{bmatrix} x_0(0) \\ x_0(1) \\ x_0(2) \\ x_0(3) \end{bmatrix} \tag{10-9}$$

The method of factorization is based on the theory of the FFT algorithm and will be developed in Chapter 11. For the present, it suffices to show that multiplication of the two square matrices of (10-9) yields the square matrix of (10-6) with the exception that rows 1 and 2 have been interchanged (The rows are numbered 0, 1, 2, and 3). Note that this interchange has been taken into account in (10-9) by rewriting the column vector $\boldsymbol{X(n)}$; let the row-interchanged vector be denoted by

$$\overline{\boldsymbol{X(n)}} = \begin{bmatrix} X(0) \\ X(2) \\ X(1) \\ X(3) \end{bmatrix} \tag{10-10}$$

Repeating, the reader should verify that Eq. (10-9) yields (10-6) with the interchanged rows as noted. This factorization is the key to the efficiency of the FFT algorithm.

Having accepted the fact that (10-9) is correct, although the results are *scrambled*, one should then examine the number of multiplications required to compute the equation. First let

$$\begin{bmatrix} x_1(0) \\ x_1(1) \\ x_1(2) \\ x_1(3) \end{bmatrix} = \begin{bmatrix} 1 & 0 & W^0 & 0 \\ 0 & 1 & 0 & W^0 \\ 1 & 0 & W^2 & 0 \\ 0 & 1 & 0 & W^2 \end{bmatrix} \begin{bmatrix} x_0(0) \\ x_0(1) \\ x_0(2) \\ x_0(3) \end{bmatrix} \tag{10-11}$$

That is, column vector $\boldsymbol{x_1(k)}$ is equal to the product of the two matrices on the right in Eq. (10-9).

Element $x_1(0)$ is computed by one complex multiplication and one complex addition (W^0 is not reduced to unity in order to develop a generalized result).

$$x_1(0) = x_0(0) + W^0 x_0(2) \tag{10-12}$$

Element $x_1(1)$ is also determined by one complex multiplication and addition. Only one complex addition is required to compute $x_1(2)$. This follows from the fact that $W^0 = -W^2$; hence

$$\begin{aligned} x_1(2) &= x_0(0) + W^2 x_0(2) \\ &= x_0(0) - W^0 x_0(2) \end{aligned} \tag{10-13}$$

where the complex multiplication $W^0 x_0(2)$ has already been computed in the determination of $x_1(0)$ (Eq. 10-12). By the same reasoning, $x_1(3)$ is computed by only one complex addition and no multiplications. The intermediate vector $\mathbf{x_1}(\boldsymbol{k})$ is then determined by four complex additions and two complex multiplications.

Let us continue by completing the computation of (10-9)

$$\begin{bmatrix} X(0) \\ X(2) \\ X(1) \\ X(3) \end{bmatrix} = \begin{vmatrix} x_2(0) \\ x_2(1) \\ x_2(2) \\ x_2(3) \end{vmatrix} = \begin{vmatrix} 1 & W^0 & 0 & 0 \\ 1 & W^2 & 0 & 0 \\ 0 & 0 & 1 & W^1 \\ 0 & 0 & 1 & W^3 \end{vmatrix} \begin{vmatrix} x_1(0) \\ x_1(1) \\ x_1(2) \\ x_1(3) \end{vmatrix} \tag{10-14}$$

Term $x_2(0)$ is determined by one complex multiplication and addition

$$x_2(0) = x_1(0) + W^0 x_1(1) \tag{10-15}$$

Element $x_2(1)$ is computed by one addition because $W^0 = -W^2$. By similar reasoning, $x_2(2)$ is determined by one complex multiplication and addition, and $x_2(3)$ by only one addition.

Computation of $\overline{\mathbf{X}(\boldsymbol{n})}$ by means of Eq. (10-9) requires a total of four complex multiplications and eight complex additions. Computation of $\mathbf{X}(\boldsymbol{n})$ by (10-4) requires sixteen complex multiplications and twelve complex additions. Note that the matrix factorization process introduces zeros into the factored matrices and, as a result, reduces the required number of multiplications. For this example, the matrix factorization process reduced the number of multiplications by a factor of two. Since computation time is largely governed by the required number of multiplications, we see the reason for the efficiency of the FFT algorithm.

For $N = 2^\gamma$ the FFT algorithm is then simply a procedure for factoring an $N \times N$ matrix into γ matrices (each $N \times N$) such that each of the factored matrices has the special property of minimizing the number of complex multiplications and additions. If we extend the results of the previous example, we note that the FFT requires $N\gamma/2 = 4$ *complex* multiplications and $N\gamma = 8$ *complex* additions, whereas the direct method [Eq. (10-4)] requires N^2 *complex* multiplications and $N(N - 1)$ *complex* additions. If we assume that computing time is proportional to the number of multiplications, then the approximate ratio of direct to FFT computing time is given by

$$\frac{N^2}{N\gamma/2} = \frac{2N}{\gamma} \tag{10-16}$$

which for $N = 1024 = 2^{10}$ is a computational reduction of more than 200 to 1. Figure 10-1 illustrates the relationship between the number of multiplications required using the FFT algorithm compared with the number of multiplications using the direct method.

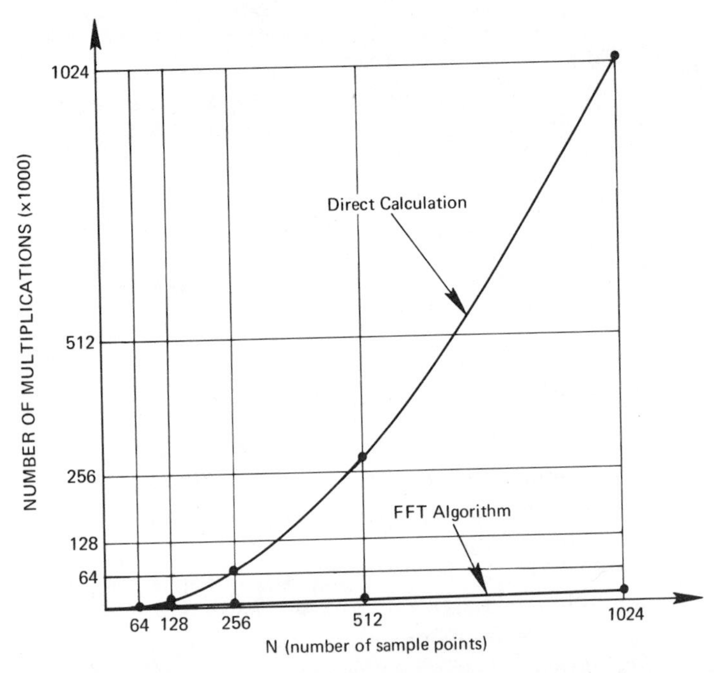

Figure 10-1. Comparison of multiplications required by direct calculation and FFT algorithm.

The matrix factoring procedure does introduce one discrepancy. Recall that the computation (10-9) yields $\overline{X(n)}$ instead of $X(n)$; that is

$$\overline{X(n)} = \begin{bmatrix} X(0) \\ X(2) \\ X(1) \\ X(3) \end{bmatrix} \text{ instead of } X(n) = \begin{bmatrix} X(0) \\ X(1) \\ X(2) \\ X(3) \end{bmatrix} \tag{10-17}$$

This rearrangement is inherent in the matrix factoring process and is a minor problem because it is straightforward to generalize a technique for *unscrambling* $\overline{X(n)}$ to obtain $X(n)$.

Rewrite $\overline{X(n)}$ by replacing argument n with its binary equivalent

$$\begin{bmatrix} X(0) \\ X(2) \\ X(1) \\ X(3) \end{bmatrix} \quad \text{becomes} \quad \begin{bmatrix} X(00) \\ X(10) \\ X(01) \\ X(11) \end{bmatrix} \tag{10-18}$$

Observe that if the binary arguments of (10-18) are *flipped* or *bit reversed* (i.e., 01 becomes 10, 10 becomes 01, etc.) then

$$\overline{X(n)} = \begin{bmatrix} X(00) \\ X(10) \\ X(01) \\ X(11) \end{bmatrix} \quad \text{flips to} \quad \begin{bmatrix} X(00) \\ X(01) \\ X(10) \\ X(11) \end{bmatrix} = X(n) \tag{10-19}$$

It is straightforward to develop a generalized result for unscrambling the FFT.

For N greater than 4, it is cumbersome to describe the matrix factorization process analogous to Eq. (10-9). For this reason we interpret (10-9) in a graphical manner. Using this graphical formulation we can describe sufficient generalities to develop a flow graph for a computer program.

10-3 SIGNAL FLOW GRAPH

We convert Eq. (10-9) into the signal flow graph illustrated in Fig. 10-2. As shown, we represent the data vector or array $x_0(k)$ by a vertical column of nodes on the left of the graph. The second vertical array of nodes is the vector $x_1(k)$ computed in Eq. (10-11), and the next vertical array corresponds to the vector $x_2(k) = \overline{X(n)}$, Eq. (10-14). In general, there will be γ computational arrays where $N = 2^\gamma$.

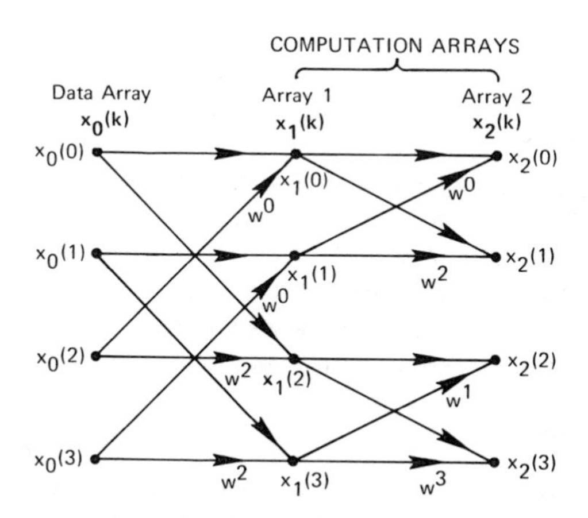

Figure 10-2. FFT signal flow graph, $N = 4$.

The signal flow graph is interpreted as follows. Each node is entered by two solid lines representing *transmission paths* from previous nodes. A path transmits or brings a quantity from a node in one array, multiplies the quantity by W^p, and inputs the result into the node in the next array. Factor W^p appears near the arrowhead of the transmission path; absence of this factor implies that $W^p = 1$. Results entering a node from the two transmission paths are combined additively.

To illustrate the interpretation of the signal flow graph, consider node $x_1(2)$ in Fig. 10-2. According to the rules for interpreting the signal flow graph,

$$x_1(2) = x_0(0) + W^2 x_0(2) \tag{10-20}$$

which is simply Eq. (10-13). Each node of the signal flow graph is expressed similarly.

The signal flow graph is then a concise method for representing the computations required in the factored matrix FFT algorithm (10-9). Each computational column of the graph corresponds to a factored matrix; γ vertical arrays of N points each ($N = 2^\gamma$) are required. Utilization of this graphical presentation allows us to easily describe the matrix factoring process for large N.

We show in Fig. 10-3 the signal flow graph for $N = 16$. With a flow graph of this size, it is possible to develop general properties concerning the matrix factorization process and thus provide a framework for developing a FFT computer program flow chart.

10-4 DUAL NODES

Inspection of Fig. 10-3 reveals that in every array we can always find two nodes whose input transmission paths stem from the same pair of nodes in the previous array. For example, nodes $x_1(0)$ and $x_1(8)$ are computed in terms of nodes $x_0(0)$ and $x_0(8)$. Note that nodes $x_0(0)$ and $x_0(8)$ do not enter into the computation of any other node. We define two such nodes as a *dual node pair*.

Since the computation of a dual node pair is independent of other nodes, it is possible to perform *in-place computation*. To illustrate, note from Fig. 10-3 that we can simultaneously compute $x_1(0)$ and $x_1(8)$ in terms of $x_0(0)$ and $x_0(8)$ and return the results to the storage locations previously occupied by $x_0(0)$ and $x_0(8)$. Storage requirements are then limited to the data array $\mathbf{x}_0(k)$ only. As each array is computed, the results are returned to this array.

Dual Node Spacing

Let us now investigate the spacing (measured vertically in terms of the index k) between a dual node pair. The following discussion will refer to

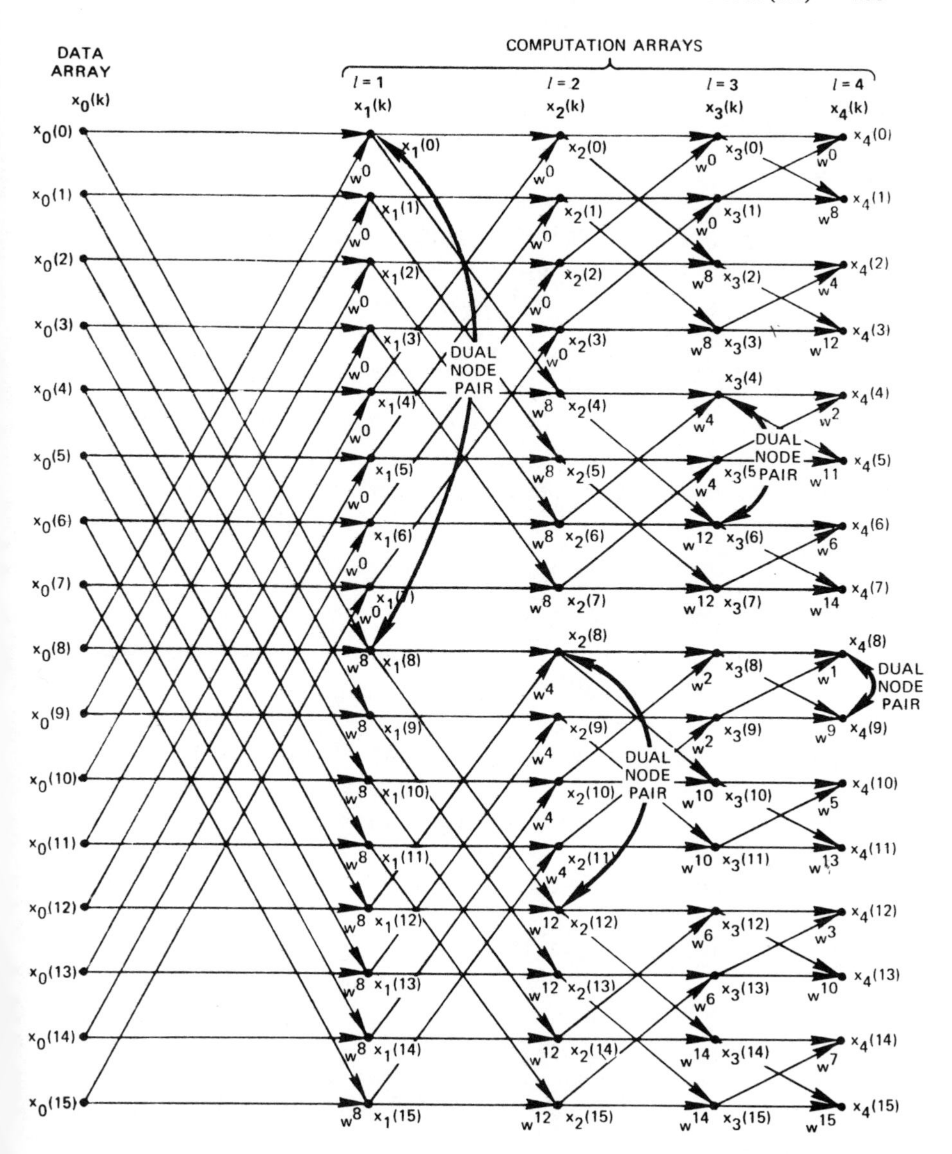

Figure 10-3. Example of dual nodes.

Fig. 10-3. First, in array $l = 1$, a dual node pair, say $x_1(0)$; $x_1(8)$, is separated by $k = 8 = N/2^l = N/2^1$. In array $l = 2$, a dual node pair, say $x_2(8)$; $x_2(12)$, is separated by $k = 4 = N/2^l = N/2^2$. Similarly, a dual node pair, $x_3(4)$; $x_3(6)$, in array $l = 3$ is separated by $k = 2 = N/2^l = N/2^3$, and in array $l = 4$, a dual node pair, $x_4(8)$; $x_4(9)$, is separated by $k = 1 = N/2^l = N/2^4$.

Generalizing these results, we observe that the spacing between dual nodes in array l is given by $N/2^l$. Thus, if we consider a particular node $x_l(k)$, then its dual node is $x_l(k + N/2^l)$. This property allows us to easily identify a dual node pair.

Dual Node Computation

The computation of a dual node pair requires only one complex multiplication. To clarify this point, consider node $x_2(8)$ and its dual $x_2(12)$, illustrated in Fig. 10-3. The transmission paths stemming from node $x_1(12)$ are multiplied by W^4 and W^{12} prior to input at nodes $x_2(8)$ and $x_2(12)$, respectively. It is important to note that $W^4 = -W^{12}$ and that only one multiplication is required since the same data $x_1(12)$ is to be multiplied by these terms. In general, if the weighting factor at one node is W^p, then the weighting factor at the dual node is $W^{p+N/2}$. Because $W^p = -W^{p+N/2}$, only one multiplication is required in the computation of a dual node pair. The computation of any dual node pair is given by the equation pair

$$\begin{aligned} x_l(k) &= x_{l-1}(k) + W^p x_{l-1}(k + N/2^l) \\ x_l(k + N/2^l) &= x_{l-1}(k) - W^p x_{l-1}(k + N/2^l) \end{aligned} \qquad (10\text{-}21)$$

In computing an array, we normally begin with node $k = 0$ and sequentially work down the array, computing the equation pair (10-21). As stated previously, the dual of any node in the lth array is always down $N/2^l$ in the array. Since the spacing is $N/2^l$, then it follows that we must *skip* after every $N/2^l$ node. To illustrate this point, consider array $l = 2$ in Fig. 10-4. If we begin with node $k = 0$, then according to our previous discussions, the dual node is located at $k = N/2^2 = 4$ which can be verified by inspection of Fig. 10-4. Proceeding down this array, we note that the dual node is always located down by 4 in the array until we reach node 4. At this point we have entered a set of nodes previously encountered; that is, these nodes are the duals for nodes $k = 0$, 1, 2, and 3. It is necessary to *skip-over* nodes $k =$ 4, 5, 6, and 7. Nodes 8, 9, 10, and 11 follow the original convention of the dual node being located 4 down in the array. In general, if we work from the top down in array l, then we will compute Eq. (10-21) for the first $N/2^l$ nodes, skip the next $N/2^l$, etc. We know to stop *skipping* when we reach a node index greater than $N - 1$.

10-5 W^p DETERMINATION

Based on the preceding discussions, we have defined the properties of each array with the exception of the value of p in Eq. (10-21). The value of p is determined by (a) writing the index k in binary form with γ bits,

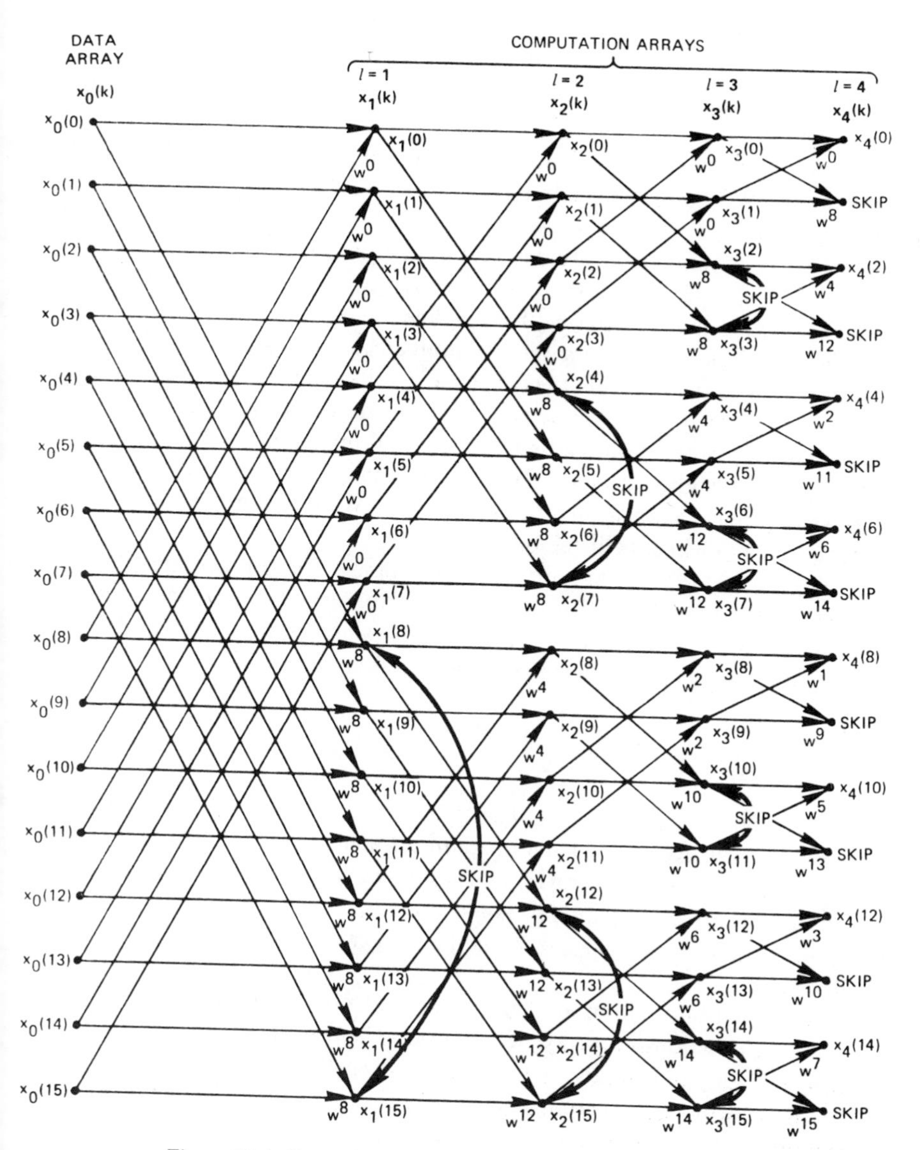

Figure 10-4. Example of nodes to be skipped when computing signal flow graph.

(b) scaling or sliding this binary number $\gamma - l$ bits to the right and filling in the newly opened bit position on the left with zeros, and (c) reversing the order of the bits. This bit-reversed number is the term p.

To illustrate this procedure, refer to Fig. 10-4 and consider node $x_3(8)$.

Since $\gamma = 4$, $k = 8$, and $l = 3$, then k in binary is 1000. We scale this number $\gamma - l = 4 - 3 = 1$ places to the right and fill in zeros; the result is 0100. We then reverse the order of the bits to yield 0010 or integer 2. The value of p is then 2.

Let us now consider a procedure for implementing this bit-reversing operation. We know that a binary number, say $a_4 a_3 a_2 a_1 \cdot$ can be written in base 10 as $a_4 \times 2^3 + a_3 \times 2^2 + a_2 \times 2^1 + a_1 \times 2^0$. The bit-reversed number which we are trying to describe is then given by $a_1 \times 2^3 + a_2 \times 2^2 + a_3 + 2^1 + a_4 \times 2^0$. If we describe a technique for determining the binary bits a_4, a_3, a_2, and a_1, then we have defined a bit-reversing operation.

Now assume that M is a binary number equal to $a_4 a_3 a_2 a_1 \cdot$. Divide M by 2, truncate, and multiply the truncated results by 2. Then compute $[a_4 a_3 a_2 a_1 \cdot - 2(a_4 a_3 a_2 \cdot)]$. If the bit a_1 is 0, then this difference will be zero because division by 2, truncation, and subsequent multiplication by 2 does not alter M. However, if the bit a_1 is 1, truncation changes the value of M and the above difference expression will be non-zero. We observe that by this technique we can determine if the bit a_1 is 0 or 1.

We can identify the bit a_2 in a similar manner. The appropriate difference expression is $[a_4 a_3 a_2 \cdot - 2(a_4 a_3 \cdot)]$. If this difference is zero, then a_2 is zero. Bits a_3 and a_4 are determined similarly. This procedure will form the basis for developing a bit-reversing computer routine in Sec. 10-7.

10-6 UNSCRAMBLING THE FFT

The final step in computing the FFT is to *unscramble* the results analogous to Eq. (10-19). Recall that the procedure for unscrambling the vector $\overline{\boldsymbol{X}(\boldsymbol{n})}$ is to write n in binary and reverse or flip the binary number. We show in Fig. 10-5 the results of this bit reversing operation; terms $x_4(k)$ and $x_4(i)$ have simply been interchanged where i is the integer obtained by bit-reversing the integer k.

Note that a situation similar to the dual node concept exists when we unscramble the output array. If we proceed down the array, interchanging $x(k)$ with the appropriate $x(i)$, we will eventually encounter a node which has previously been interchanged. For example, in Fig. 10-5, node $k = 0$ remains in its location, nodes $k = 1$, 2, and 3 are interchanged with nodes 8, 4, and 12, respectively. The next node to be considered is node 4, but this node was previously interchanged with node 2. To eliminate the possibility of considering a node that has previously been interchanged, we simply check to see if i (the integer obtained by bit-reversing k) is less than k. If so, this implies that the node has been interchanged by a previous operation. With this check we can insure a straightforward unscrambling procedure.

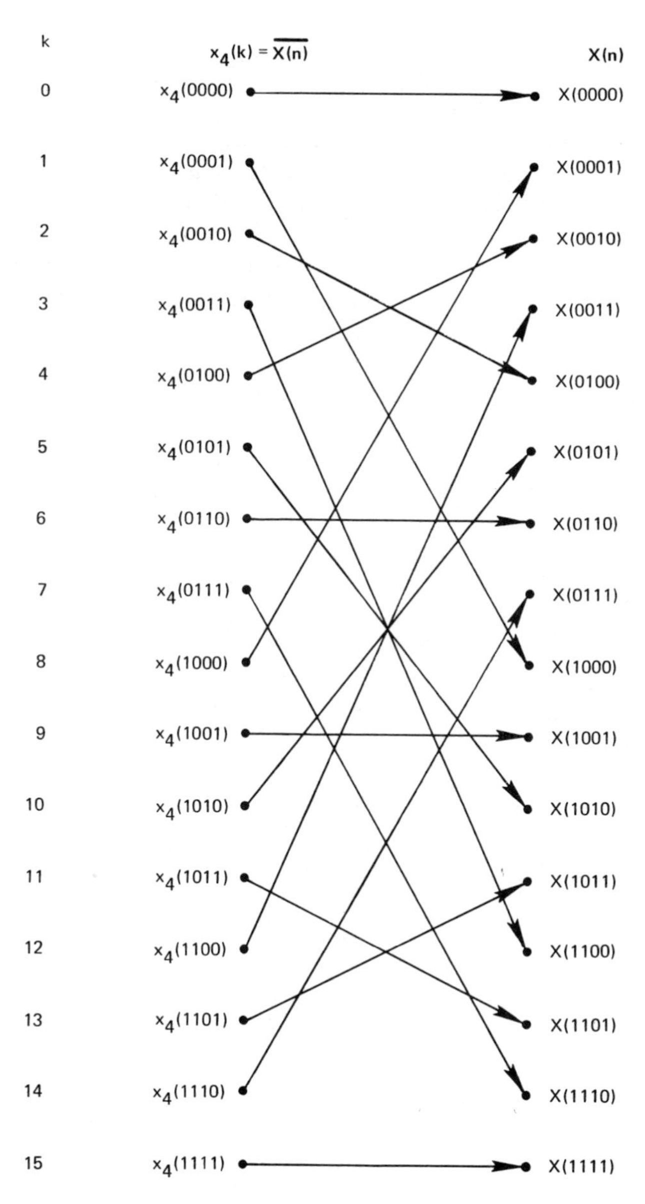

Figure 10-5. Example of bit-reversing operation for $N = 16$.

10-7 FFT COMPUTATION FLOW CHART

Using the discussed properties of the FFT signal flow graph, we can easily develop a flow chart for programming the algorithm on a digital computer. We know from the previous discussions that we first compute array $l = 1$ by starting at node $k = 0$ and working down the array. At each node k we compute the equation pair (10-21) where p is determined by the described procedure. We continue down the array computing the equation pair (10-21) until we reach a region of nodes which must be *skipped-over*. We skip over the appropriate nodes and continue until we have computed the entire array. We then proceed to compute the remaining arrays using the same procedures. Finally we unscramble the final array to obtain the desired results. Figure 10-6 illustrates a flow chart for computer programming the FFT algorithm.

Box 1 describes the necessary input data. Data vector $\boldsymbol{x}_0(\boldsymbol{k})$ is assumed to be complex and is indexed as $k = 0, 1, \ldots, N - 1$. If $\boldsymbol{x}_0(\boldsymbol{k})$ is real, then the imaginary part should be set to zero. The number of sample points N must satisfy the relationship $N = 2^\gamma$, γ integer valued.

Initialization of the various program parameters is accomplished in Box 2. Parameter l is the array number being considered. We start with array $l = 1$. The spacing between dual nodes is given by the parameter $N2$; for array $l = 1$, $N2 = N/2$ and is initialized as such. Parameter $NU1$ is the right shift required when determining the value of p in Eq. (10-21); $NU1$ is initialized to $\gamma - 1$. The index k of the array is initialized to $k = 0$; thus we will work from the top and progress down the array.

Box 3 checks to see if the array l to be computed is greater than γ. If yes, then the program branches to Box 13 to unscramble the computed results by bit inversion. If all arrays have not been computed, then we proceed to Box 4.

Box 4 sets a counter $I = 1$. This counter monitors the number of dual node pairs which have been considered. Recall from Sec. 10-4 that it is necessary to skip certain nodes in order to insure that previously considered nodes are not encountered a second time. Counter I is the control for determining when the program must skip.

Boxes 5 and 6 perform the computation of Eq. (10-21). Since k and l have been initialized to 0 and 1, respectively, the initial node considered is the first node of the first array. To determine the factor p for this node, recall that we must first scale the binary number k to the right $\gamma - l$ bits. To accomplish this, we compute the integer value of $k/2^{\gamma-l} = k/2^{NU1}$ and set the result to M as shown in Box 5. According to the procedure for determining p, we must bit reverse M where M is represented by $\gamma = NU$ bits. The func-

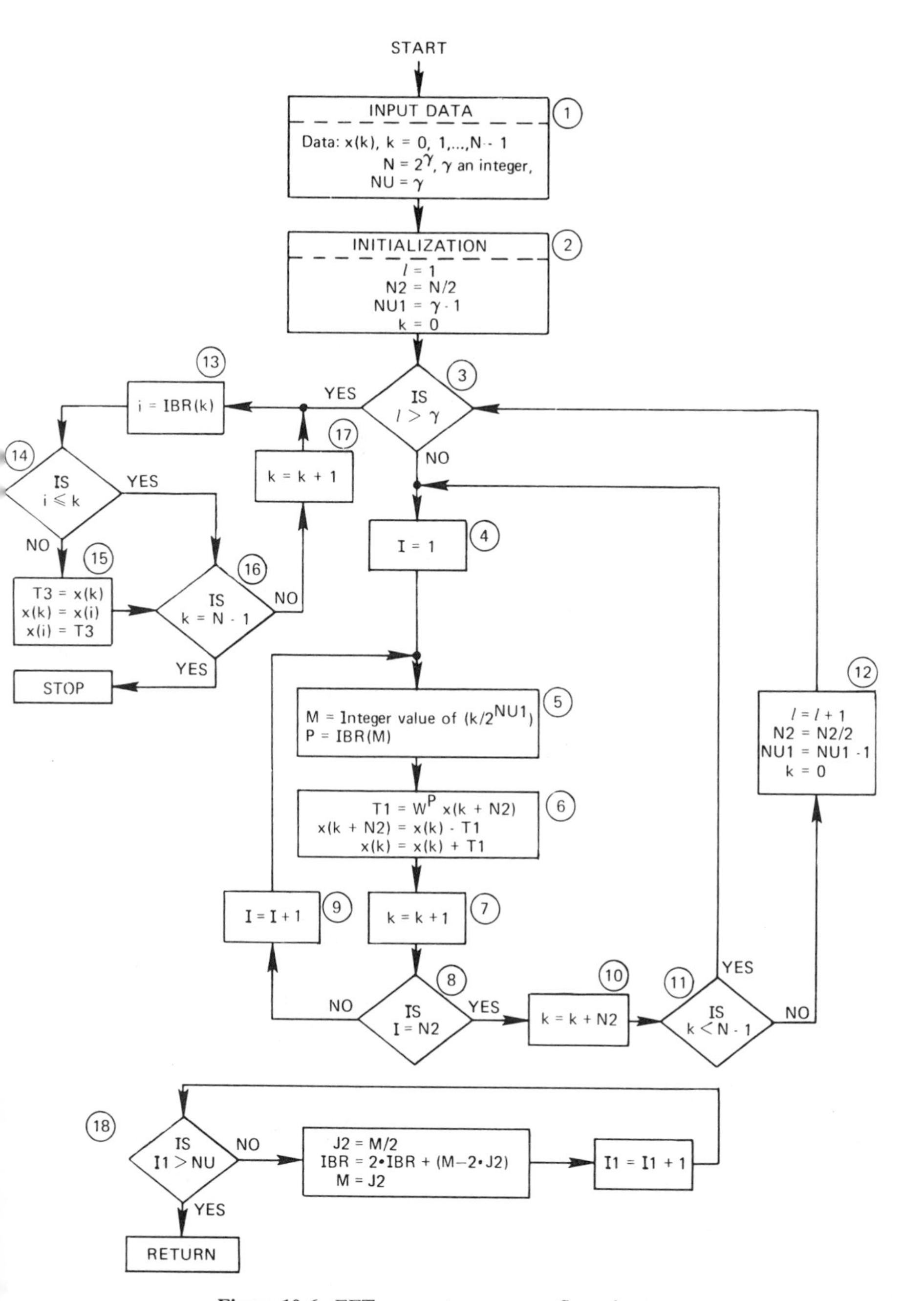

Figure 10-6. FFT computer program flow chart.

tion $IBR(M)$ denoted in Box 5 is a special function routine for bit inversion; this routine will be described later.

Box 6 is the computation of Eq. (10-21). We compute the product W^p $x(k + N2)$ and assign the result to a temporary storage location. Next we add and subtract this term according to Eq. (10-21). The result is the dual node output.

We then proceed down the array to the next node. As shown in Box 7, k is incremented by 1.

To avoid recomputing a dual node that has been considered previously, we check in Box 8 to determine if the counter I is equal to $N2$. For array 1, the number of nodes that can be considered consecutively without skipping is equal to $N/2 = N2$. Box 8 determines this condition. If I is not equal to $N2$, then we proceed down the array and increment the counter I as shown in Box 9. Recall that we have already incremented k in Box 7. Boxes 5 and 6 are then repeated for the new value of k.

If $I = N2$ in Box 8, then we know that we have reached a node previously considered. We then skip $N2$ nodes by setting $k = k + N2$. Because k has already been incremented by 1 in Box 7, it is sufficient to skip the previously considered nodes by incrementing k by $N2$.

Before we perform the required computations indicated by Boxes 5 and 6 for the new node $k = k + N2$, we must first check to see that we have not exceeded the array size. As shown in Box 11, if k is less than $N - 1$ (recall k is indexed from 0 to $N - 1$), then we reset the counter I to 1 in Box 4 and repeat Boxes 5 and 6.

If $k > N - 1$ in Box 11, we know that we must proceed to the next array. Hence, as shown in Box 12, l is indexed by 1. The new spacing $N2$ is simply $N2/2$ (recall the spacing is $N/2^l$). $NU1$ is decremented by 1 ($NU1$ is equal to $\gamma - l$.), and k is reset to zero. We then check in Box 3 to see if all arrays have been computed. If so, then we proceed to unscramble the final results. This operation is performed by Boxes 13 through 17.

Box 13 bit-reverses the integer k to obtain the integer i. Again we use the bit-reversing function $IBR(k)$ which is to be explained later. Recall that to unscramble the FFT we simply interchange $x(k)$ and $x(i)$. This manipulation is performed by the operations indicated in Box 15. However, before Box 15 is entered it is necessary to determine, as shown in Box 14, if i is less than or equal to k. This step is necessary to prohibit the altering of previously unscrambled nodes.

Box 16 determines when all nodes have been unscrambled and Box 17 is simply an index for k.

In Box 18, we describe the logic of the bit-reversing function $IBR(k)$. We have implemented the bit-reversing procedure discussed in Sec. 10-5.

When one proceeds to implement the flow graph of Fig. 10-6 into a computer program, it is necessary to consider the variables $x(k)$ and W^p as complex numbers and they must be handled accordingly.

10-8 FFT FORTRAN PROGRAM

A listing of a FORTRAN program based on the FFT algorithm flow chart in Fig. 10-6 is shown in Fig. 10-7. This program does not attempt to accomplish the ultimate in efficiency but rather is designed to acquaint the reader with the computer programming procedure of the FFT algorithm. Efficient programming results in a slight increase in computing speed. We will reference additional FFT programs in the discussions to follow.

The input to the FFT program are: XREAL, the real part of the function to be discrete Fourier transformed; XIMAG, the imaginary part; N, the number of points; and NU, where $N = 2^{NU}$. Upon completion, XREAL is the real part of the transform and XIMAG is the imaginary part of the transform. Input data is destroyed.

10-9 FFT ALGOL PROGRAM

In Fig. 10-8 we show an ALGOL program based on the flow graph of Fig. 10-6. As in the FORTRAN case, this program is not designed for maximum efficiency. Additional ALGOL programs will be referenced in future discussions.

10-10 FFT ALGORITHMS FOR REAL DATA

In applying the FFT, we often consider only real functions of time whereas the frequency functions are, in general, complex. Thus, a single computer program written to determine both the discrete transform and its inverse is written such that a complex time waveform is assumed;

$$H(n) = \frac{1}{N} \sum_{k=0}^{N-1} [h_r(k) + jh_i(k)]\, e^{-j2\pi nk/N} \tag{10-22}$$

This follows from the fact that the alternate inversion formula (8-9) is given by

$$h(k) = \frac{1}{N} \left[\sum_{n=0}^{N-1} [H_r(n) + jH_i(n)]^* \, e^{-j2\pi nk/N} \right]^* \tag{10-23}$$

and since both (10-22) and (10-23) contain the common factor $e^{-j2\pi nk/N}$, then a single computer program can be used to compute both the discrete transform and its inverse.

```
      SUBROUTINE FFT(XREAL,XIMAG,N,NU)
      DIMENSION XREAL(N),XIMAG(N)
      N2=N/2
      NU1=NU-1
      K=Ø
      DO 1ØØ L=1,NU
1Ø2   DO 1Ø1 I=1,N2
      P=IBITR(K/2**NU1,NU)
      ARG=6.283185*P/FLOAT(N)
      C=COS(ARG)
      S=SIN(ARG)
      K1=K+1
      K1N2=K1+N2
      TREAL=XREAL(K1N2)*C+XIMAG(K1N2)*S
      TIMAG=XIMAG(K1N2)*C-XREAL(K1N2)*S
      XREAL(K1N2)=XREAL(K1)-TREAL
      XIMAG(K1N2)=XIMAG(K1)-TIMAG
      XREAL(K1)=XREAL(K1)+TREAL
      XIMAG(K1)=XIMAG(K1)+TIMAG
1Ø1   K=K+1
      K=K+N2
      IF(K.LT.N) GO TO 1Ø2
      K=Ø
      NU1=NU1-1
1ØØ   N2=N2/2
      DO 1Ø3 K=1,N
      I=IBITR(K-1,NU)+1
      IF(I.LE.K) GO TO 1Ø3
      TREAL=XREAL(K)
      TIMAG=XIMAG(K)
      XREAL(K)=XREAL(I)
      XIMAG(K)=XIMAG(I)
      XREAL(I)=TREAL
      XIMAG(I)=TIMAG
1Ø3   CONTINUE
      RETURN
      END
      FUNCTION IBITR(J,NU)
      J1=J
      IBITR=Ø
      DO 2ØØ I=1,NU
      J2=J1/2
      IBITR=IBITR*2+(J1-2*J2)
2ØØ   J1=J2
      RETURN
      END
```

Figure 10-7. FFT FORTRAN computer subroutine.

```
PROCEDURE FFT(XREAL,XIMAG,N,NU);
VALUE N,NU;
REAL ARRAY XREAL[0],XIMAG[0];
INTEGER N,NU;
BEGIN INTEGER N2,NU1,I,L,K;
      REAL TREAL,TIMAG,P,ARG,C,S;
      LABEL LBL;
      INTEGER PROCEDURE BITREV(J,NU);
      VALUE J,NU;
      INTEGER J,NU;
      BEGIN INTEGER I,J1,J2,K;
            J1:=J;
            K:=0;
            FOR I:=1 STEP 1 UNTIL NU DO
            BEGIN J2:=J1 DIV 2;
                  K:=K*2+(J1-2*J2);
                  J1:=J2 END;
            BITREV:=K END OF PROCEDURE BITREV;
      N2:=N DIV 2;
      NU1:=NU-1;
      K:=0;
      FOR L:=1 STEP 1 UNTIL NU DO
      BEGIN
            FOR I:=1 STEP 1 UNTIL N2 DO
            BEGIN P:=BITREV(K DIV 2**NU1,NU);
                  ARG:=6.283185*P/N;
                  C:=COS(ARG);
                  S:=SIN(ARG);
                  TREAL:=XREAL[K+N2]*C+XIMAG[K+N2]*S;
                  TIMAG:=XIMAG[K+N2]*C-XREAL[K+N2]*S;
                  XREAL[K+N2]:=XREAL[K]-TREAL;
                  XIMAG[K+N2]:=XIMAG[K]-TIMAG;
                  XREAL[K]:=XREAL[K]+TREAL;
                  XIMAG[K]:=XIMAG[K]+TIMAG;
                  K:=K+1 END;
            K:=K+N2;
            IF K<N THEN GO TO LBL;
            K:=0;
            NU1:=NU1-1;
            N2:=N2 DIV 2 END;
      FOR K:=0,K+1 WHILE K<N DO
      BEGIN I:=BITREV(K,NU);
            IF I>K THEN
            BEGIN TREAL:=XREAL[K];
                  TIMAG:=XIMAG[K];
                  XREAL[K]:=XREAL[I];
                  XIMAG[K]:=XIMAG[I];
                  XREAL[I]:=TREAL;
                  XIMAG[I]:=TIMAG END END END OF PROCEDURE FFT;
```

Figure 10-8. FFT ALGOL computer subroutine.

If the time function being considered is real, we must set to zero the imaginary part of the complex time function in (10-22). This approach is inefficient in that the computer program will still perform the multiplications involving $jh_i(k)$ in Eq. (10-23) even though $jh_i(k)$ is zero.

In this section we will develop two techniques for using this imaginary part of the complex time function to more efficiently compute the FFT of real functions.

FFT of Two Real Functions Simultaneously

It is desired to compute the discrete Fourier transform of the real time functions $h(k)$ and $g(k)$ from the complex function

$$y(k) = h(k) + jg(k) \tag{10-24}$$

That is, $y(k)$ is constructed to be the sum of two real functions where one of these real functions is taken to be imaginary. From the linearity property (8-1), the discrete Fourier transform of $y(k)$ is given by

$$\begin{aligned} Y(n) &= H(n) + jG(n) \\ &= [H_r(n) + jH_i(n)] + j[G_r(n) + jG_i(n)] \\ &= [H_r(n) - G_i(n)] + j[H_i(n) + G_r(n)] \\ &= R(n) + jI(n) \end{aligned} \tag{10-25}$$

By means of the frequency domain equivalent of (8-15) we decompose both $R(n)$, the real part of $Y(n)$, and $I(n)$, the imaginary part of $Y(n)$, into even and odd components

$$\begin{aligned} Y(n) = &\left(\frac{R(n)}{2} + \frac{R(N-n)}{2}\right) + \left(\frac{R(n)}{2} - \frac{R(N-n)}{2}\right) \\ &+ j\left(\frac{I(n)}{2} + \frac{I(N-n)}{2}\right) + j\left(\frac{I(n)}{2} - \frac{I(N-n)}{2}\right) \end{aligned} \tag{10-26}$$

From Eqs. (8-21) and (8-22)

$$\begin{aligned} H(n) &= R_e(n) + jI_0(n) \\ &= \left(\frac{R(n)}{2} + \frac{R(N-n)}{2}\right) + j\left(\frac{I(n)}{2} - \frac{I(N-n)}{2}\right) \end{aligned} \tag{10-27}$$

Similarly from (8-23) and (8-24)

$$jG(n) = R_0(n) + jI_e(n)$$

or

$$\begin{aligned} G(n) &= I_e(n) - jR_0(n) \\ &= \left(\frac{I(n)}{2} + \frac{I(N-n)}{2}\right) - j\left(\frac{R(n)}{2} - \frac{R(N-n)}{2}\right) \end{aligned} \tag{10-28}$$

Thus, if the real and imaginary parts of the discrete transform of a complex time function are decomposed according to Eqs. (10-27) and (10-28), then the simultaneous discrete transform of two real time functions can be accomplished. As is easily seen, this technique results in a *two-times* capability with only the requirement for *sorting* out the results. For ease of reference, the necessary steps to simultaneously compute the FFT of two real functions are listed in Fig. 10-9.

1. Functions $h(k)$ and $g(k)$ are real $\quad k = 0, 1, \ldots, N-1$
2. Form the complex function
$$y(k) = h(k) + jg(k) \qquad k = 0, 1, \ldots, N-1$$
3. Compute
$$Y(n) = \sum_{k=0}^{N-1} y(k) e^{-j2\pi nk/N}$$
$$= R(n) + jI(n) \qquad n = 0, 1, \ldots, N-1$$
where $R(n)$ and $I(n)$ are the real and imaginary parts of $Y(n)$, respectively.
4. Compute
$$H(n) = \left[\frac{R(n)}{2} + \frac{R(N-n)}{2}\right] + j\left[\frac{I(n)}{2} - \frac{I(N-n)}{2}\right]$$
$$G(n) = \left[\frac{I(n)}{2} + \frac{I(N-n)}{2}\right] - j\left[\frac{R(n)}{2} - \frac{R(N-n)}{2}\right]$$
$$n = 0, 1, \ldots, N-1$$
where $H(n)$ and $G(n)$ are the discrete transforms of $h(k)$ and $g(k)$, respectively.

Figure 10-9. Computation procedure for simultaneous discrete Fourier transform of two real functions.

Transform of 2N Samples with an N Sample Transform

The imaginary part of the complex time function can also be used to compute more efficiently the discrete transform of a single real time function. Consider a function $x(k)$ which is described by $2N$ samples. It is desired to compute the discrete transform of this function using Eq. (10-22). That is, we wish to break the $2N$ point function $x(k)$ into two N sample functions. Function $x(k)$ cannot simply be divided into half; instead we divide $x(k)$ as follows:

$$\begin{aligned} h(k) &= x(2k) \\ g(k) &= x(2k+1) \end{aligned} \qquad k = 0, 1, \ldots, N-1 \tag{10-29}$$

That is, function $h(k)$ is equal to the even numbered samples of $x(k)$, and $g(k)$ is equal to the odd numbered samples. [Note that $h(k)$ and $g(k)$ are not the even and odd function decomposition of $x(k)$ which was described in

(8-17)]. Eq. (10-22) can then be written as

$$\begin{aligned} X(n) &= \sum_{k=0}^{2N-1} x(k)\, e^{-j2\pi nk/2N} \\ &= \sum_{k=0}^{N-1} x(2k)\, e^{-j2\pi n(2k)/2N} + \sum_{k=0}^{N-1} x(2k+1)\, e^{-j2\pi n(2k+1)/2N} \\ &= \sum_{k=0}^{N-1} x(2k)\, e^{-j2\pi nk/N} + e^{-j\pi n/N} \sum_{k=0}^{N-1} x(2k+1)\, e^{-j2\pi nk/N} \\ &= \sum_{k=0}^{N-1} h(k)\, e^{-j2\pi nk/N} + e^{-j\pi n/N} \sum_{k=0}^{N-1} g(k)\, e^{-j2\pi nk/N} \\ &= H(n) + e^{-j\pi n/N}\, G(n) \end{aligned} \tag{10-30}$$

To efficiently compute $H(n)$ and $G(n)$, use the previously discussed technique. Let

$$y(k) = h(k) + jg(k) \tag{10-31}$$

then

$$Y(n) = R(n) + jI(n)$$

From Eqs. (10-27) and (10-28)

$$\begin{aligned} H(n) &= R_e(n) + jI_0(n) \\ G(n) &= I_e(n) - jR_0(n) \end{aligned} \tag{10-32}$$

Substitution of (10-32) into (10-30) yields

$$\begin{aligned} X(n) &= R_e(n) + jI_0(n) + e^{-j\pi n/N}[I_e(n) - jR_0(n)] \\ &= \left[R_e(n) + \cos\left(\frac{\pi n}{N}\right) I_e(n) - \sin\left(\frac{\pi n}{N}\right) R_0(n)\right] \\ &\quad + j\left[I_0(n) - \sin\left(\frac{\pi n}{N}\right) I_e(n) - \cos\left(\frac{\pi n}{N}\right) R_0(n)\right] \\ &= X_r(n) + jX_i(n) \end{aligned} \tag{10-33}$$

Hence, the real part of the $2N$ sample function $x(k)$ is

$$\begin{aligned} X_r(n) = \left[\frac{R(n)}{2} + \frac{R(N-n)}{2}\right] + \cos\frac{\pi n}{N}\left[\frac{I(n)}{2} + \frac{I(N-n)}{2}\right] \\ - \sin\frac{\pi n}{N}\left[\frac{R(n)}{2} - \frac{R(N-n)}{2}\right] \end{aligned} \tag{10-34}$$

and similarly the imaginary part is

$$\begin{aligned} X_i(n) = \left[\frac{I(n)}{2} - \frac{I(N-n)}{2}\right] - \sin\frac{\pi n}{N}\left[\frac{I(n)}{2} + \frac{I(N-n)}{2}\right] \\ - \cos\frac{\pi n}{N}\left[\frac{R(n)}{2} - \frac{R(N-n)}{2}\right] \end{aligned} \tag{10-35}$$

Thus, the imaginary part of the complex time function can be used advantageously to compute the transform of a function defined by $2N$ samples by using a discrete transform which sums only over N values. We normally speak of this computation as performing a $2N$-point transform by means of a N-point transform. For reference, an outline of the computation approach is given in Fig. 10-10. This technique as well as the one described previously is used repeatedly in applications of the FFT. FORTRAN and ALGOL programs for computing the FFT with real input data are given in [7] and [8].

1. Function $x(k)$ is real $\quad k = 0, 1, \ldots, 2N - 1$
2. Divide $x(k)$ into two functions

$$\begin{aligned} h(k) &= x(2k) \\ g(k) &= x(2k+1) \end{aligned} \qquad k = 0, 1, \ldots, N-1$$

3. Form the complex function

$$y(k) = h(k) + jg(k) \qquad k = 0, 1, \ldots, N-1$$

4. Compute

$$\begin{aligned} Y(n) &= \sum_{k=0}^{N-1} y(k) e^{-j2\pi nk/N} \\ &= R(n) + jI(n) \qquad n = 0, 1, \ldots, N-1 \end{aligned}$$

where $R(n)$ and $I(n)$ are the real and imaginary parts of $Y(n)$, respectively.

5. Compute

$$\begin{aligned} X_r(n) &= \left[\frac{R(n)}{2} + \frac{R(N-n)}{2}\right] + \cos\frac{\pi n}{N}\left[\frac{I(n)}{2} + \frac{I(N-n)}{2}\right] \\ &\quad - \sin\frac{\pi n}{N}\left[\frac{R(n)}{2} - \frac{R(N-n)}{2}\right] \qquad n = 0, 1, \ldots, N-1 \\ X_i(n) &= \left[\frac{I(n)}{2} - \frac{I(N-n)}{2}\right] - \sin\frac{\pi n}{N}\left[\frac{I(n)}{2} + \frac{I(N-n)}{2}\right] \\ &\quad - \cos\frac{\pi n}{N}\left[\frac{R(n)}{2} - \frac{R(N-n)}{2}\right] \qquad n = 0, 1, \ldots, N-1 \end{aligned}$$

where $X_r(n)$ and $X_i(n)$ are respectively the real and imaginary parts of the $2N$ point discrete transform of $x(k)$.

Figure 10-10. Computation procedure for discrete Fourier transform of a $2N$ point function by means of an N point transform.

PROBLEMS

10-1. Let $x_0(k) = k$; $k = 0, 1, 2$ and 3. Compute Eq. (10-1) and note the total number of multiplications and additions. Repeat the calculation following the procedure outlined by Eqs. (10-6) through (10-14) and again note the total number of multiplications and additions. Compare your results.

10-2. It has been shown that the matrix factoring procedure introduces scrambled results. For the case $N = 8$, 16 and 32 show the order of $X(n)$ which results from the scrambling.

10-3. It is desired to convert Eq. (10-9) into a signal flow graph for the case $N = 8$.
a. How many computation arrays are there?
b. Define dual nodes for this case. What is the dual node spacing for each array? Give a general expression and then identify each node with its duals for each array.
c. Write the equation pair (21) for each node for array 1. Repeat for the other arrays.
d. Determine W^p for each node and substitute these values into the equation determined in part c.
e. Draw the signal flow graph for this case.
f. Show how to unscramble the results of the last computational array.
g. Illustrate on the signal flow graph the concept of node skipping.

10-4. Verify the computer program flow chart illustrated in Fig. 10-6 by mentally observing that each of the arrays determined in Problem 10-3 is correctly computed.

10-5. Relate each statement of the FORTRAN (ALGOL) program illustrated in Fig. 10-7 (10-8) with the computer program flow chart shown in Fig. 10-6.

PROJECTS

The following projects will require access to a digital computer.

10-6. Write an FFT computer program based on the flow chart illustrated in Fig. 10-6. The program should be capable of accepting complex time functions and performing the inverse transform using the alternate inversion formula. Call this program FFT.

10-7. Let $h(t) = e^{-t}$; $t > 0$. Sample $h(t)$ with $T = 0.01$ and $N = 1024$. Compute the discrete Fourier transform of $h(k)$ with both FFT and DFT. Compare computing times.

10-8. Demonstrate the technique illustrated in Fig. 10-10 on the function defined in Problem 10-7. Let $2N = 1024$.

10-9. Let $h(k)$ be defined according to Problem 10-7. Let

$$g(k) = \cos \frac{2\pi k}{1024} \qquad k = 0, \ldots, 1023$$

Simultaneously compute the discrete Fourier transform of $h(k)$ and $g(k)$ using the procedure described in Fig. 10-9.

REFERENCES

1. Bergland, G. D., "A guided tour of the fast Fourier transform," *IEEE Spectrum* (July 1969) Vol. 6, No. 7, pp. 41–52.

2. Brigham, E. O., and R. E. Morrow, "The fast Fourier transform," *IEEE Spectrum* (December 1967), Vol. 4, pp. 63–70.

3. Cooley, J. W., and J. W. Tukey, "An algorithm for machine calculation of complex Fourier series," *Math. Computation* (April 1965), Vol. 19, pp. 297–301.

4. G-AE Subcommittee on Measurement Concepts, "What is the fast Fourier transform?" *IEEE Trans. Audio and Electroacoustics* (June 1967), Vol. AU–15, pp. 45–55. also *Proc. IEEE* (October 1967), Vol. 55, pp. 1664–1674.

5. Gentleman, W. M., "Matrix multiplication and fast Fourier transforms," *Bell System Tech. J.* (July–August 1968), Vol. 47, pp. 1099–1103.

6. Gentleman, W. M., and G. Sande, "Fast Fourier transforms for fun and profit," *AFIPS Proc.*, 1966 Fall Joint Computer Conf., Vol. 29, pp. 563–678, Washington, D. C.: Spartan, 1966.

7. *IBM Applications Program*, System/360. Scientific Subroutine Package (360A-CM-03X), Version II, 1966.

8. Singleton, R. C., "Algol Procedures for the Fast Fourier Transform," *Communications of the ACM* (Nov. 1968), Vol. 11, No. 11, pp. 773–776.

9. Theilheimer, F., "A matrix version of the fast Fourier transform," *IEEE Trans. Audio and Electroacoustics* (June 1969), Vol. AU–17, No. 2, pp. 158–161.

11

THEORETICAL DEVELOPMENT OF THE BASE 2 FFT ALGORITHM

In Sec. 10-2 we pursued a matrix argument to develop an understanding of why the FFT is an efficient algorithm. We then constructed a signal flow graph which described the algorithm for any $N = 2^\gamma$. In this chapter, we relate each of these developments to a theoretical basis. First we will develop a theoretical proof of the algorithm for the case $N = 4$. We then extend these arguments to the case $N = 8$. The reason for these developments for specific cases is to establish the notation which we use in the final derivation of the algorithm for the case $N = 2^\gamma$, γ integer valued.

11-1 DEFINITION OF NOTATION

A necessary evil of most theoretical developments is the introduction of new and unfamiliar notation. In the case of the FFT algorithm, the simplicity achieved as a result of the notation change is worth the effort.

Consider the discrete Fourier transform relationship (10-1)

$$X(n) = \sum_{k=0}^{N-1} x_0(k) W^{nk} \qquad n = 0, 1, \ldots, N-1 \tag{11-1}$$

where we have set $W = e^{-j2\pi/N}$. It is desirable to represent the integers n and k as binary numbers; that is, if we assume $N = 4$, then $\gamma = 2$ and we can represent k and n as 2-bit binary numbers,

$$k = 0, 1, 2, 3 \qquad \text{or} \qquad k = (k_1, k_0) = 00, 01, 10, 11$$

$$n = 0, 1, 2, 3 \qquad \text{or} \qquad n = (n_1, n_0) = 00, 01, 10, 11$$

A compact method of writing k and n is

$$k = 2k_1 + k_0 \qquad n = 2n_1 + n_0 \tag{11-2}$$

where k_0, k_1, n_0, and n_1 can take on the values of 0 and 1 only. Equation (11-2) is simply the method of writing a binary number as its base 10 equivalent.

Using the representation (11-2), we can rewrite (11-1) for the case $N = 4$ as

$$X(n_1, n_0) = \sum_{k_0=0}^{1} \sum_{k_1=0}^{1} x_0(k_1, k_0)\, W^{(2n_1+n_0)(2k_1+k_0)} \tag{11-3}$$

Note that the single summation in (11-1) must now be replaced by γ summations in order to enumerate all the bits of the binary representation of k.

11-2 FACTORIZATION OF W^p

Now consider the W^p term. Since $W^{a+b} = W^a W^b$, then

$$\begin{aligned} W^{(2n_1+n_0)(2k_1+k_0)} &= W^{(2n_1+n_0)2k_1} W^{(2n_1+n_0)k_0} \\ &= [W^{4n_1k_1}]\, W^{2n_0k_1}\, W^{(2n_1+n_0)k_0} \\ &= W^{2n_0k_1}\, W^{(2n_1+n_0)k_0} \end{aligned} \tag{11-4}$$

Note that the term in brackets is equal to unity since

$$W^{4n_1k_1} = [W^4]^{n_1k_1} = [e^{-j2\pi 4/4}]^{n_1k_1} = [1]^{n_1k_1} = 1 \tag{11-5}$$

Thus, Eq. (11-3) can be written in the form

$$X(n_1, n_0) = \sum_{k_0=0}^{1} \left[\sum_{k_1=0}^{1} x_0(k_1, k_0)\, W^{2n_0k_1} \right] W^{(2n_1+n_0)k_0} \tag{11-6}$$

This equation represents the foundation of the FFT algorithm. To demonstrate this point, let us consider each of the summations of (11-6) individually. First rewrite the summation in brackets as

$$x_1(n_0, k_0) = \sum_{k_1=0}^{1} x_0(k_1, k_0)\, W^{2n_0k_1} \tag{11-7}$$

Enumerating the equations represented by (11-7) we obtain

$$\begin{aligned} x_1(0, 0) &= x_0(0, 0) + x_0(1, 0)W^0 \\ x_1(0, 1) &= x_0(0, 1) + x_0(1, 1)W^0 \\ x_1(1, 0) &= x_0(0, 0) + x_0(1, 0)W^2 \\ x_1(1, 1) &= x_0(0, 1) + x_0(1, 1)W^2 \end{aligned} \tag{11-8}$$

If we rewrite (11-8) in matrix notation we have

$$\begin{bmatrix} x_1(0, 0) \\ x_1(0, 1) \\ x_1(1, 0) \\ x_1(1, 1) \end{bmatrix} = \begin{bmatrix} 1 & 0 & W^0 & 0 \\ 0 & 1 & 0 & W^0 \\ 1 & 0 & W^2 & 0 \\ 0 & 1 & 0 & W^2 \end{bmatrix} \begin{bmatrix} x_0(0, 0) \\ x_0(0, 1) \\ x_0(1, 0) \\ x_0(1, 1) \end{bmatrix} \tag{11-9}$$

Note that (11-9) is exactly the factored matrix equation (10-11), developed in Sec. 10-2, with the index k written in binary notation. Thus, the inner summation of (11-6) specifies the first of the factored matrices for the example developed in Sec. 10-2 or, equivalently, the array $l = 1$ of the signal flow graph illustrated in Fig. 10-2.

Similarly, if we write the outer summation of (11-6) as

$$x_2(n_0, n_1) = \sum_{k_0=0}^{1} x_1(n_0, k_0)\, W^{(2n_1+n_0)k_0} \tag{11-10}$$

and enumerate the results in matrix form we obtain

$$\begin{bmatrix} x_2(0,0) \\ x_2(0,1) \\ x_2(1,0) \\ x_2(1,1) \end{bmatrix} = \begin{bmatrix} 1 & W^0 & 0 & 0 \\ 1 & W^2 & 0 & 0 \\ 0 & 0 & 1 & W^1 \\ 0 & 0 & 1 & W^3 \end{bmatrix} \begin{bmatrix} x_1(0,0) \\ x_1(0,1) \\ x_1(1,0) \\ x_1(1,1) \end{bmatrix} \tag{11-11}$$

which is Eq. (10-14). Thus, the outer summation of (11-6) determines the second of the factored matrices of the example in Sec. 10-2.

From Eqs. (11-6) and (11-10) we have

$$X(n_1, n_0) = x_2(n_0, n_1) \tag{11-12}$$

That is, the final results $x_2(n_0, n_1)$ as obtained from the outer sum are in bit-reversed order with respect to the desired values $X(n_1, n_0)$. This is simply the scrambling which results from the FFT algorithm.

If we combine Eqs. (11-7), (11-10), and (11-12),

$$\begin{aligned} x_1(n_0, k_0) &= \sum_{k_1=0}^{1} x_0(k_1, k_0)\, W^{2n_0k_1} \\ x_2(n_0, n_1) &= \sum_{k_0=0}^{1} x_1(n_0, k_0)\, W^{(2n_1+n_0)k_0} \\ X(n_1, n_0) &= x_2(n_0, n_1) \end{aligned} \tag{11-13}$$

then the set (11-13) represents the original Cooley-Tukey [5] formulation of the FFT algorithm for $N = 4$. We term these equations recursive in that the second is computed in terms of the first.

EXAMPLE 11-1

To illustrate further the notation associated with the Cooley-Tukey formulation of the FFT, consider Eq. (11-1) for the case $N = 2^3 = 8$. For this case

$$\begin{aligned} n &= 4n_2 + 2n_1 + n_0 \qquad n_i = 0 \text{ or } 1 \\ k &= 4k_2 + 2k_1 + k_0 \qquad k_i = 0 \text{ or } 1 \end{aligned} \tag{11-14}$$

and (11-1) becomes

$$X(n_2, n_1, n_0) = \sum_{k_0=0}^{1} \sum_{k_1=0}^{1} \sum_{k_2=0}^{1} x_0(k_2, k_1, k_0) W^{(4n_2+2n_1+n_0)(4k_2+2k_1+k_0)} \tag{11-15}$$

Rewriting W^p we obtain

$$W^{(4n_2+2n_1+n_0)(4k_2+2k_1+k_0)} = W^{(4n_2+2n_1+n_0)(4k_2)} W^{(4n_2+2n_1+n_0)(2k_1)} \times W^{(4n_2+2n_1+n_0)(k_0)} \tag{11-16}$$

We note that since $W^8 \triangleq [e^{j2\pi/8}]^8 = 1$, then

$$\begin{aligned} W^{(4n_2+2n_1+n_0)(4k_2)} &= [W^{8(2n_2k_2)}][W^{8(n_1k_2)}] W^{4n_0k_2} = W^{4n_0k_2} \\ W^{(4n_2+2n_1+n_0)(2k_1)} &= [W^{8(n_2k_1)}] W^{(2n_1+n_0)(2k_1)} = W^{(2n_1+n_0)(2k_1)} \end{aligned} \tag{11-17}$$

Hence, Eq. (11-15) can be rewritten as

$$X(n_2, n_1, n_0) = \sum_{k_0=0}^{1} \sum_{k_1=0}^{1} \sum_{k_2=0}^{1} x_0(k_2, k_1, k_0) W^{4n_0k_2} \times W^{(2n_1+n_0)(2k_1)} W^{(4n_2+2n_1+n_0)(k_0)} \tag{11-18}$$

If we let

$$x_1(n_0, k_1, k_0) = \sum_{k_2=0}^{1} x_0(k_2, k_1, k_0) W^{4n_0k_2} \tag{11-19}$$

$$x_2(n_0, n_1, k_0) = \sum_{k_1=0}^{1} x_1(n_0, k_1, k_0) W^{(2n_1+n_0)(2k_1)} \tag{11-20}$$

$$x_3(n_0, n_1, n_2) = \sum_{k_0=0}^{1} x_2(n_0, n_1, k_0) W^{(4n_2+2n_1+n_0)(k_0)} \tag{11-21}$$

$$X(n_2, n_1, n_0) = x_3(n_0, n_1, n_2) \tag{11-22}$$

then we have determined the required matrix factorization or equivalently, the

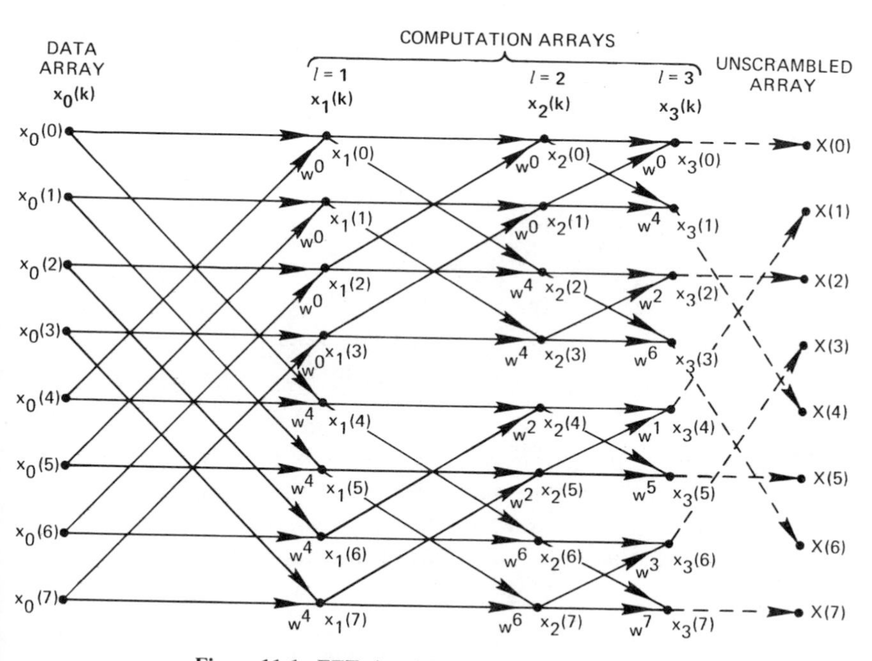

Figure 11-1. FFT signal flow graph for $N = 8$.

signal flow graph for $N = 8$. The signal flow graph as developed from Eqs. (11-19), (11-20), (11-21), and (11-22) is shown in Fig. 11-1.

11-3 DERIVATION OF THE COOLEY-TUKEY ALGORITHM FOR $N = 2^\gamma$

Previous discussions have considered the Cooley-Tukey algorithm for the special cases of $N = 4$ and $N = 8$. We will now develop the general results for $N = 2^\gamma$, γ integer valued.

When $N = 2^\gamma$, n and k can be represented in binary form as

$$\begin{aligned} n &= 2^{\gamma-1}n_{\gamma-1} + 2^{\gamma-2}n_{\gamma-2} + \cdots + n_0 \\ k &= 2^{\gamma-1}k_{\gamma-1} + 2^{\gamma-2}k_{\gamma-2} + \cdots + k_0 \end{aligned} \tag{11-23}$$

Using this representation we can rewrite (11-1) as

$$X(n_{\gamma-1}, n_{\gamma-2}, \ldots, n_0) = \sum_{k_0=0}^{1} \sum_{k_1=0}^{1} \cdots \sum_{k_{\gamma-1}=0}^{1} x(k_{\gamma-1}, k_{\gamma-2}, \ldots, k_0) W^p \tag{11-24}$$

where

$$p = (2^{\gamma-1}n_{\gamma-1} + 2^{\gamma-2}n_{\gamma-2} + \cdots + n_0)(2^{\gamma-1}k_{\gamma-1} + 2^{\gamma-2}k_{\gamma-2} + \cdots + k_0) \tag{11-25}$$

Since $W^{a+b} = W^a W^b$, we rewrite W^p as

$$\begin{aligned} W^p = {} & W^{(2^{\gamma-1}n_{\gamma-1}+2^{\gamma-2}n_{\gamma-2}+\cdots+n_0)(2^{\gamma-1}k_{\gamma-1})} W^{(2^{\gamma-1}n_{\gamma-1}+2^{\gamma-2}n_{\gamma-2}+\cdots+n_0)(2^{\gamma-2}k_{\gamma-2})} \\ & \times \cdots W^{(2^{\gamma-1}n_{\gamma-1}+2^{\gamma-2}n_{\gamma-2}+\cdots+n_0)k_0} \end{aligned} \tag{11-26}$$

Now consider the first term of (11-26);

$$\begin{aligned} W^{(2^{\gamma-1}n_{\gamma-1}+2^{\gamma-2}n_{\gamma-2}+\cdots+n_0)(2^{\gamma-1}k_{\gamma-1})} &= [W^{2^\gamma(2^{\gamma-2}n_{\gamma-1}k_{\gamma-1})}][W^{2^\gamma(2^{\gamma-3}n_{\gamma-2}k_{\gamma-1})}] \\ & \quad \times \cdots [W^{2^\gamma(n_1k_{\gamma-1})}] W^{2^{\gamma-1}(n_0k_{\gamma-1})} \\ &= W^{2^{\gamma-1}(n_0k_{\gamma-1})} \end{aligned} \tag{11-27}$$

since

$$W^{2^\gamma} = W^N = [e^{-j2\pi/N}]^N = 1 \tag{11-28}$$

Similarly, the second term of (11-26) yields

$$\begin{aligned} W^{(2^{\gamma-1}n_{\gamma-1}+2^{\gamma-2}n_{\gamma-2}+\cdots+n_0)(2^{\gamma-2}k_{\gamma-2})} &= [W^{2^\gamma(2^{\gamma-3}n_{\gamma-1}k_{\gamma-2})}][W^{2^\gamma(2^{\gamma-4}n_{\gamma-2}k_{\gamma-2})}] \\ & \quad \times \cdots W^{2^{\gamma-1}(n_1k_{\gamma-2})} W^{2^{\gamma-2}(n_0k_{\gamma-2})} \\ &= W^{(2n_1+n_0)2^{\gamma-2}k_{\gamma-2}} \end{aligned} \tag{11-29}$$

Note that as we progress through the terms of (11-26), we add another factor which does not cancel by the condition $W^{2^\gamma} = 1$. This process continues until we reach the last term in which there is no cancellation.

Using these relationships, Eq. (11-24) can be rewritten as

$$\begin{aligned} X(n_{\gamma-1}, n_{\gamma-2}, \ldots, n_0) = {} & \sum_{k_0=0}^{1} \sum_{k_1=0}^{1} \cdots \sum_{k_{\gamma-1}=0}^{1} x_0(k_{\gamma-1}, k_{\gamma-2}, \ldots, k_0) \\ & \times W^{2^{\gamma-1}(n_0k_{\gamma-1})} W^{(2n_1+n_0)2^{\gamma-2}k_{\gamma-2}} \cdots \\ & \times W^{(2^{\gamma-1}n_{\gamma-1}+2^{\gamma-2}n_{\gamma-2}+\cdots+n_0)k_0} \end{aligned} \tag{11-30}$$

Performing each of the summations separately and labeling the intermediate results, we obtain

$$x_1(n_0, k_{\gamma-2}, \ldots, k_0) = \sum_{k_{\gamma-1}=0}^{1} x_0(k_{\gamma-1}, k_{\gamma-2}, \ldots, k_0)W^{2^{\gamma-1}(n_0 k_{\gamma-1})}$$

$$x_2(n_0, n_1, k_{\gamma-3}, \ldots, k_0) = \sum_{k_{\gamma-2}=0}^{1} x_1(n_0, k_{\gamma-2}, \ldots, k_0)W^{(2n_1+n_0)2^{\gamma-2}k_{\gamma-2}}$$

$$\vdots$$

$$x_\gamma(n_0, n_1, \ldots, n_{\gamma-1}) = \sum_{k_0=0}^{1} x_{\gamma-1}(n_0, n_1, \ldots, k_0)W^{(2^{\gamma-1}n_{\gamma-1}+2^{\gamma-2}n_{\gamma-2}+\cdots+n_0)k_0}$$

$$X(n_{\gamma-1}, n_{\gamma-2}, \ldots, n_0) = x_\gamma(n_0, n_1, \ldots, n_{\gamma-1}) \tag{11-31}$$

This set of recursive equations represents the original Cooley-Tukey formulation of the FFT, $N = 2^\gamma$. Recall that the direct evaluation of an N point transform requires approximately N^2 complex multiplications. Now consider the number of multiplications required to compute the relationships (11-31). There are γ summation equations which each represent N equations. Each of the latter equations contains two *complex* multiplications; however the first multiplication of each equation is actually a multiplication by unity. This follows from the fact that the first multiplication is always of the form $W^{ak_{\gamma-i}}$, where $k_{\gamma-i} = 0$. Thus only $N\gamma$ *complex* multiplications are required. Noting Problem 11-5, it can be shown that in the computation of an array, there occurs the relationship $W^p = -W^{p+N/2}$; the number of multiplications can be reduced by another factor of 2. The number of *complex* multiplications for $N = 2^\gamma$ is then $N\gamma/2$. Similarly, one can reason that there are $N\gamma$ *complex* additions.

Extension of these results to the case for N chosen according to other criteria than $N = 2^\gamma$ will be discussed in Chapter 12. Prior to these discussions we will develop the canonic forms of the FFT.

11-4 CANONIC FORMS OF THE FFT

Up to this point in our discussion we have considered the specific algorithm originally developed by Cooley and Tukey [5]. However, there exist many variations of the algorithm which are, in a sense, *canonic*. Each particular algorithm variation is formulated to exploit a particular property of the data being transformed or the machine being used to perform the computation. Most of these variants are based either on the Cooley-Tukey or the Sande-Tukey algorithm [7]. In this section, we describe the most often used forms of the FFT.

Cooley-Tukey Algorithm

As stated previously, our discussions have considered only the Cooley-Tukey algorithm. In particular, recall from Eq. (11-13) that for $N = 4$, the

FFT algorithm can be written as

$$\begin{aligned} x_1(n_0, k_0) &= \sum_{k_1=0}^{1} x_0(k_1, k_0)\, W^{2n_0k_1} \\ x_2(n_0, n_1) &= \sum_{k_0=0}^{1} x_1(n_0, k_0)\, W^{(2n_1+n_0)k_0} \end{aligned} \tag{11-32}$$

and illustrated graphically as shown by the signal flow graph of Fig. 11-2(a). We observe from the tree graph that this form of the algorithm can be computed *in-place*; that is, a dual node pair can be computed and the results stored in the original data storage locations. Further, we observe that with this form of the algorithm the input data is in natural order and the output data is in scrambled order.

Recall from the discussions in Sec. 10-5 that to determine p, the power of W, it is necessary to bit-reverse the node index k. We thus describe this form of the algorithm as requiring the powers of W in bit-reversed order. This is contrasted to other canonic forms of the FFT, to be discussed next, which require the powers of W to be in natural order. FORTRAN and ALGOL programs for computing the FFT via the signal flow graph in Fig. 11-2(a) are given in [9] and [10], respectively.

If one so desires, it is possible to rearrange the signal flow graph shown in Fig. 11-2(a) in order that the input data is in *scrambled* order and the output data is in natural order. To accomplish this rearrangement, simply interchange nodes (01) and (10) in each array of Fig. 11-2(a), carrying the inputs with that node to each respective node when interchanging. The resulting signal flow graph is illustrated in Fig. 11-2(b). We observe from the flow graph that computation of this algorithm can be done *in place* and that the powers of W necessary to perform the computation occur in natural order.

Mathematically, the signal flow graph of Fig. 11-2(b) is described by Eqs. (11-32), but with the change that the input data $x_0(k_1, k_0)$ is scrambled prior to computing the arrays. An ALGOL program for computing the FFT by means of Fig. 11-2(b) is given in [10].

These two algorithms are often referred to in the literature by the term *decimation in time*. This terminology arises from the fact that alternate derivations of the algorithm [6] are structured to appeal to the concept of sample rate reduction or throwing away samples; thus, the term decimation in time.

Sande-Tukey Algorithm

Another distinct form of the FFT is due to Sande [7]. To develop this form, let $N = 4$ and write

$$X(n_1, n_0) = \sum_{k_0=0}^{1} \sum_{k_1=0}^{1} x_0(k_1, k_0)\, W^{(2n_1+n_0)(2k_1+k_0)} \tag{11-33}$$

In contrast to the Cooley-Tukey approach, we separate the components of n instead of the components of k.

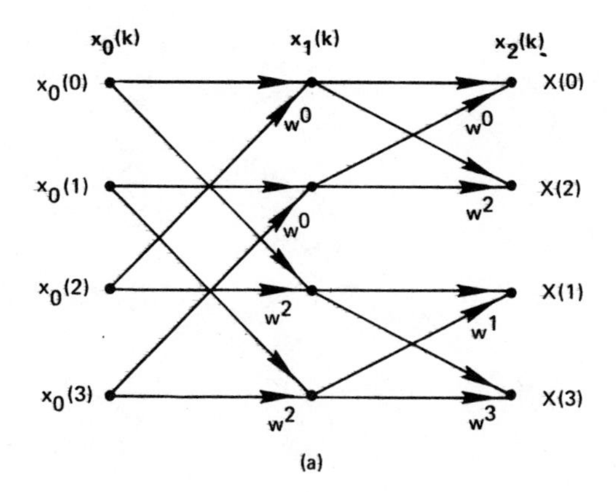

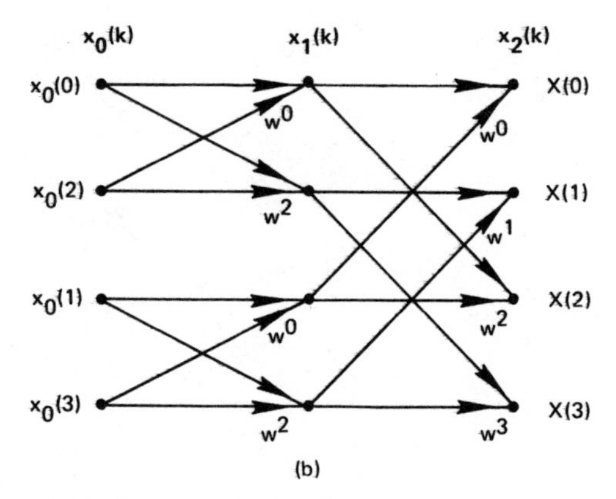

Figure 11-2. Cooley-Tukey FFT algorithm signal flow graphs: (a) data naturally ordered; (b) data in inverse order.

$$\begin{aligned}W^{(2n_1+n_0)}W^{(2k_1+k_0)} &= W^{(2n_1)(2k_1+k_0)}W^{n_0(2k_1+k_0)}\\&= [W^{4n_1k_1}]W^{2n_1k_0}W^{n_0(2k_1+k_0)}\\&= W^{2n_1k_0}W^{n_0(2k_1+k_0)}\end{aligned} \tag{11-34}$$

where $W^4 = 1$.

Thus Eq. (11-33) can be written as

$$X(n_1, n_0) = \sum_{k_0=0}^{1}\left[\sum_{k_1=0}^{1} x_0(k_1, k_0)\,W^{2n_0k_1}\,W^{n_0k_0}\right]W^{2n_1k_0} \tag{11-35}$$

If we define the intermediate computational steps, then

$$\begin{aligned}x_1(n_0, k_0) &= \sum_{k_1=0}^{1} x_0(k_1, k_0)\,W^{2n_0k_1}\,W^{n_0k_0}\\x_2(n_0, n_1) &= \sum_{k_0=0}^{1} x_1(n_0, k_0)\,W^{2n_1k_0}\\X(n_1, n_0) &= x_2(n_0, n_1)\end{aligned} \tag{11-36}$$

The signal flow graph describing Eqs. (11-36) is shown in Fig. 11-3(a).

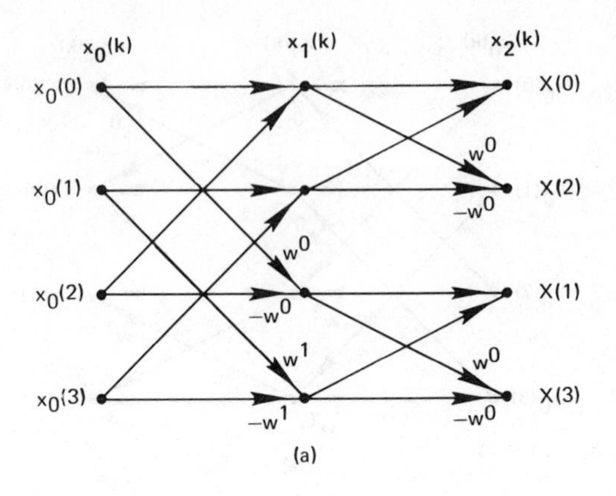

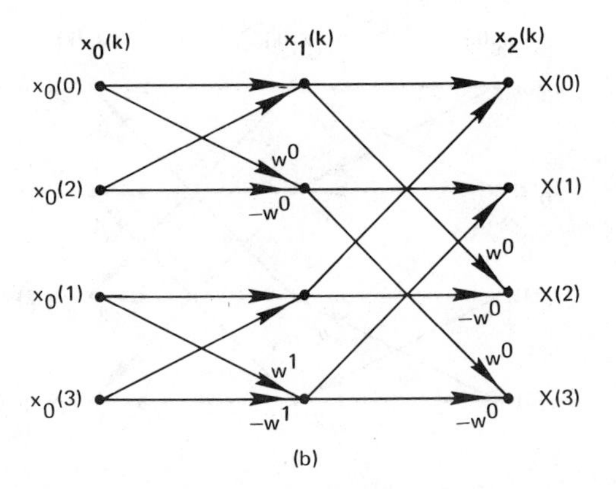

Figure 11-3. Sande-Tukey FFT algorithm signal flow graphs: (a) data naturally ordered; (b) data in inverse order.

We note that the input data is in natural order, the output data is in scrambled order, and the powers of W occur in natural order. A FORTRAN program for computing the FFT according to Fig. 11-3(a) is given by [11].

To develop a signal flow graph which yields results in natural order, we proceed as in the Cooley-Tukey case and interchange node (01) and (10) in Fig. 11-3(a). The resulting signal flow graph is shown in Fig. 11-3(b). The input data is now in bit-reversed order and the powers of W now occur in bit-reversed order. Reference [12] describes a FFT program which is organized according to the signal flow graph of Fig. 11-3(d).

These two forms of the FFT algorithm are also known by the terms *decimation in frequency* where the reasoning for the terminology is analogous to that for *decimation in time*.

Summary

For ease of reference, all four of these variants of the FFT algorithm are illustrated in Fig. 11-4 for the case $N = 8$. We may choose among these

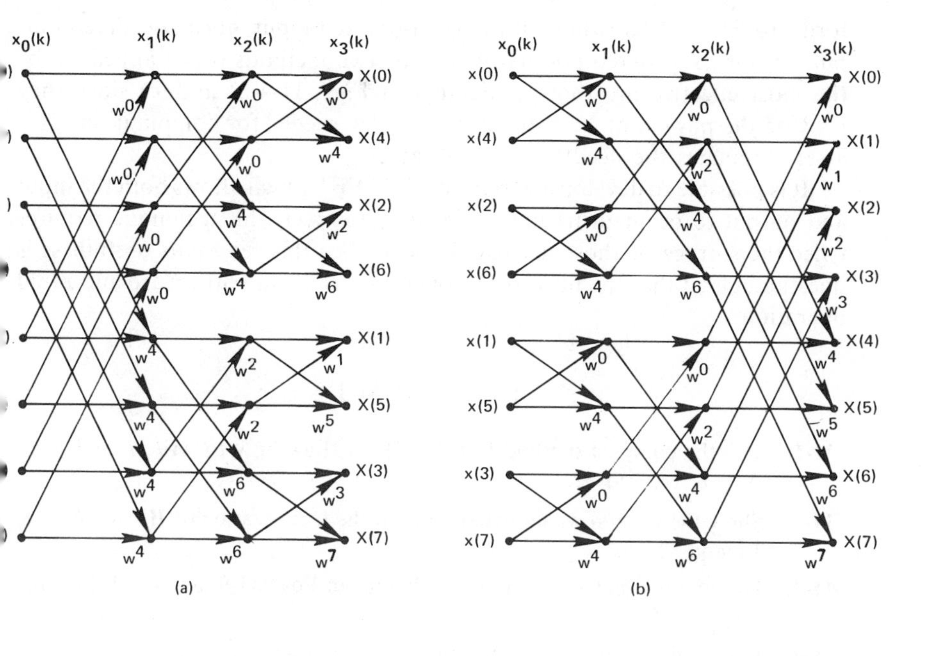

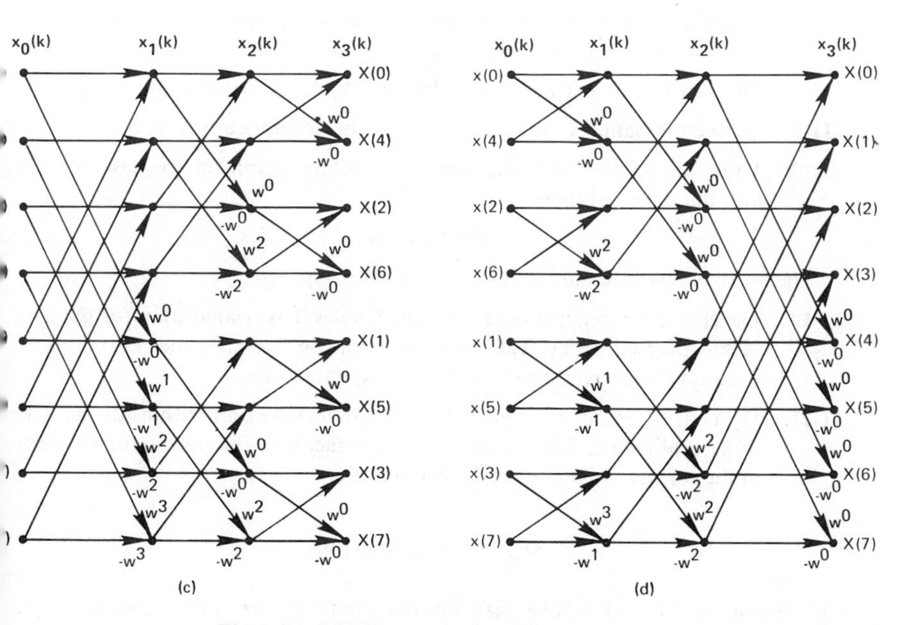

Figure 11-4. FFT canonic signal flow graphs.

forms to find an algorithm with normal-ordered input, normal-ordered output, or normal-ordered powers of W. The two methods which appear to be the most effective are those illustrated in Figs. 11-4(b) and (c) since they provide the powers of W in the correct order needed for computation. This asset eliminates the need for storage tables.

It is possible to develop a variation of the FFT which allows both the input and output to occur in natural order [6]. However, this technique requires twice the storage of those discussed above. For this reason its usefulness is questionable in that the bit-reversal operation is a straightforward and rapid operation.

PROBLEMS

11-1. Use the compact method [Eq. (11-2)] to show how to write k and n for $N = 8$, 16, and 32.

11-2. Show the term(s) which equal unity in the factorization of W^p for $N = 8$, 16, and 32.

11-3. Develop the matrix equations analogous to Eqs. (11-9) and (11-11) for the case $N = 8$.

11-4. Repeat Example 11-1 for the case $N = 16$.

11-5. For each array in the Equation set (11-31) show the occurrence of the relationship $W^p = -W^{p+N/2}$ and, hence, the reduction by 2 in the number of required multiplications.

11-6. Draw the Cooley-Tukey algorithm signal flow graph for the case $N = 16$ and the data in bit-reversed order.

11-7. Derive the Sande-Tukey algorithm equations for the case $N = 8$.

11-8. Derive the Sande-Tukey algorithm for the general case $N = 2^\gamma$.

11-9. Draw the Sande-Tukey algorithm signal flow graph for the case $N = 16$ and the data in bit-reversed order.

PROJECTS

These projects will require access to a digital computer:

11-10. Develop a computer program for the Cooley-Tukey algorithm for the data in bit-reversed order. Take advantage of the fact that the powers of W occur in the correct order needed for computation.

11-11. Develop a computer program for the Sande-Tukey algorithm for the data in natural order. Take advantage of the fact that the powers of W occur in the correct order needed for computation.

REFERENCES

1. BERGLAND, G. D., "The fast Fourier transform recursive equations for arbitrary length records," *Math. Computation* (April 1967), Vol. 21, pp. 236–238.

2. ———, "A guided tour of the fast Fourier transform," *IEEE Spectrum* (July 1969), Vol. 6, No. 7, pp. 41–52.

3. BRIGHAM, E. O., and R. E. MORROW, "The fast Fourier transform," *IEEE Spectrum* (December 1967), Vol. 4, pp. 63–70.

4. COOLEY, J. W., P. A. W. LEWIS, and P. D. WELCH, "The fast Fourier transform algorithm and its applications." IBM Corp., Research Paper RC–1743, February 9, 1967.

5. COOLEY, J. W., and J. W. TUKEY, "An algorithm for machine calculation of complex Fourier series," *Math. Computation* (April 1965), Vol. 19, pp. 297–301.

6. G-AE Subcommittee on Measurement Concepts, "What is the fast Fourier transform?" *IEEE Trans. Audio and Electroacoustics* (June 1967), Vol. AU–15, pp. 45–55. Also *Proc. IEEE* (October 1967), Vol. 55, pp. 1664–1674.

7. GENTLEMAN, W. M., and G. SANDE, "Fast Fourier transforms for fun and profit," *AFIPS Proc.*, 1966 Fall Joint Computer Conf., Vol. 29, pp. 563–678, Washington, D. C.: Spartan, 1966.

8. SINGLETON, R. C., "On computing the fast Fourier transform," *Commun. ACM* (October 1957), Vol. 10, pp. 647–654.

9. *IBM Applications Program*, System/360, Scientific Subroutine Package (360A-CM-03X), Version II, 1966.

10. SINGLETON, R. C., "ALGOL procedures for the Fast Fourier Transform," *Communications of the ACM* (Nov. 1968), Vol. 11, No. 11, pp. 773–776.

11. KAHANER, D. K., "Matrix Description of the Fast Fourier Transform," *IEEE Trans. on Audio and Electroacoustics* (Dec. 1970), Vol. AU-18, No. 4, pp. 442–450.

12. COOLEY, J. W., P. A. LEWIS, and P. D. WELCH, "The fast Fourier Transform and its Applications," *IEEE Trans. on Education* (March 1969), Vol. 12, No. 1.

12

FFT ALGORITHMS FOR ARBITRARY FACTORS

In the discussions to this point we have assumed that the number of points N to be Fourier transformed satisfy the relationship $N = 2^\gamma$, γ integer valued. As we saw, this base 2 algorithm resulted in a tremendous computation time savings; however, the constraint $N = 2^\gamma$ can be rather restrictive. In this section we develop FFT algorithms which remove this assumption. We will show that significant time savings can be obtained as long as N is highly composite; that is, $N = r_1 r_2 \ldots r_m$ where r_i is an integer.

To develop the FFT algorithm for arbitrary factors, we will first consider case $N = r_1 r_2$. This approach allows us to develop the notation required in the proof for the general case. Examples for the base 4 and base "4 + 2" algorithms will be used to further develop the case $N = r_1 r_2$. The Cooley-Tukey and Sande-Tukey algorithms for the case $N = r_1 r_2 \ldots r_m$ will then be developed.

12-1 FFT ALGORITHM FOR $N = r_1 r_2$

Assume that the number of points N satisfies the relationship $N = r_1 r_2$, where r_1 and r_2 are integer valued. To derive the FFT algorithm for this case we first express the n and k indices in Eq. (11-1) as

$$\begin{aligned} n &= n_1 r_1 + n_0 \qquad n_0 = 0, 1, \ldots, r_1 - 1 \qquad n_1 = 0, 1, \ldots, r_2 - 1, \\ k &= k_1 k_2 + k_0 \qquad k_0 = 0, 1, \ldots, r_2 - 1 \qquad k_1 = 0, 1, \ldots, r_1 - 1 \end{aligned} \tag{12-1}$$

We observe that this method of writing the indices allows us to give a unique representation of each decimal integer. Using (12-1) we can rewrite Eq. (11-1) as

$$X(n_1, n_0) = \sum_{k_0=0}^{r_2-1} \left[\sum_{k_1=0}^{r_1-1} x_0(k_1, k_0) \, W^{nk_1r_2} \right] W^{nk_0} \tag{12-2}$$

Rewriting $W^{nk_1r_2}$ we obtain

$$\begin{aligned} W^{nk_1r_2} &= W^{(n_1r_1+n_0)k_1r_2} \\ &= W^{r_1r_2n_1k_1} W^{n_0k_1r_2} \\ &= [W^{r_1r_2}]^{n_1k_1} W^{n_0k_1r_2} \\ &= W^{n_0k_1r_2} \end{aligned} \tag{12-3}$$

where we have used the fact that $W^{r_1r_2} = W^N = 1$.

From (12-3) we rewrite the inner sum of (12-2) as a new array

$$x_1(n_0, k_0) = \sum_{k_1=0}^{r_1-1} x_0(k_1, k_0) \, W^{n_0k_1r_2} \tag{12-4}$$

If we expand the terms W^{nk_0} the outer loop can be written as

$$x_2(n_0, n_1) = \sum_{k_0=0}^{r_2-1} x_1(n_0, k_0) \, W^{(n_1r_1+n_0)k_0} \tag{12-5}$$

The final result can be written as

$$X(n_1, n_0) = x_2(n_0, n_1) \tag{12-6}$$

Thus, as in the base 2 algorithm, the results are in reversed order.

Equations (12-4), (12-5), and (12-6) are the defining FFT algorithm relationships for the case $N = r_1r_2$. To further illustrate this particular algorithm, consider the following examples.

Base 4 Algorithm for $N = 16$

Let us consider the case $N = r_1r_2 = 4 \times 4 = 16$; that is we will develop the base 4 algorithm for the case $N = 16$. Using Eq. (12-1) we represent the variables n and k in Eq. (11-1) in a base 4 or quaternary number system

$$\begin{aligned} n &= 4n_1 + n_0 \qquad n_1, n_0 = 0, 1, 2, 3 \\ k &= 4k_1 + k_0 \qquad k_1, k_0 = 0, 1, 2, 3 \end{aligned} \tag{12-7}$$

Equation (12-2) then becomes

$$X(n_1, n_0) = \sum_{k_0=0}^{3} \left[\sum_{k_1=0}^{3} x_0(k_1, k_0) \, W^{4nk_1} \right] W^{nk_0} \tag{12-8}$$

Rewriting W^{4nk_1} we obtain

$$\begin{aligned} W^{4nk_1} &= W^{4(4n_1+n_0)k_1} \\ &= W^{16n_1k_1} W^{4n_0k_1} \\ &= [W^{16}]^{n_1k_1} W^{4n_0k_1} \\ &= W^{4n_0k_1} \end{aligned} \tag{12-9}$$

The term in brackets is equal to unity since $W^{16} = 1$.

Substitution of (12-9) into (12-4) yields the inner sum of the base 4 algorithm

$$x_1(n_0, k_0) = \sum_{k_1=0}^{3} x_0(k_1, k_0)W^{4n_0k_1} \tag{12-10}$$

From (12-5) the outer sum is

$$x_2(n_0, n_1) = \sum_{k_0=0}^{3} x_1(n_0, k_0)W^{(4n_1+n_0)k_0} \tag{12-11}$$

and from (12-6) the base 4 algorithm results are given by

$$X(n_1, n_0) = x_2(n_0, n_1) \tag{12-12}$$

Equations (12-10), (12-11), and (12-12) define the base 4 algorithm for the case $N = 16$. Based on these equations, we can develop a base 4 signal flow graph.

Base 4 Signal Flow Graph for $N = 16$

From the defining relationships (12-10) and (12-11) we observe that there are $\gamma = 2$ computational arrays and there are four inputs to each node. The inputs to node $x_1(n_0, k_0)$ are $x_0(0, k_0)$, $x_0(1, k_0)$, $x_0(2, k_0)$, and $x_0(3, k_0)$. That is, the four inputs to a node i in array l are those nodes in array $l - 1$ whose subscripts differ from i only in the $(\gamma - l)$th quaternary digit.

We show in Fig. 12-1 an abbreviated signal flow graph for the base 4 $N = 16$ algorithm. To alleviate confusion, only representative transmission paths are shown and all W^p factors have been omitted. W^p factors can be determined from Eqs. (12-10) and (12-11). Each pattern of transmission paths shown is applied sequentially to successive nodes until all nodes have been considered. Figure 12-1 also illustrates the unscrambling procedure for the base 4 algorithm.

Enumeration of Eqs. (12-10) and (12-11) reveals that the base 4 algorithm requires approximately 30 percent fewer multiplications than the base 2 algorithm. We will develop in detail these relationships in Sec. 12-4.

Base "4 + 2" Algorithm for $N = 8$

Let us now consider the case $N = r_1 r_2 = 4 \times 2 = 8$. This case represents the simplest form of the base "4 + 2" algorithm. Base "4 + 2" implies that we compute as many arrays as possible with a base 4 algorithm and then compute a base 2 array.

To develop the "4 + 2" algorithm, we first substitute $r_1 = 4$ and $r_2 = 2$ into Eq. (12-1);

$$\begin{aligned} n &= 4n_1 + n_0 \qquad & n_0 &= 0, 1, 2, 3 \qquad & n_1 &= 0, 1 \\ k &= 2k_1 + k_0 \qquad & k_0 &= 0, 1 \qquad & k_1 &= 0, 1, 2, 3 \end{aligned} \tag{12-13}$$

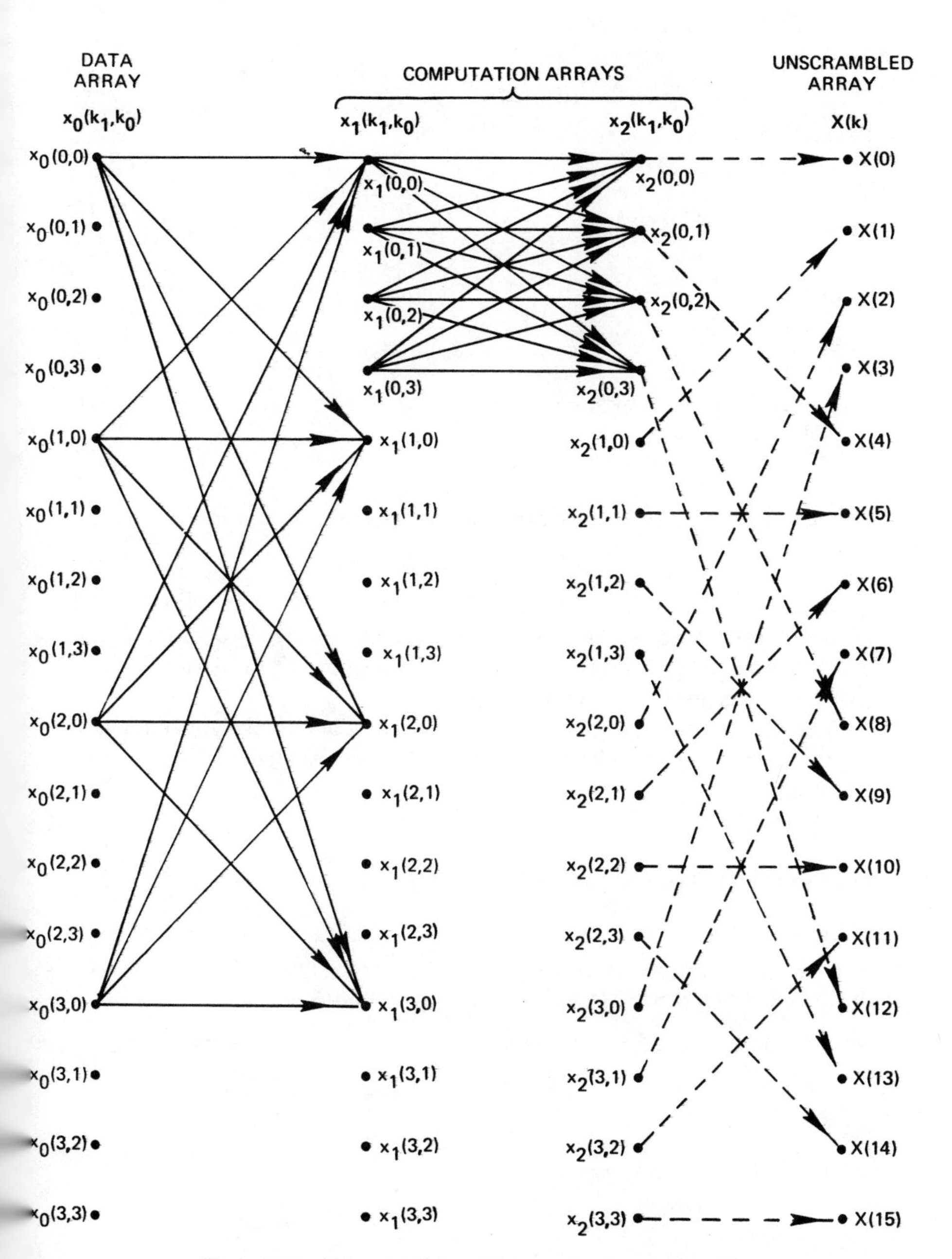

Figure 12-1. Abbreviated signal flow graph—base 4, $N = 16$.

Equation (12-2) then becomes

$$X(n_1, n_0) = \sum_{k_0=0}^{1}\left[\sum_{k_1=0}^{3} x_0(k_1, k_0)W^{2nk_1}\right]W^{nk_0} \tag{12-14}$$

Expanding W^{2nk_1} we obtain

$$\begin{aligned} W^{2nk_1} &= W^{2(4n_1+n_0)k_1} \\ &= [W^8]^{n_1k_1}W^{2n_0k_1} \\ &= W^{2n_0k_1} \end{aligned} \tag{12-15}$$

With (12-15) the inner sum of (12-14) becomes

$$x_1(n_0, k_0) = \sum_{k_1=0}^{3} x_0(k_1, k_0)W^{2n_0k_1} \tag{12-16}$$

The outer loop can be written as

$$x_2(n_0, n_1) = \sum_{k_0=0}^{1} x_1(n_0, k_0)W^{(4n_1+n_0)k_0} \tag{12-17}$$

and the unscrambling is accomplished according to the relationship

$$X(n_1, n_0) = x_2(n_0, n_1) \tag{12-18}$$

Equations (12-16), (12-17), and (12-18) represent the base "4 + 2" FFT algorithm for $N = 8$. We observe that Eq. (12-16) is a base 4 iteration on the data array and Eq. (12-17) is a base 2 iteration on array $l = 1$. An abbreviated signal flow graph is illustrated in Fig. 12-2.

The "4 + 2" algorithm is more efficient than the base 2 algorithm and is equally restrictive in the choice of N. We will now develop an FFT algorithm for N highly composite.

12-2 COOLEY TUKEY ALGORITHM FOR $N = r_1r_2 \ldots r_m$

Assume that the number of points to be discretely transformed satisfies $N = r_1r_2 \ldots r_m$, where $r_1, r_2, \ldots, r_m$ are integer valued. We first express the indices n and k in a variable radix representation

$$\begin{aligned} n &= n_{m-1}(r_1r_2 \ldots r_{m-1}) + n_{m-2}(r_1r_2 \ldots r_{m-2}) + \cdots + n_1r_1 + n_0 \\ k &= k_{m-1}(r_2r_3 \ldots r_m) + k_{m-2}(r_3r_4 \ldots r_m) + \cdots + k_1r_m + k_0 \end{aligned} \tag{12-19}$$

where

$$\begin{aligned} n_{i-1} &= 0, 1, 2, \ldots, r_i - 1 \qquad && 1 \leq i \leq m \\ k_i &= 0, 1, 2, \ldots, r_{m-i} - 1 \qquad && 0 \leq i \leq m - 1 \end{aligned}$$

We can now rewrite Eq. (11-1) as

$$X(n_{m-1}, n_{m-2}, \ldots, n_1, n_0) = \sum_{k_0}\sum_{k_1}\cdots\sum_{k_{m-1}} x_0(k_{m-1}, k_{m-2}, \ldots, k_0)W^{nk} \tag{12-20}$$

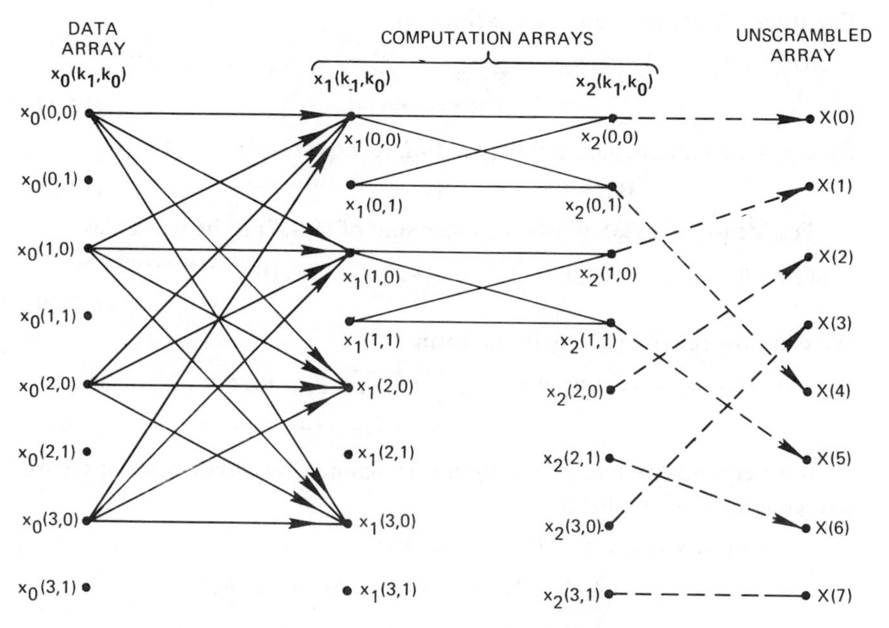

Figure 12-2. Abbreviated signal flow graph—base "4 & 2," $N = 8$.

where $\sum_{k_i}$ represents a summation over all $k_i = 0, 1, 2, \ldots, r_{m-i} - 1; 0 \leq i \leq m - 1$. Note that

$$W^{nk} = W^{n[k_{m-1}(r_2 r_3 \cdots r_m) + \cdots + k_0]} \tag{12-21}$$

and the first term of the summation expands to

$$\begin{aligned} W^{nk_{m-1}(r_2 r_3 \cdots r_m)} &= W^{[n_{m-1}(r_1 r_2 \cdots r_{m-1}) + \cdots + n_0][k_{m-1}(r_2 r_3 \cdots r_m)]} \\ &= [W^{r_1 r_2 \cdots r_m}]^{[n_{m-1}(r_2 r_3 \cdots r_{m-1}) + \cdots + n_1]k_{m-1}} W^{n_0 k_{m-1}(r_2 \cdots r_m)} \end{aligned} \tag{12-22}$$

Because $W^{r_1 r_2 \cdots r_m} = W^N = 1$ then Eq. (12-22) may be written as

$$W^{nk_{m-1}(r_2 r_3 \cdots r_m)} = W^{n_0 k_{m-1}(r_2 \cdots r_m)} \tag{12-23}$$

and hence Eq. (12-21) becomes

$$W^{nk} = W^{n_0 k_{m-1}(r_2 \cdots r_m)} W^{n[k_{m-2}(r_3 \cdots r_m) + \cdots + k_0]} \tag{12-24}$$

Equation (12-20) can now be rewritten as

$$\begin{aligned} X(n_{m-1}, n_{m-2}, \ldots, n_1, n_0) = \sum_{k_0} \sum_{k_1} \cdots [\sum_{k_{m-1}} x_0(k_{m-1}, k_{m-2}, \ldots, k_0) \\ \times W^{n_0 k_{m-1}(r_2 \cdots r_m)}] W^{n[k_{m-2}(r_3 \cdots r_m) + \cdots + k_0]} \end{aligned} \tag{12-25}$$

Note that the inner sum is over k_{m-1} and is only a function of the variables n_0 and $k_{m-2}, \ldots, k_0$. Thus, we define a new array as

$$x_1(n_0, k_{m-2}, \ldots, k_0) = \sum_{k_{m-1}} x_0(k_{m-1}, \ldots, k_0) W^{n_0 k_{m-1}(r_2 \cdots r_m)} \tag{12-26}$$

Equation (12-25) can now be rewritten as

$$X(n_{m-1}, n_{m-2}, \ldots, n_1, n_0) = \sum_{k_0} \sum_{k_1} \cdots \sum_{k_{m-2}} x_1(n_0, k_{m-2}, \ldots, k_0) \times W^{n[k_{m-2}(r_3 \cdots r_m) + \cdots + k_0]} \tag{12-27}$$

By arguments analogous to those leading to Eq. (12-23), we obtain

$$W^{nk_{m-2}(r_3 r_4 \cdots r_m)} = W^{(n_1 r_1 + n_0)k_{m-2}(r_3 r_4 \cdots r_m)} \tag{12-28}$$

The identity (12-28) allows the inner sum of (12-27) to be written as

$$x_2(n_0, n_1, k_{m-3}, \ldots, k_0) = \sum_{k_{m-2}} x_1(n_0, k_{m-2}, \ldots, k_0) W^{(n_1 r_1 + n_0)k_{m-2}(r_3 r_4 \cdots r_m)} \tag{12-29}$$

We can now rewrite (12-27) in the form

$$X(n_{m-1}, n_{m-2}, \ldots, n_1, n_0) = \sum_{k_0} \sum_{k_1} \cdots \sum_{k_{m-3}} x_2(n_0, n_1, k_{m-3}, \ldots, k_0) \times W^{n[k_{m-3}(r_4 r_5 \cdots r_m) + \cdots + k_0]} \tag{12-30}$$

If we continue reducing (12-30) in this manner, we obtain a set of recursive equations of the form

$$x_i(n_0, n_1, \ldots, n_{i-1}, k_{m-i-1}, \ldots, k_0) = \sum_{k_{m-i}} x_{i-1}(n_0, n_1, \ldots, n_{i-2}, k_{m-i}, \ldots, k_0) \times W^{[n_{i-1}(r_1 r_2 \cdots r_{i-1}) + \cdots + n_0]k_{m-i}(r_{i+1} \cdots r_m)} \qquad i = 1, 2, \ldots, m \tag{12-31}$$

Expression (12-31) is valid provided we define $(r_{i+1} \ldots r_m) = 1$ for $i > m - 1$ and $k_{-1} = 0$.

The final results are given by

$$X(n_{m-1}, \ldots, n_0) = x_m(n_0, \ldots, n_{m-1}) \tag{12-32}$$

Expression (12-31) is an extension due to Bergland [1] of the original Cooley-Tukey [9] algorithm. We note that there are N elements in array $\boldsymbol{x}_1$, each requiring r_1 operations (one complex multiplication and one complex addition), giving a total of Nr_1 operations to obtain $\boldsymbol{x}_1$. Similarly, it takes Nr_2 operations to calculate $\boldsymbol{x}_2$ from $\boldsymbol{x}_1$. Thus, the computation of $\boldsymbol{x}_m$ requires $N(r_1 + r_2 + \ldots + r_m)$ operations. This bound does not take into account the symmetries of the complex exponential which can be exploited. These symmetries will be discussed at length in Sec. 12-4.

FORTRAN and ALGOL FFT computer programs for $N = r_1 r_2 \ldots r_m$ are given in [13] and [14], respectively.

12-3 SANDE-TUKEY ALGORITHM FOR $N = r_1 r_2 \ldots r_m$

As discussed in Sec. 11-4, an alternate algorithm results if we expand the nk product in terms of n instead of k. If we represent

$$W^{nk} = W^{n_0 k} W^{[n_1 r_1 + n_2(r_1 r_2) + \cdots + n_{m-1}(r_1 \cdots r_{m-1})]k} \tag{12-33}$$

then the inner sum analogous to Eq. (12-26) can be written as

$$x_1'(n_0, k_{m-2}, \ldots, k_0) = \sum_{k_{m-1}} x_0(k_{m-1}, \ldots, k_0) W^{n_0[k_{m-1}(r_2 r_3 \cdots r_m) + \cdots + k_0]} \tag{12-34}$$

Similar results can be obtained for $\boldsymbol{x}_2', \ldots, \boldsymbol{x}_m'$. The general form of these recursive equations is

$$\begin{aligned} x_i'(n_0, n_1, \ldots, n_{i-1}, k_{m-i-1}, \ldots, k_0) &= \sum_{k_{m-i}} x_{i-1}'(n_0, n_1, \ldots, n_{i-2}, k_{m-i}, \ldots, k_0) \\ &\quad \times W^{n_{i-1}[k_{m-i}(r_{i+1} \cdots r_m) + \cdots + k_0](r_1 \cdots r_{i-1})} \\ & \qquad i = 1, 2, \ldots, m \end{aligned} \tag{12-35}$$

As before, the final result given by

$$X(n_{m-1}, \ldots, n_0) = x_m'(n_0, \ldots, n_{m-1}) \tag{12-36}$$

The set of recursive equations (12-35) represents the general form of the Sande-Tukey algorithm. In Sec. 12-4 we will develop the more efficient computational form for these equations.

12-4 TWIDDLE FACTOR FFT ALGORITHMS

In the previous section, we briefly alluded to the fact that further efficiencies in FFT algorithms can be achieved if the symmetries of the sine and cosine functions are exploited. In this section we first describe an example of a base 4 algorithm with *twiddle factors* for $N = 16$. It is shown that the introduction of these *twiddle factors* results in a more efficient algorithm. Next, we develop the general form of the Cooley-Tukey and Sande-Tukey *twiddle factor* algorithms. Finally, we describe in detail the computational efficiencies of these algorithms for base 2, base 4, base 8, and base 16.

Base 4, N = 16, Twiddle Factor Algorithm

Recall from Eqs. (12-10) and (12-11) that the recursive equations for the base 4 FFT algorithm for $N = 16$ are given by

$$\begin{aligned} x_1(n_0, k_0) &= \sum_{k_1=0}^{3} x_0(k_1, k_0) W^{4n_0k_1} \\ x_2(n_0, n_1) &= \sum_{k_0=0}^{3} x_1(n_0, k_0) W^{(4n_1+n_0)k_0} \\ X(n_1, n_0) &= x_2(n_0, n_1) \end{aligned} \tag{12-37}$$

To illustrate the *twiddle factor* concept let us rewrite (12-37) as

$$X(n_1, n_0) = \sum_{k_0=0}^{3} \left[\sum_{k_1=0}^{3} x_0(k_1, k_0) W^{4n_0k_1} \right] W^{4n_1k_0} W^{n_0k_0} \tag{12-38}$$

Note that the term $W^{n_0k_0}$ has been arbitrarily grouped with the outer sum and could have just as easily been grouped with the inner sum. By regrouping, (12-38) becomes

$$X(n_1, n_0) = \sum_{k_0=0}^{3} \left[\left\{ \sum_{k_1=0}^{3} x_0(k_1, k_0) W^{4n_0k_1} \right\} W^{n_0k_0} \right] W^{4n_1k_0} \tag{12-39}$$

or in recursive form

$$x_1(n_0, k_0) = \left[\sum_{k_1=0}^{3} x_0(k_1, k_0) W^{4n_0k_1} \right] W^{n_0k_0} \tag{12-40}$$

$$x_2(n_0, n_1) = \left[\sum_{k_0=0}^{3} x_1(n_0, k_0) W^{4n_1k_0} \right] \tag{12-41}$$

$$X(n_1, n_0) = x_2(n_0, n_1) \tag{12-42}$$

The form of the algorithm given by (12-40) exploits the symmetries of the sine and cosine function. To illustrate this point, consider the term $W^{4n_0k_1}$ in brackets in Eq. (12-40). Because $N = 16$, then

$$W^{4n_0k_1} = (W^4)^{n_0k_1} = (e^{-j2\pi(4)/16})^{n_0k_1} = (e^{-j\pi/2})^{n_0k_1} \tag{12-43}$$

Thus $W^{4n_0k_1}$ only takes on the values $\pm i$ and ± 1 depending on the integer n_0k_1. As a result the 4-point transform in brackets in Eq. (12-40) can be evaluated without multiplications. These results are then *referenced* or *twiddled* [10] by the factor $W^{n_0k_0}$ which is outside the brackets in Eq. (12-40). Note that by similar arguments Eq. (12-41) can be evaluated without multiplications.

We see that the total computations required to evaluate the base 4 algorithm appear to have been reduced by this regrouping. Before identifying exactly the required number of operations, let us first develop a general formulation of this regrouping concept.

Cooley-Tukey and Sande-Tukey Twiddle Factor Algorithms

The original Cooley-Tukey formulation is given by the set of recursive Eqs. (12-31). If we regroup these equations the first array takes the form

$$\tilde{x}_1(n_0, k_{m-2}, \ldots, k_0) = [\sum_{k_{m-1}} x_0(k_{m-1}, \ldots, k_0) W^{n_0k_{m-1}(N/r_1)}] W^{(n_0k_{m-2})(r_3\cdots r_m)} \tag{12-44}$$

and the succeeding equations are given by

$$\begin{aligned}\tilde{x}_i(n_0, n_1, \ldots, n_{i-1}, k_{m-i-1}, \ldots, k_0) &= [\sum_{k_{m-i}} \tilde{x}_{i-1}(n_0, \ldots, n_{i-2}, k_{m-i}, \ldots, k_0) W^{n_{i-1}k_{m-i}(N/r_i)}] \\ &\quad \times W^{(n_{i-1}(r_1r_2\cdots r_{i-1})+\cdots+n_1r_1+n_0)k_{m-i-1}(r_{i+2}\cdots r_m)}\end{aligned} \tag{12-45}$$

We have used the notation $\tilde{x}$ to indicate that these results have been obtained by *twiddling*. Equation (12-40) is valid for $i = 1, 2, \ldots, m$ if we interpret the case $i = 1$ in the sense of (12-44) and if we define $(r_{i+2} \ldots r_m) = 1$ for $i > m - 2$ and $k_{-1} = 0$.

Similarly, if we regroup the Sande-Tukey equations (12-35) we obtain

$$\begin{aligned}\tilde{x}'_i(n_0, n_1, \ldots, n_{i-1}, k_{m-i-1}, \ldots, k_0) \\ = \Big[\sum_{k_{m-i}} \tilde{x}'_{i-1}(n_0, \ldots, n_{i-2}, k_{m-i}, \ldots, k_0) W^{n_{i-1}k_{m-i}(N/r_i)}\Big] \\ \times W^{n_{i-1}(k_{m-i-1}(r_{i+2}\cdots r_m)+\cdots+k_1 r_m+k_0)(r_1 r_2\cdots r_{i-1})}\end{aligned} \quad (12\text{-}46)$$

Note that the bracketed terms of (12-45) and (12-46) are identical. Actually, Eq. (12-46) only represents a slightly different manner in writing the Eqs. (12-35) whereas (12-45) represents a regrouping of the W^p factors in (12-31).

Each iteration of both (12-45) and (12-46) requires the evaluation of an r_i-point Fourier transform followed by a referencing or twiddling operation. The importance of this formulation is that the bracketed r_i-point Fourier transforms can be computed with a minimum number of multiplications. For example, if $r_i = 8$ (that is, a base 8 transform), then the W^p factor in brackets only takes on the values ± 1, $\pm j$, $\pm e^{j\pi/4}$, and $\pm e^{-j\pi/4}$. Since the first two factors require no multiplications and the product of a complex number and either of the last 2 factors requires only 2 real multiplications, a total of only 4 real multiplications are required in evaluating each 8-point transform.

As we see, the twiddle factor algorithms allow us to take advantage of the properties of the sine and cosine functions. We will now examine in detail the number of operations required to evaluate these algorithms for various number base systems.

12-5 COMPUTATIONS REQUIRED BY BASE 2, BASE 4, BASE 8, AND BASE 16 ALGORITHMS

Let us consider the case $N = 2^{12} = 4096$. The real number of multiplications and additions required to evaluate the recursive Eqs. (12-45) and (12-46) is given in Table 12-1. This summary of operations was first reported by Bergland [2]. In counting the number of multiplications and additions it is assumed that each of the twiddling operations requires one complex multiplication except when the multiplier is W^0.

TABLE 12-1 OPERATIONS REQUIRED IN COMPUTING BASE 2, BASE 4, BASE 8, AND BASE 16 FFT ALGORITHMS FOR $N = 4096$

Algorithm	Number of Real Multiplications	Number of Real Additions
Base 2	81,924	139,266
Base 4	57,348	126,978
Base 8	49,156	126,978
Base 16	48,132	125,442

TABLE 12-2† COMPARISON OF ARITHMETIC OPERATIONS REQUIRED FOR BASE 2, BASE 4, BASE 8, AND BASE 16 ALGORITHMS

Algorithm	Required Computation for	Real Multiplications	Real Additions
Base 2 algorithm for $N = 2^{\gamma}$ $\gamma = 0, 1, 2, \ldots$	Evaluating $(N/2)\gamma$, 2 term Fourier transforms.	0	$2N\gamma$
	Referencing	$((\gamma/2 - 1)N + 1)(4)$	$((\gamma/2 - 1)N + 1)(2)$
	Complete analysis	$(2\gamma - 4)N + 4$	$(3\gamma - 2)N + 2$
Base 4 algorithm for $N = (2^2)^{\gamma/2}$ $\gamma/2 = 0, 1, 2, \ldots$	Evaluating $(N/4)(\gamma/2)$, 4 term Fourier transforms.	0	$2N\gamma$
	Referencing	$((3\gamma/8 - 1)N + 1)(4)$	$((3\gamma/8 - 1)N + 1)(2)$
	Complete analysis	$(1.5\gamma - 4)N + 4$	$(2.75\gamma - 2)N + 2$
Base 8 algorithm for $N = (2^3)^{\gamma/3}$ $\gamma/3 = 0, 1, 2, \ldots$	Evaluating $(N/8)(\gamma/3)$, 8 term Fourier transforms.	$N\gamma/6$	$13N\gamma/6$
	Referencing	$((7\gamma/24 - 1)N + 1)(4)$	$((7\gamma/24 - 1)N + 1)(2)$
	Complete analysis	$(1.333\gamma - 4)N + 4$	$(2.75\gamma - 2)N + 2$
Base 16 algorithm for $N = (2^4)^{\gamma/4}$ $\gamma/4 = 0, 1, 2, \ldots$	Evaluating $(N/16)(\gamma/4)$, 16 term Fourier transforms.	$3N\gamma/8$	$9N\gamma/4$
	Referencing	$((15\gamma/64 - 1)N + 1)(4)$	$((15\gamma/64 - 1)N + 1)(2)$
	Complete analysis	$(1.3125\gamma - 4)N + 4$	$(2.71875\gamma - 2)N + 2$

†Reprinted by permission of G. D. Bergland, "A Fast Fourier Transform Algorithm Using Base 8 Iterations," *Math. Computation*, Vol. 22 (April 1968): 275–279.

Table 12-2 gives the required number of operations for N any power of 2. This table is also due to Bergland [2]. It is assumed that each algorithm performs as many operations as possible in the most efficient manner.

It is readily apparent from Table 12-2 that performing as many high base iterations as possible reduces the total required computations. However, as the base of the algorithm increases, the algorithm becomes involved. Base 4 and base 8 algorithms appear to be the most efficient while at the same time relatively easy to compute.

12-6 SUMMARY OF FFT ALGORITHMS

The number of variations of the FFT algorithm appears to be almost limitless. Each version has been formulated to exploit properties of the data being analyzed, properties of the computer, or special-purpose FFT hardware being used. However, the majority of these different algorithms are based on the Cooley-Tukey or Sande-Tukey algorithms which we have discussed. In this section we will briefly summarize some of these variations of the basic FFT algorithms.

FFT Algorithm for N Arbitrary

An FFT algorithm for the case N arbitrary was contributed by Bluestein [6]. To describe this algorithm, we write the discrete Fourier transform

$$X(n) = \sum_{k=0}^{N-1} x(k)\, W^{nk} \tag{12-47}$$

where $W = e^{-j2\pi/N}$. Equation (12-47) can be rewritten as

$$\begin{aligned} X(n) &= \sum_{k=0}^{N-1} x(k)\, W^{nk+[(k^2-k^2+n^2-n^2)/2]} \\ &= W^{n^2/2}\left\{\sum_{k=0}^{N-1} [W^{k^2/2}\, x(k)]\, W^{-(n-k)^2/2}\right\} \end{aligned} \tag{12-48}$$

If we let $y(k) = W^{k^2/2}x(k)$ and $h(n-k) = W^{-(n-k)^2/2}$ then Eq. (12-48) can be rewritten as

$$X(n) = W^{n^2/2}\left\{\sum_{k=0}^{N-1} y(k)\, h(n-k)\right\} \tag{12-49}$$

Equation (12-49) is in the form of a convolution.

We will show in Chapter 13 that the most efficient way to evaluate an equation of the form of (12-49) is to *augment* functions $y(k)$ and $h(k)$ with sufficient zeros so that they are of length N', where N' is a highly factorable number, and apply the FFT. Using this procedure we can evaluate the FFT for any value of N by multiplying the result of the convolution by the factor $W^{n^2/2}$.

Restructured FFT Algorithm for Real Data

In Sec. 10-10 we discussed a technique for computing the discrete Fourier transform of a $2N$-point real function with an N-point transform. An alternate approach is to restructure the FFT algorithm by eliminating those computations which lead to redundant results [3]. A base 8 FFT computer program for implementing this technique can be found in Ref. [4].

FFT Algorithms for Large Data Arrays

In general, the number of points N to be Fourier transformed is limited only by the memory size of the computer. As we increase the size of N beyond core memory size, it is necessary to store data in slower memory devices such as a drum, disk, or tape. It is then necessary to compute separate transforms and then combine these individual transforms in a fashion similar to that employed when computing the discrete transform of $2N$ real samples using only an N-point transform. Discussions and computer programs for performing the FFT on large data arrays can be found in Refs. [7], [8] and [12].

PROBLEMS

12-1. Derive the FFT algorithm for $N = r_1 r_2$ for the case where the components of n are separated, i.e., the Sande-Tukey algorithm.

12-2. Develop the signal flow graph for the base 4 Sande-Tukey algorithm for $N = 16$.

12-3. Develop the Sande-Tukey base "4 + 2" algorithm equation for the case $N = 8$.

12-4. Develop fully the Sande-Tukey algorithm for the case $N = r_1, r_2, \ldots, r_m$.

12-5. Develop the Cooley-Tukey base 8 algorithm for the case $N = 64$.

12-6. Repeat Problem 12-5 for the Sande-Tukey algorithm.

12-7. Let $N = 16$. Develop the Sande-Tukey twiddle factor algorithm equations.

12-8. Let $N = 8$. Is there an advantage in using twiddle factors in computing the FFT by the Cooley-Tukey base 2 algorithm? Verify your conclusions by demonstrating the required number of multiplications in each case.

12-9. Repeat Problem 12-8 for the base "4 + 2" algorithm.

PROJECTS

These projects will require access to a digital computer.

12-10. Develop a FFT computer program for a base "4 + 2" Cooley-Tukey algorithm and bit-reversed data.

12-11. Develop a FFT computer program for a base "4 + 2" Sande-Tukey algorithm and the data in natural order.

12-12. Develop a FFT computer program for a base "8 + 4 + 2" Sande-Tukey algorithm with data in natural order. The program should maximize the number of base 8 computations, then maximize the number of base 4 computations.

REFERENCES

1. BERGLAND, G. D., "The fast Fourier transform recursive equations for arbitrary length records," *Math. Computation* (April 1967), Vol. 21, pp. 236–238.
2. ———, "A fast Fourier transform algorithm using base eight iterations," *Math Computation* (April 1968), Vol. 22, pp. 275–279.
3. ———, "A fast Fourier transform algorithm for real-valued series," *Commun. ACM* (October 1968) Vol. 11, pp. 703–710.
4. ———, "A radix-eight fast Fourier transform subroutine for real-valued series," *IEEE Trans. Audio and Electroacoustics* (June 1969), Vol. AU–17, No. 2, pp. 138–144.
5. ———, "A guided tour of the fast Fourier transform," *IEEE Spectrum*, (July 1969), Vol. 6, No. 7, pp. 41–52.
6. BLUESTEIN, L. I., "A linear filter approach to the computation of the discrete Fourier transform," *1968 Nerem Rec.*, pp. 218–219.
7. BRENNER, N. M., "Fast Fourier transform of externally stored data," *IEEE Trans. Audio and Electroacoustics* (June 1969), Vol. AU-17, 128–132.
8. BUIJS, H. L., "Fast Fourier transformation of large arrays of data," *Applied Optics* (January 1969), Vol. 8, 211–212.
9. COOLEY, J. W., and J. W. TUKEY, "An algorithm for machine calculation of complex Fourier series," *Math. Computation* Vol. 19, pp. 297–301.
10. GENTLEMAN, W. M., and G. SANDE, "Fast Fourier transforms for fun and profit," *1966 Fall Joint Computer Conf., AFIPS Proc.*, Vol. 29, pp. 563–678, Washington, D. C.: Spartan, 1966.
11. RADAR, C. M., "Discrete Fourier transforms when the number of data samples is prime," *Proc. IEEE* (Letters) (June 1968), Vol. 56 pp. 1107–1108.
12. SINGLETON, R. C., "A method for computing the fast Fourier transform with auxiliary memory and limited high-speed storage," *IEEE Trans. Audio and Electroacoustics* (June 1967), Vol. AU-15, pp. 91–98.
13. ———, "An Algol procedure for the fast Fourier transform with arbitrary factors," *Comm. ACM* (Nov. 1968), Vol. 11, No. 11, 776–779.
14. ———, "An algorithm for computing the mixed radix fast Fourier transform," *IEEE Trans. Audio and Electroacoustics* (June 1969), Vol. AU-17, No. 2, pp. 93–103.

13

FFT CONVOLUTION AND CORRELATION

Applications of the FFT such as digital filtering, power spectrum analysis, simulation, system analysis, communication theory, etc., are based, in general, on a specific implementation of the convolution or correlation integral or on using the FFT as an approximation to the continuous Fourier transform. If we explore in detail these two FFT uses, we have developed the fundamental basics for general FFT applications.

In Chapter 9, we described the application of the discrete transform to the computation of the continuous Fourier transform. Because the FFT is simply a procedure for rapidly computing the discrete Fourier transform, we have already explored the fundamentals of this basic FFT application.

It remains to describe the techniques for applying the FFT to the computation of the convolution and correlation integrals. As discussed in Chapter 6, both of these integrals can be computed by means of the discrete Fourier transform. We expect that it is impractical to use the frequency domain relationships to compute either convolution or correlation because of the apparent number of increased multiplications required. However, with the tremendous increase in computational speed which can be achieved using the FFT, it is advantageous to use frequency domain analysis.

In this chapter we develop the techniques for using the FFT for high speed convolution and correlation.

13-1 FFT CONVOLUTION OF FINITE DURATION WAVEFORMS

The discrete convolution relationship is given by Eq. (7-1) as

$$y(k) = \sum_{i=0}^{N-1} x(i)\, h(k-i) \tag{13-1}$$

where both $x(k)$ and $h(k)$ are periodic functions with period N. As discussed in Chapter 7, discrete convolution, if correctly performed, will produce a replica of the continuous convolution provided that both the functions $x(t)$ and $h(t)$ are of finite duration. We now extend that discussion to include efficient computation by means of the FFT.

Consider the finite duration or *aperiodic* waveforms $x(t)$ and $h(t)$ illustrated in Fig. 13-1(a). Continuous convolution of these functions is also shown. By means of discrete convolution, it is desired to produce a replica of the continuous convolution. Recall from Chapter 7 that discrete convolution requires that we sample both $x(t)$ and $h(t)$ and form periodic functions with period N as illustrated in Fig. 13-1(b). The resulting discrete convolution [Fig. 13-1(c)] is periodic; however, each period is a replica of the desired finite duration or aperiodic. Scaling constant T (sample period) has been introduced to obtain results comparable with continuous convolution. Note that since both $x(t)$ and $h(t)$ are shifted from the origin, a large N is required to produce a period sufficiently large to eliminate the *overlap* or *end effect* described in Chapter 7. Computationally, the discrete convolution of

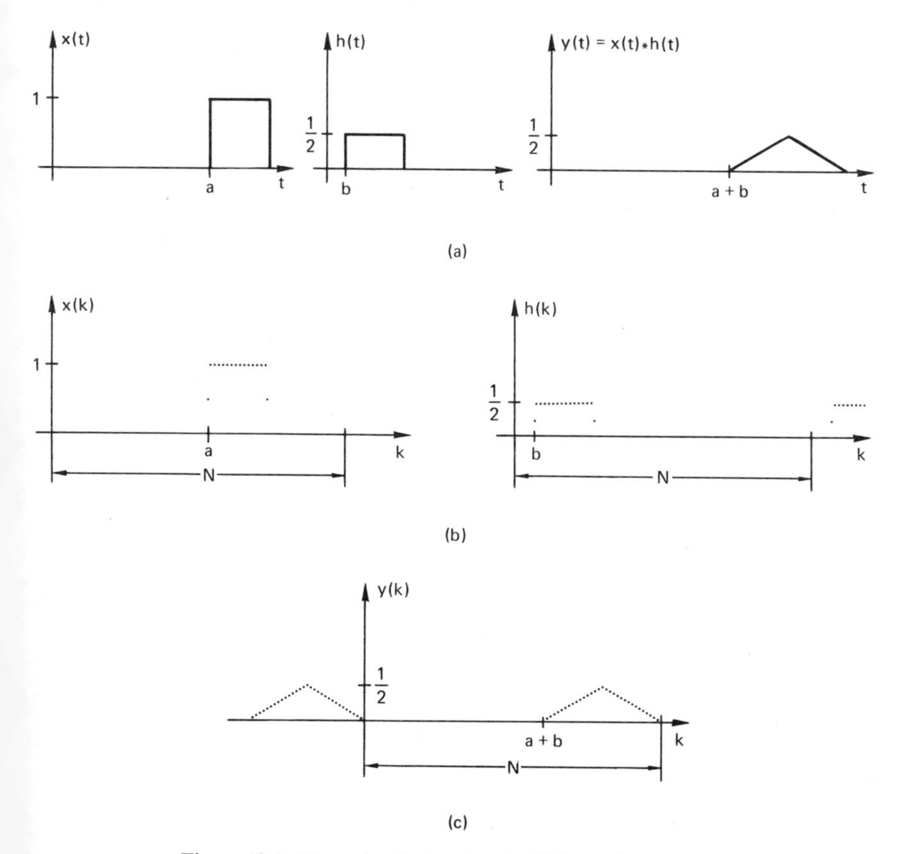

Figure 13-1. Example illustrating inefficient discrete convolution.

Fig. 13-1(c) is very inefficient because of the large number of zeros produced in the interval $[0, a + b]$ as illustrated in Fig. 13-1(c). To perform the discrete convolution more efficiently we simply restructure the data.

Restructuring the Data

As illustrated in Fig. 13-2, we have shifted each of the sampled functions [Fig. 13-1(b)] to the origin; from Eq. (7-4) we chose the period $N > P + Q - 1$ to eliminate *overlap* effects. Since we ultimately desire to use the FFT to perform the convolution, we also require that $N = 2^\gamma$, γ integer valued; we have assumed that a radix 2 algorithm will be used. Our results are easily extended for the case of other algorithms.

Functions $x(k)$ and $h(k)$ are required to have a period N satisfying

$$\begin{aligned} &N > P + Q - 1 \\ &N = 2^\gamma \qquad \gamma \text{ integer valued} \end{aligned} \tag{13-2}$$

Discrete convolution for this choice of N is illustrated in Fig. 13-2(b); the results differ from that of Fig. 13-1(c) only in a shift of origin. But this shift is known *a priori*. From Fig. 13-1(a) the shift of the convolution $y(t)$ is simply the sum of the shifts of the functions being convolved. Consequently, no information is lost if we shift each function to the origin prior to convolution.

To compute the identical waveform Fig. 13-2(b) by means of the FFT, we first shift both $x(t)$ and $h(t)$ to the origin. Let the shifts of $x(t)$ and $h(t)$ be a and b, respectively. Both functions are then sampled. Next, N is chosen to satisfy (13-2). The resulting sampled periodic functions are defined by the relationships

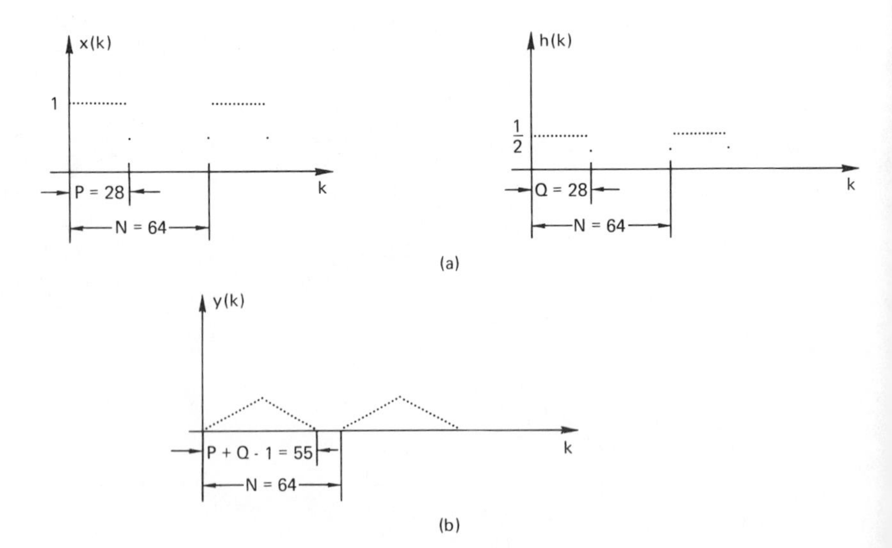

Figure 13-2. Discrete convolution of restructured data.

$$
\begin{aligned}
x(k) &= x(kT+a) && k = 0, 1, \ldots, P-1 \\
x(k) &= 0 && k = P, P+1, \ldots, N-1 \\
h(k) &= h(kT+b) && k = 0, 1, \ldots, Q-1 \\
h(k) &= 0 && k = Q, Q+1, \ldots, N-1
\end{aligned}
\tag{13-3}
$$

The same notation is used to emphasize that our discussions should assume only sampled periodic functions shifted to the origin. We now compute the discrete convolution by means of the discrete convolution theorem (7-8). The discrete Fourier transforms of $x(k)$ and $h(k)$ are computed;

$$X(n) = \sum_{k=0}^{N-1} x(k)\, e^{-j2\pi nk/N} \tag{13-4}$$

$$H(n) = \sum_{k=0}^{N-1} h(k)\, e^{-j2\pi nk/N} \tag{13-5}$$

Next the product

$$Y(n) = X(n)\, H(n) \tag{13-6}$$

is formed, and finally we compute the inverse discrete transform of $Y(n)$ and obtain the discrete convolution $y(k)$;

$$y(k) = \frac{1}{N} \sum_{n=0}^{N-1} Y(n)\, e^{j2\pi nk/N} \tag{13-7}$$

Note that the single discrete convolution Eq. (13-1) has now been replaced by Eqs. (13-4), (13-5), (13-6), and (13-7). This gives rise to the term *the long way around.* However, because of the computing efficiency of the FFT algorithm, these four equations define in fact *a shortcut by the long way around.*

Computing Speed of FFT Convolution

Evaluation of the N samples of the convolution result $y(k)$ by means of Eq. (13-1) requires a computation time proportional to N^2, the number of multiplications. From Sec. 11-3, the computation time of the FFT is proportional to $N \log_2 N$; computation time of Eqs. (13-4), (13-5), and (13-6), is then proportional to $3\, N \log_2 N$ and computation time of Eq. (13-7) is proportional to N. It is generally faster to use the FFT and Eq. (13-4) through (13-7) to compute the discrete convolution rather than computing Eq. (13-1) directly.

Exactly how much faster the FFT approach is than the conventional approach depends not only on the number of points but also on the details of the FFT and convolution programs being employed. To indicate the point at which FFT convolution is faster and the time savings which can be obtained by means of FFT convolution, we have observed as a function of N the time required to compute (13-1) by both the direct and FFT approach; data was computed on a GE 630 computer. Results of this simulation are given in Table 13-1. As shown, with our computer programs it is faster to employ the FFT for convolution if N exceeds 64. In Sec. 13-4 we will describe a technique for reducing the FFT computing time by an additional factor of 2; as a result the breakeven point will be for $N = 32$.

TABLE 13-1 COMPUTING TIMES IN SECONDS

N	Direct Method	FFT Method	Speed Factor
16	0.0008	0.003	0.27
32	0.003	0.007	0.43
64	0.012	0.015	0.8
128	0.047	0.033	1.4
256	0.19	0.073	2.6
512	0.76	0.16	4.7
1024	2.7	0.36	7.5
2048	11.0	0.78	14.1
4096	43.7	1.68	26.0

1. Let $x(t)$ and $h(t)$ be finite-length functions shifted from the origin by a and b, respectively.
2. Shift $x(t)$ and $h(t)$ to the origin and sample
$$x(k) = x(kT + a) \qquad k = 0, 1, \ldots, P - 1$$
$$h(k) = h(kT + b) \qquad k = 0, 1, \ldots, Q - 1$$
3. Choose N to satisfy the relationships
$$N \geq P + Q - 1$$
$$N = 2^{\gamma} \qquad \gamma \text{ integer valued}$$
where P is the number of samples defining $x(t)$ and Q is the number of samples defining $h(t)$.
4. Augment with zeros the sampled functions of step (2)
$$x(k) = 0 \qquad k = P, P + 1, \ldots, N - 1$$
$$h(k) = 0 \qquad k = Q, Q + 1, \ldots, N - 1$$
5. Compute the discrete transform of $x(k)$ and $h(k)$
$$X(n) = \sum_{k=0}^{N-1} x(k) e^{-j2\pi nk/N}$$
$$H(n) = \sum_{k=0}^{N-1} h(k) e^{-j2\pi nk/N}$$
6. Compute the product
$$Y(n) = X(n)H(n)$$
7. Compute the inverse discrete transform using the forward transform
$$y(k) = \sum_{n=0}^{N-1} \left[\frac{1}{N} Y^*(n)\right] e^{-j2\pi nk/N}$$

Figure 13-3. Computation procedure for FFT convolution of finite-length functions.

A step-by-step computation procedure for applying the FFT to convolution of discrete functions is given in Fig. 13-3. Note that we have used the alternate inversion formula (8-9) in step (7). Also, these results must be scaled by T if comparison with continuous results is desired.

13-2 FFT CORRELATION OF FINITE DURATION WAVEFORMS

Application of the FFT to discrete correlation is very similar to FFT convolution. As a result, our discussion on correlation will only point out the differences in the two techniques.

Consider the discrete correlation relationship,

$$z(k) = \sum_{i=0}^{N-1} h(i)\, x(k+i) \tag{13-8}$$

where both $x(k)$ and $h(k)$ are periodic functions with period N. Fig. 13-4(a) illustrates the same periodic functions $x(k)$ and $h(k)$ considered in Fig. 13-1(b). Correlation of these two functions according to (13-8) is shown in Fig. 13-4(b). Scaling factor T has been introduced for ease of comparison with continuous results. Note from Fig. 13-4(b) that the shift from the origin of the resultant correlation function is given by the difference between the leading edge of $x(k)$ and the trailing edge of $h(k)$. As in our convolution example, the correlation computation illustrated in Fig. 13-4(b) is inefficient because of the number of zeros included in the N points defining one period

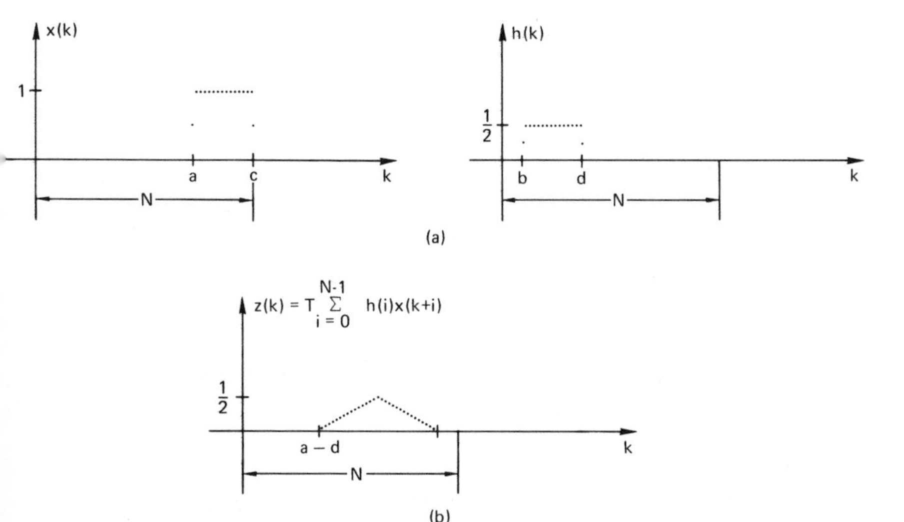

Figure 13-4. Example illustrating inefficient discrete correlation.

of the periodic correlation function. Restructuring of the data is again the solution we choose for efficient computation.

If we shift both functions to the origin as shown in Fig. 13-5(a), then the resulting correlation is as illustrated in Fig. 13-5(b). Although the correlation waveform is correct, it must be *unraveled* before it is meaningful. We can remedy this situation by restructing the waveform $x(k)$ as shown in Fig. 13-5(c). For this condition, the resulting correlation waveform is as illustrated in Fig. 13-5(d). This is the desired waveform with the exception of a known time shift.

To apply the FFT to the computation of (13-8), we choose the period N to satisfy the relationships

$$\begin{aligned} &N \geq P + Q - 1 \\ &N = 2^{\gamma} \qquad \gamma \text{ integer valued} \end{aligned} \tag{13-9}$$

We shift and sample $x(t)$ as follows:

$$\begin{aligned} x(k) &= 0 & k &= 0, 1, \ldots, N - P \\ x(k) &= x[kT + a] & k &= N - P + 1, N - P + 2, \ldots, N - 1 \end{aligned} \tag{13-10}$$

That is, we shift the P samples of $x(k)$ to the extreme right of the N samples defining a period. Function $h(t)$ is shifted and sampled according to the relations

$$\begin{aligned} h(k) &= h(kT + b) & k &= 0, 1, \ldots, Q - 1 \\ h(k) &= 0 & k &= Q, Q + 1, \ldots, N - 1 \end{aligned} \tag{13-11}$$

Based on the discrete correlation theorem (7-13), we compute the following

$$X(n) = \sum_{k=0}^{N-1} x(k)\, e^{-j2\pi nk/N} \tag{13-12}$$

$$H(n) = \sum_{k=0}^{N-1} h(k)\, e^{-j2\pi nk/N} \tag{13-13}$$

$$Z(n) = X(n)\, H^*(n) \tag{13-14}$$

$$z(k) = \frac{1}{N} \sum_{n=0}^{N-1} Z(n)\, e^{j2\pi nk/N} \tag{13-15}$$

The resulting $z(k)$ is identical to the illustration of Fig. 13-5(d).

Computing times of Eqs. (13-12) through (13-15) are essentially the same as the convolution Eqs. (13-4) through (13-7) and the results of the previous section are applicable. The computations leading to (13-15) are outlined in Fig. 13-6 for easy reference.

Because the application of the FFT to discrete convolution and correlation is so similar we will limit the remainder of the chapter to a discussion of FFT convolution.

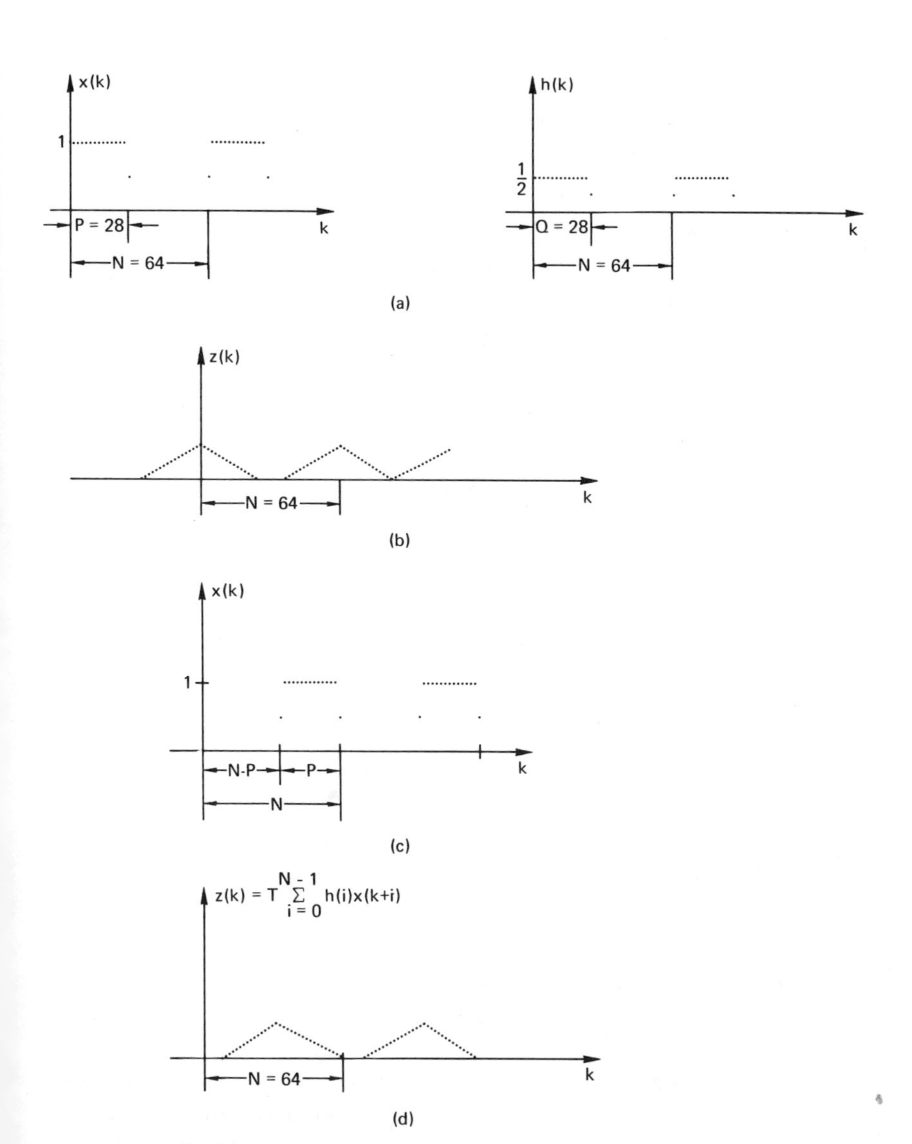

Figure 13-5. Discrete correlation of restructured data.

1. Let $x(t)$ and $h(t)$ be finite-length functions shifted from the origin by a and b, respectively.
2. Let P be the number of samples defining $x(t)$ and Q be the number of samples defining $h(t)$.
3. Choose N to satisfy the relationships

$$N \geq P + Q - 1$$

$$N = 2^{\gamma} \qquad \gamma \text{ integer valued}$$

4. Define $x(k)$ and $h(k)$ as follows:

$$\begin{aligned} x(k) &= 0 & k &= 0, 1, \ldots, N - P \\ x(k) &= x(kT + a) & k &= N - P + 1, N - P + 2, \ldots, N - 1 \\ h(k) &= h(kT + b) & k &= 0, 1, \ldots, Q - 1 \\ h(k) &= 0 & k &= Q, Q + 1, \ldots, N - 1 \end{aligned}$$

5. Compute the discrete transform of $x(k)$ and $h(k)$

$$X(n) = \sum_{k=0}^{N-1} x(k) e^{-j2\pi nk/N}$$

$$H(n) = \sum_{k=0}^{N} h(k) e^{-j2\pi nk/N}$$

6. Change the sign of the imaginary part of $H(n)$ to obtain $H^*(n)$.
7. Compute the product

$$Z(n) = X(n)H^*(n)$$

8. Compute the inverse transform using the forward transform

$$z(k) = \sum_{n=0}^{N-1} \left(\frac{1}{N} Z^*(n)\right) e^{-j2\pi nk/N}$$

Figure 13-6. Computation procedure for FFT correlation of finite-length functions.

13-3 FFT CONVOLUTION OF AN INFINITE AND A FINITE DURATION WAVEFORM

We have discussed to this point only the class of functions for which both $x(t)$ and $h(t)$ are of finite duration. Further we have assumed that $N = 2^{\gamma}$ was sufficiently small so that the number of samples did not exceed our computer memory. When either of these two assumptions is false it is necessary to use the concept of *sectioning*.

Consider the waveforms $x(t)$, $h(t)$, and their convolution $y(t)$ as illustrated in Fig. 13-7. We assume that $x(t)$ is of infinite duration or that the number of samples representing $x(t)$ exceeds the memory of the computer. As a result,

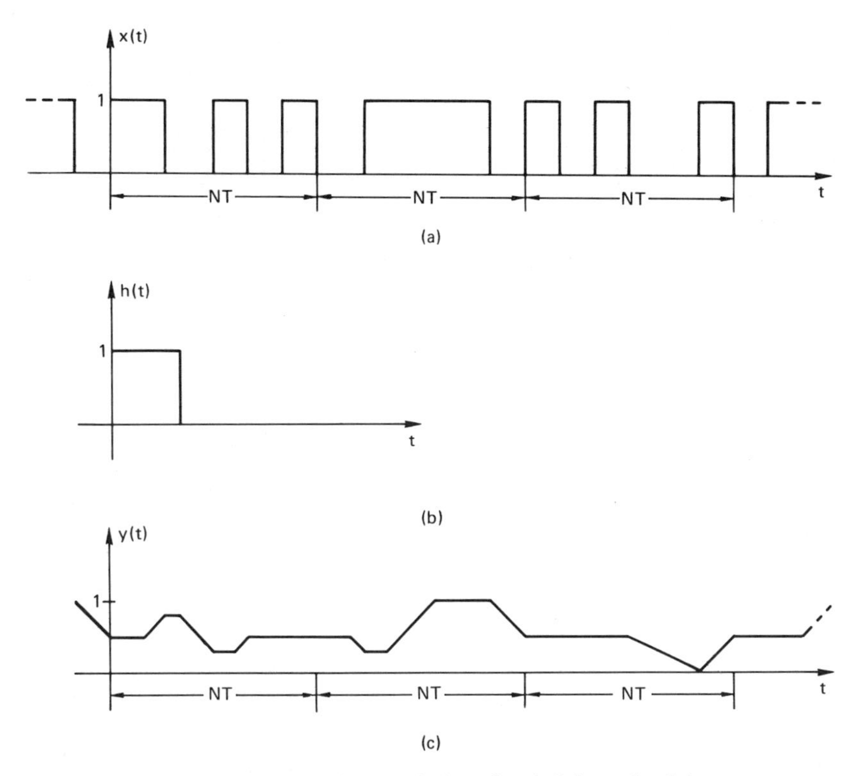

Figure 13-7. Example convolution of an infinite and a finite duration waveform.

it is necessary to decompose $x(t)$ into sections and compute the discrete convolution as many smaller convolutions. Let NT be the time duration of each section of $x(t)$ to be considered; these sections are illustrated in Fig. 13-7(a). As shown in Fig. 13-8(a), we form the periodic sampled function $x(k)$ where a period is defined by the first section of $x(t)$; $h(t)$ is sampled and zeros are added to obtain the same period. Convolution $y(k)$ of these functions is also illustrated in Fig. 13-8(a). Note that we do not show the first $Q - 1$ points of the discrete convolution; these samples are incorrect because of the *end effect*. Recall from Sec. 7-3 for $h(k)$ defined by Q samples that the first $Q - 1$ samples of $y(k)$ have no relationship to the desired continuous convolution and should be discarded.

In Fig. 13-8(b) we illustrate the discrete convolution of the second section of duration NT illustrated in Fig. 13-7(a). As described in Sec. 13-1, we have shifted this section to the origin for purposes of efficient convolution. The section is then sampled and forced to be periodic; functions $h(k)$ and the resulting convolution $y(k)$ are also shown. Again the first $Q - 1$ samples of the convolution function are deleted because of the *end effect*.

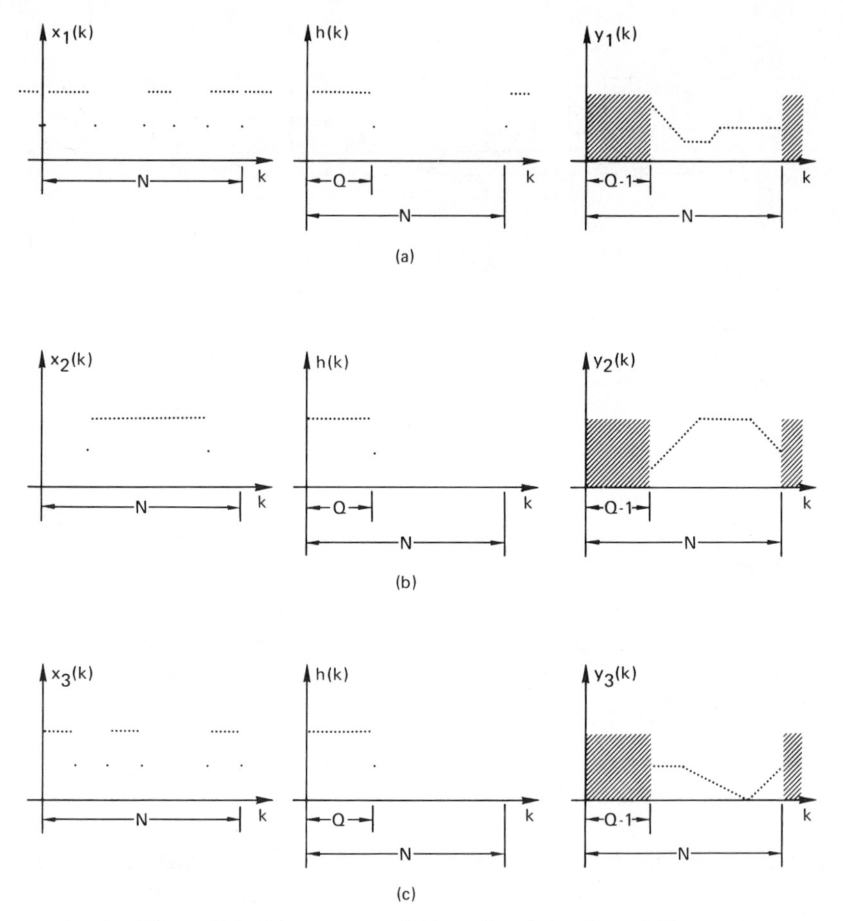

Figure 13-8. Discrete convolution of each section of Fig. 13-7(a).

The final section of $x(t)$ is shifted to the origin and sampled as illustrated in Fig. 13-8(c); discrete convolution results with the first $Q - 1$ samples deleted are also shown.

Each of the discrete convolution sections of Figs. 13-8(a), (b), and (c) is reconstructed in Figs. 13-9(a), (b), and (c), respectively. We have replaced the shift from the origin which was removed for efficient convolution. Note that with the exception of the *holes* created by the addition of these sectioned results, Fig. 13-9(d) approximates closely the desired continuous convolution of Fig. 13-9(e). By simply overlapping the sections of $x(t)$ by a duration $(Q - 1)T$ we can eliminate these holes entirely.

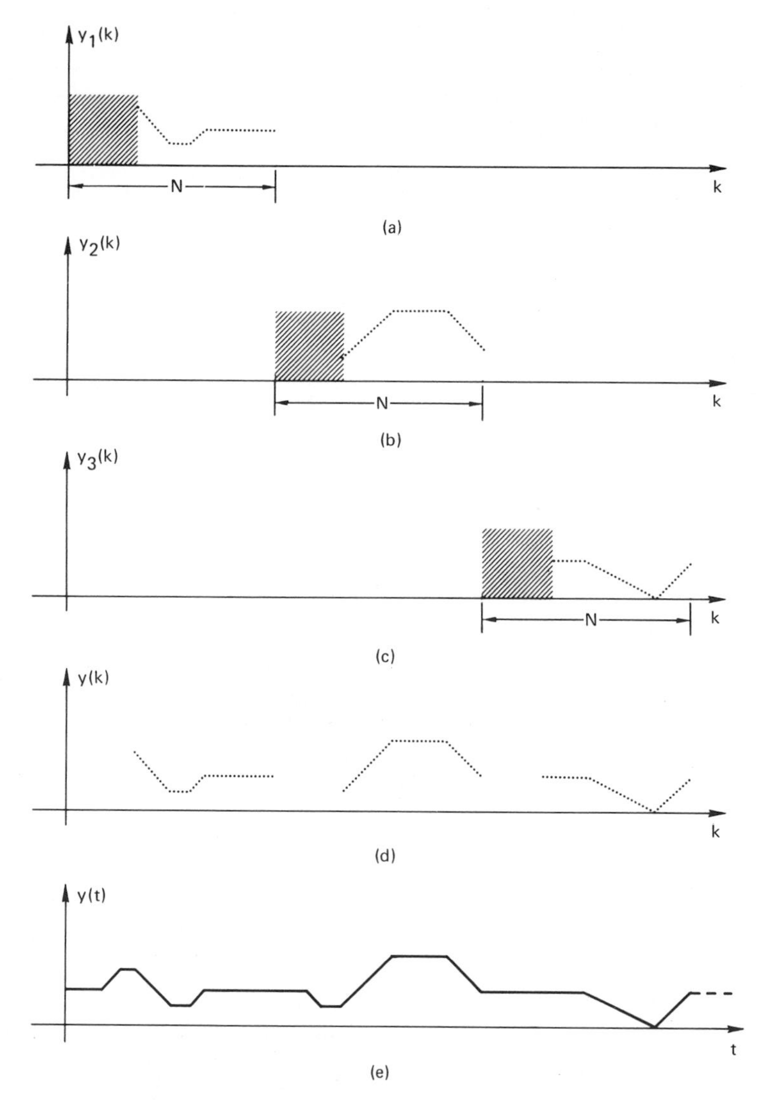

Figure 13-9. Reconstructed results of the discrete convolutions of Fig. 13-8.

Overlap-Save Sectioning

In Fig. 13-10(a) we show the identical waveform $x(t)$ of Fig. 13-7(a). However, note that the sections of $x(t)$ are now overlapped by $(Q-1)T$, the duration of the function $h(t)$ minus T.

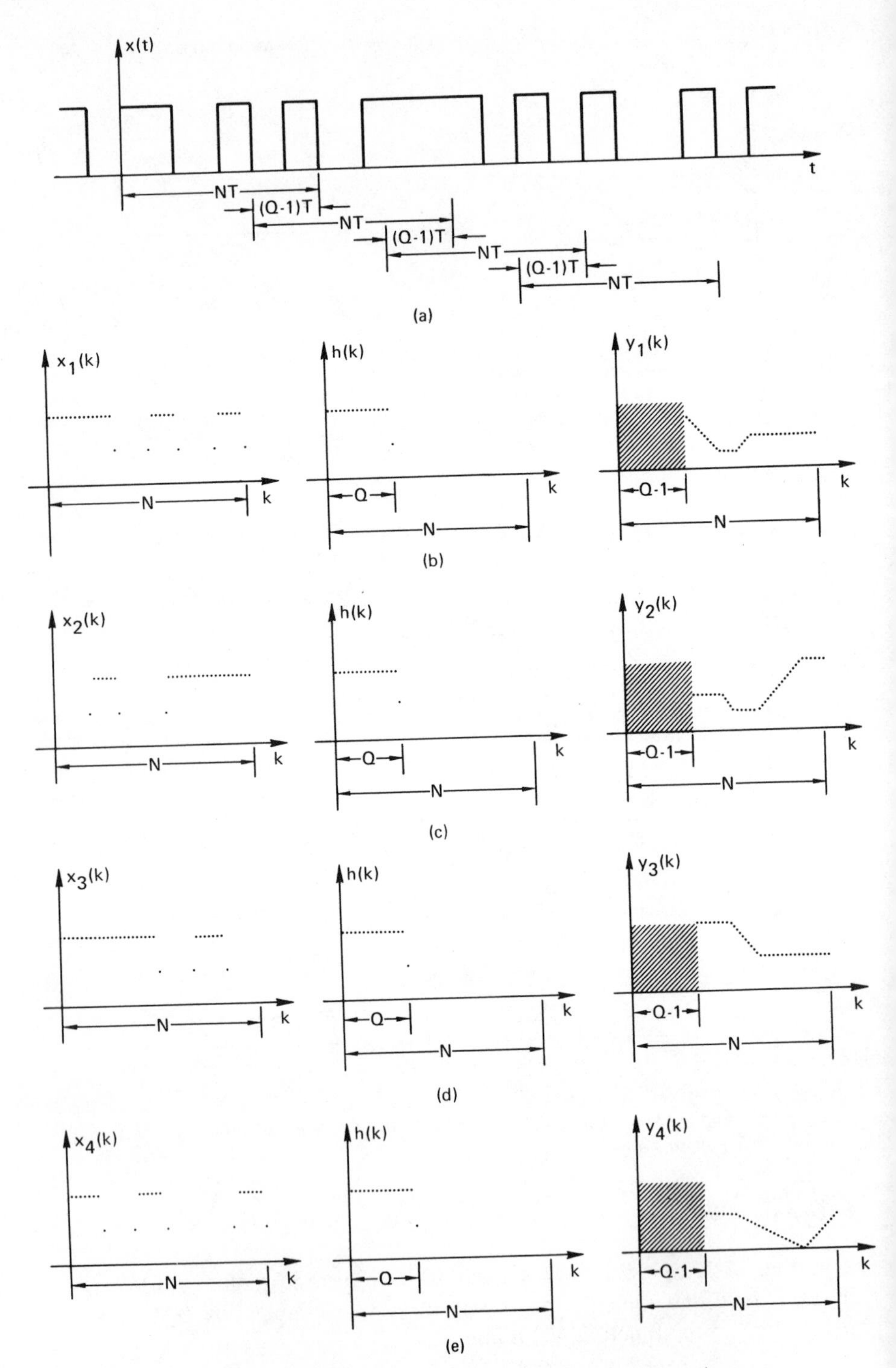

Figure 13-10. Discrete convolution of overlapped sections of data.

We shift each section of $x(t)$ to the origin, sample the section and form a periodic function. Figs. 13-10(b), (c), (d), and (e) illustrate the discrete convolution resulting from each section. Note that as a result of the overlap, additional sections are necessary. The first $Q - 1$ samples of each section are again eliminated because of the end effect.

As illustrated in Fig. 13-11, we add each of the sections of the discrete correlation. The appropriate shift has been added to each of the sections. We do not have *holes* as before because the end effect occurs for a duration of the convolution which was computed by the previous section. Combination of each of the sections yields over the entire range the desired continuous convolution [Fig. 13-7(c)]. The only *end effect* which cannot be compensated is the first one, as illustrated. All illustrations have been scaled by the factor T for convenience of comparison with continuous results. It remains to specify mathematically the relationships which have been developed graphically.

Refer to the illustration of Fig. 13-10(a). Note that we chose the first section to be of duration NT. To use the FFT, we require that

$$N = 2^{\gamma} \qquad \gamma \text{ integer valued} \tag{13-16}$$

and obviously, we require $N > Q$ (the optimum choice of N is discussed later). We form the sampled periodic function

$$x_1(k) = x(kT) \qquad k = 0, 1, \ldots, N-1$$

and by means of the FFT compute

$$X_1(n) = \sum_{k=0}^{N-1} x_1(k)\, e^{-j2\pi nk/N} \tag{13-17}$$

Next we take the Q sample values defining $h(t)$ and assign zero to the remaining samples to form a periodic function with period N,

$$h(k) = \begin{cases} h(kT) & k = 0, 1, \ldots, Q-1 \\ 0 & k = Q, Q+1, \ldots, N-1 \end{cases} \tag{13-18}$$

If $h(t)$ is not shifted to the origin, as illustrated in Fig. 13-7(b), then $h(t)$ is first shifted to the origin and Eq. (13-18) is applied. Using the FFT we compute

$$H(n) = \sum_{k=0}^{N-1} h(k)\, e^{-j2\pi nk/N} \tag{13-19}$$

and then the product

$$Y_1(n) = X_1(n)\, H(n) \tag{13-20}$$

Finally we compute the inverse discrete transform of $Y_1(n)$,

$$y_1(k) = \frac{1}{N} \sum_{n=0}^{N-1} Y_1(n)\, e^{j2\pi nk/N} \tag{13-21}$$

and because of the end effect, delete the first $Q - 1$ samples of $y(k)$; $y(0)$,

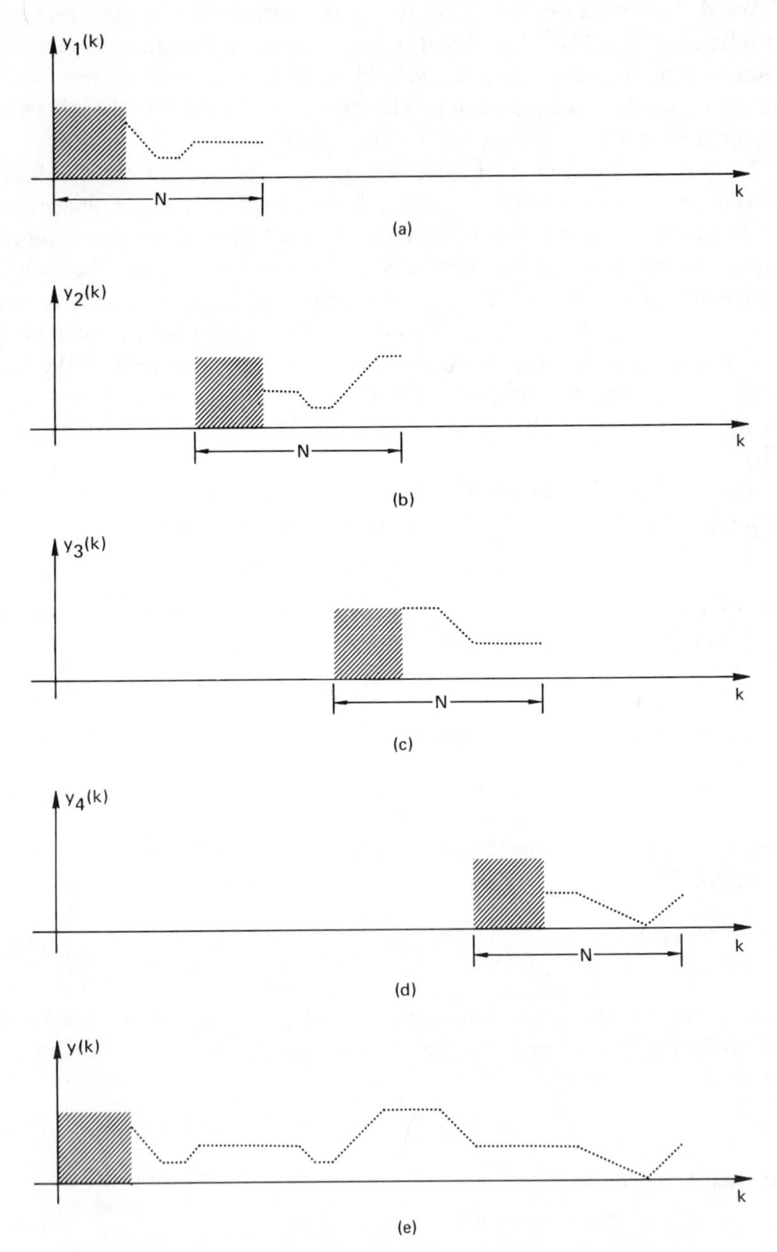

Figure 13-11. Reconstructed results of the discrete convolutions of Fig. 13-10.

$y(1), \ldots, y(Q-2)$. The remaining samples are identical to those illustrated in Fig. 13-11(a) and should be saved for future combination.

The second section of $x(t)$, illustrated in Fig. 13-10(a), is shifted to the origin and sampled

$$x_2(k) = x[(k + [N - Q + 1])T] \qquad k = 0, 1, \ldots, N-1 \tag{13-22}$$

Equations (13-19) through (13-21) are then repeated. From Eq. (13-19), the frequency function $H(n)$ has previously been determined and need not be recomputed. Multiplication as indicated by (13-20) and subsequent inverse transformation as indicated in (13-21) yields the waveform $y_2(k)$, illustrated in Fig. 13-11(b). Again, the first $Q-1$ samples of $y_2(k)$ are deleted because of the end effect. All remaining sectioned convolution results are determined similarly.

The method of combining the sectioned results is as illustrated in Fig. 13-11(e);

$$\begin{aligned}
&y(k) \text{ undefined} && k = 0, 1, \ldots, Q-2 \\
&y(k) = y_1(k) && k = Q-1, Q, \ldots, N-1 \\
&y(k+N) = y_2(k+Q-1) && k = 0, 1, \ldots, N-Q \\
&y(k+2N) = y_3(k+Q-1) && k = 0, 1, \ldots, N-Q \\
&y(k+3N) = y_4(k+Q-1) && k = 0, 1, \ldots, N-Q
\end{aligned} \tag{13-23}$$

The term *select-saving* or *overlap-save* is the term given in the literature [3, 5] for this technique of sectioning.

Overlap-Add Sectioning

An alternate technique for sectioning has been termed the *overlap-add* [3, 5] method. Consider the illustrations of Fig. 13-12. We assume that the finite-length function $x(t)$ is of a duration such that the samples representing $x(t)$ will exceed the memory of our computer. As a result, we show the sections of length $(N-Q)T$ as illustrated in Fig. 13-12(a). The desired convolution is illustrated in Fig. 13-12(c). To implement this technique, we first sample the first section of Fig. 13-12(a); these samples are illustrated in Fig. 13-13(a). The samples are augmented with zeros to form one period of a periodic function. In particular, we choose $N = 2^\gamma$, $N-Q$ samples of the function $x(t)$,

$$x_1(k) = x(kT) \qquad k = 0, 1, \ldots, N-Q \tag{13-24}$$

and $Q-1$ zero values,

$$x_1(k) = 0 \qquad k = N-Q+1, \ldots, N-1 \tag{13-25}$$

Note that the addition of $Q-1$ zeros insures that there will be no end effect. Function $h(t)$ is sampled to form a function $h(k)$ with period N as illustrated; the resulting convolution is also shown.

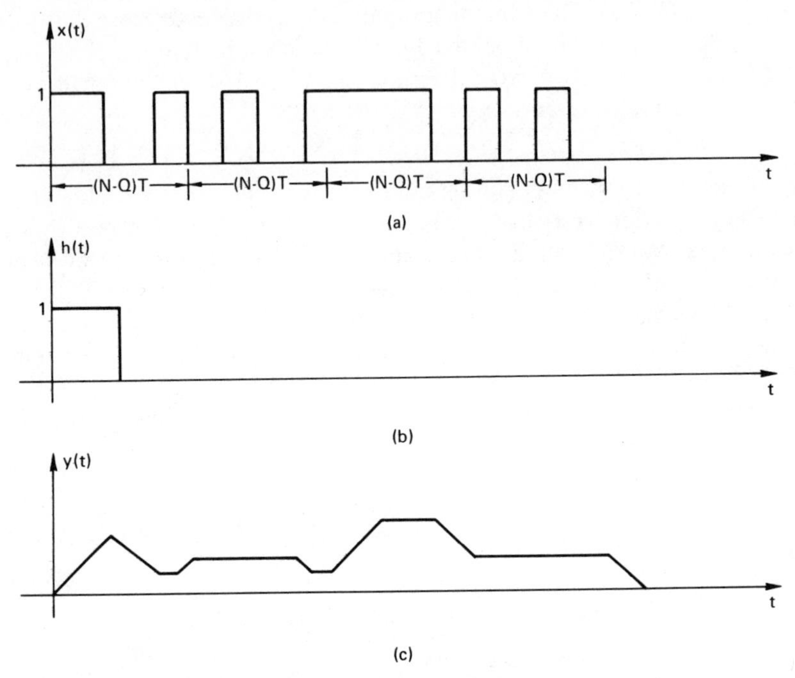

Figure 13-12. Example illustrating proper sectioning for applying overlap–add discrete convolution.

The second section of $x(t)$, illustrated in Fig. 13-12(a), is shifted to zero and then sampled;

$$\begin{aligned} x_2(k) &= x[(k + N - Q + 1)T] \qquad && k = 0, \ldots, N - Q \\ &= 0 && k = N - Q + 1, \ldots, N - 1 \end{aligned} \tag{13-26}$$

As before, we add $Q - 1$ zeros to the sampled function $x(t)$. Convolution with $h(k)$ yields the function $y_2(k)$, as illustrated in Fig. 13-13(b). Convolution of each of the additional sequences is obtained similarly; results are illustrated in Figs. 13-13(c) and (d).

We now combine these sectioned convolution results as illustrated in Fig. 13-14. Each section has been shifted to the appropriate value. Note that the resulting addition yields a replica of the desired convolution. The trick of this technique is to add sufficient zeros to eliminate any end effects. These convolution results are then *overlapped* and added at identically those samples where zeros were added. This gives rise to the term *overlap-add*.

Computing Speed of FFT Sectioned Convolution

In both of the sectioning techniques which we have described, the choice of N seems to be rather arbitrary as long as $N = 2^\gamma$. This choice will deter-

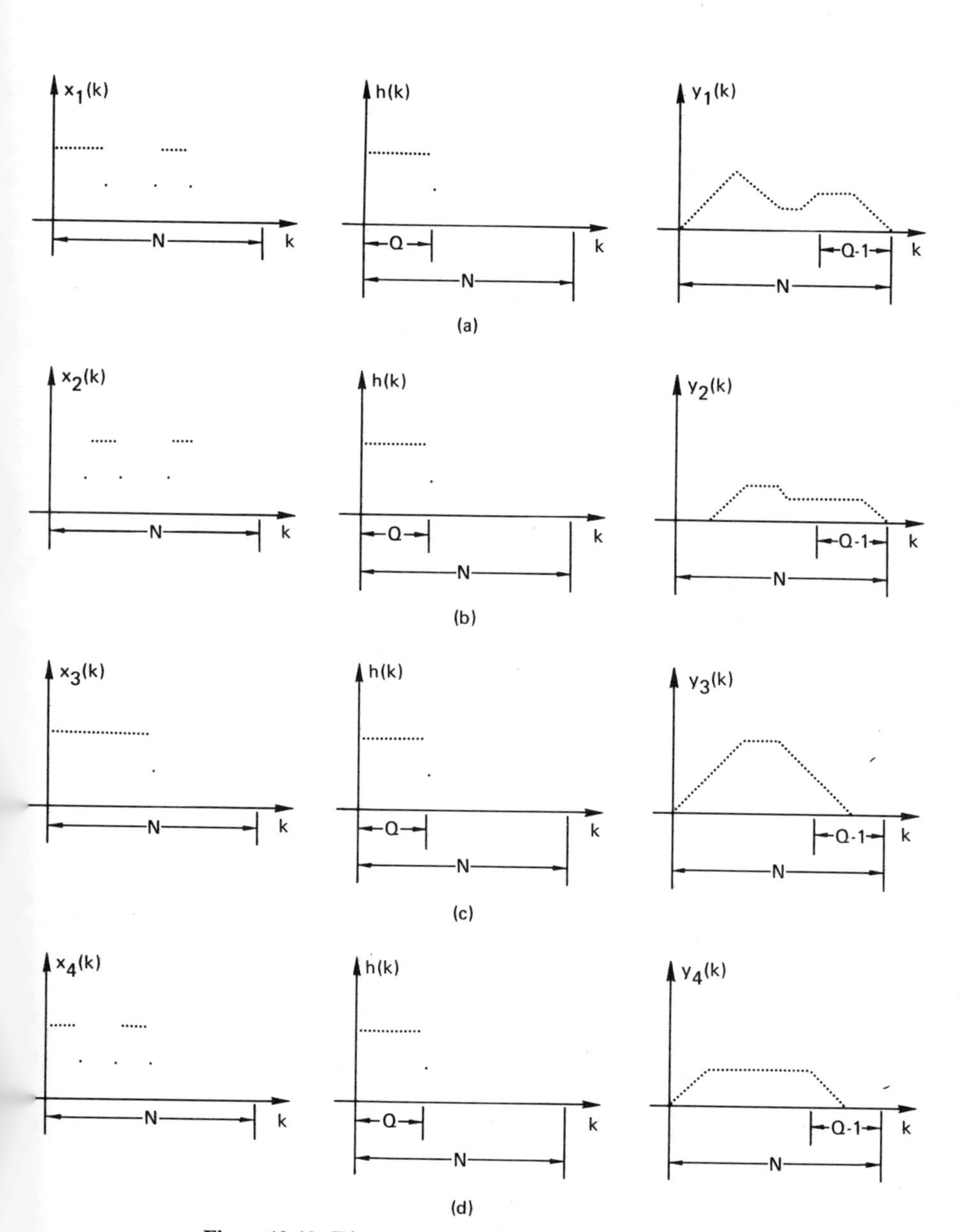

Figure 13-13. Discrete convolution of each section of Fig. 13-12.

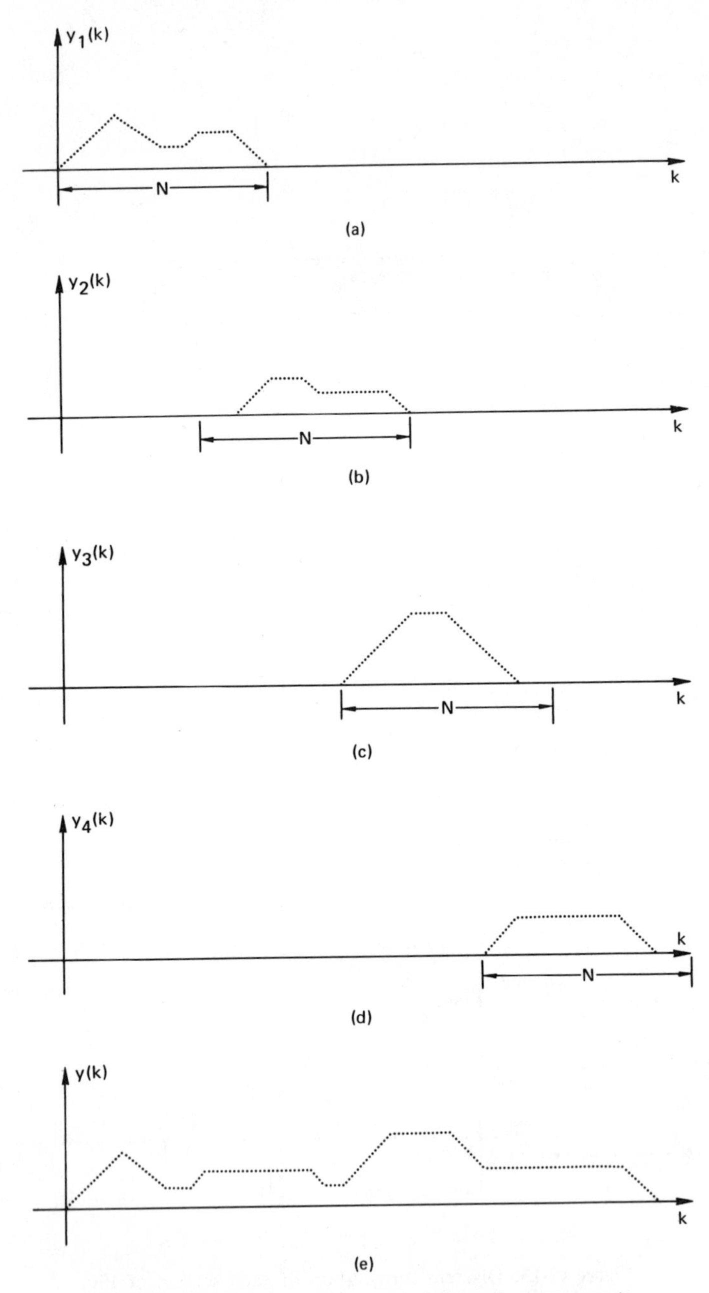

Figure 13-14. Reconstructed results of the discrete convolutions of Fig. 13-13.

mine the number of sections which must be computed, and thus the computing time. If an M-point convolution is desired, then approximately $M/(N - Q + 1)$ sections must be computed. If it is assumed that M is sufficiently greater than $(N - Q + 1)$, then the time required to compute $H(n)$ via the FFT can be ignored. Each section requires a forward and an inverse transform; hence the FFT must be repeated approximately $2M/(N - Q + 1)$ times. We have experimentally determined the optimum value of N; the results of this investigation are given in Table 13-2. One can depart from the values of N shown without greatly increasing the computing time.

TABLE 13-2 OPTIMUM VALUE OF N FOR FFT CONVOLUTION

Q	N	γ
≥ 10	32	5
11–19	64	6
20–29	128	7
30–49	256	8
50–99	512	9
100–199	1024	10
200–299	2048	11
300–599	4096	12
600–999	8192	13
1000–1999	16,384	14
2000–3999	32,768	15

We describe the step-by-step computational procedure for the *select saving* and the *overlap-add* method of sectioning in Figs. 13-15 and 13-16, respectively. Both algorithms are approximately equivalent with respect to computational efficiency.

If the functions $x(t)$ and $h(t)$ are real then we can use additional techniques to more efficiently compute the FFT. We describe in the next section exactly how this is accomplished.

13-4 EFFICIENT FFT CONVOLUTION

We have to this point in the discussion considered that the functions being convolved were real functions of time. As a result, we have not been utilizing the full capabilities of the FFT. In particular, the FFT algorithm is designed for complex input functions; thus if we only consider real functions, then the imaginary part of the algorithm is being wasted. In this section, we described how to divide a single real waveform into two parts, calling one part real, one part imaginary, and how to compute the convolution in one-half the

1. Refer to Figs. 13-10 and 13-11 for a graphical interpretation of the algorithm.
2. Let Q be the number of samples representing $h(t)$.
3. Choose N according to Table 13-2.
4. Form the sampled periodic function $h(k)$

$$\begin{aligned} h(k) &= h(kT) \qquad k = 0, 1, \ldots, Q-1 \\ &= 0 \qquad k = Q, Q+1, \ldots, N-1 \end{aligned}$$

5. Compute the discrete Fourier transform of $h(k)$

$$H(n) = \sum_{k=0}^{N-1} h(k) e^{-j2\pi nk/N}$$

6. Form the sampled periodic function

$$x_i(k) = x(kT) \qquad k = 0, 1, \ldots, N-1$$

7. Compute the discrete Fourier transform of $x_i(k)$

$$X_i(n) = \sum_{k=0}^{N-1} x_i(k) e^{-j2\pi nk/N}$$

8. Compute the product

$$Y_i(n) = X_i(n) H(n)$$

9. Compute the inverse discrete transform of $Y_i(n)$

$$y_i(k) = \sum_{n=0}^{N-1} \left(\frac{1}{N} Y_i(n)\right)^{\#} e^{-j2\pi nk/N}$$

10. Delete samples $y_i(0), y_i(1), \ldots, y_i(Q-2)$, and save the remaining samples.
11. Repeat steps 6-10 until all sections are computed.
12. Combine the sectioned results by the relationships

$$\begin{aligned} y(k) &\text{ undefined} \qquad && k = 0, 1, \ldots, Q-2 \\ y(k) &= y_i(k) && k = Q-1, Q, \ldots, N-1 \\ y(k+N) &= y_2(k+Q-1) && k = 0, 1, \ldots, N-Q \\ y(k+2N) &= y_3(k+Q-1) && k = 0, 1, \ldots, N-Q \\ &\vdots \end{aligned}$$

Figure 13-15. Computation procedure for FFT convolution: select–savings method.

normal FFT computing time. Alternately, our technique can be used to convolve two signals with an identical function simultaneously.

Consider the real periodic sampled functions $g(k)$ and $s(k)$. It is desired to convolve simultaneously these two functions with the real function $h(k)$ by means of the FFT. We accomplish this task by applying the technique of efficient discrete transforms which was discussed in Sec. 10-10. First we

1. Refer to Figs. 13-13 and 13-14 for a graphical interpretation of the algorithm.
2. Let Q be the number of samples representing $h(t)$.
3. Choose N according to Table 13-2.
4. Form the sampled periodic function $h(k)$

$$\begin{aligned} h(k) &= h(kT) \qquad k = 0, 1, \ldots, Q-1 \\ &= 0 \qquad k = Q, Q+1, \ldots, N-1 \end{aligned}$$

5. Compute the discrete Fourier transform of $h(k)$

$$H(n) = \sum_{k=0}^{N-1} h(k) e^{-j2\pi nk/N}$$

6. Form the sampled periodic function

$$\begin{aligned} x_i(k) &= x(kT) \qquad k = 0, 1, \ldots, N-Q \\ &= 0 \qquad k = N-Q+1, \ldots, N-1 \end{aligned}$$

7. Compute the discrete Fourier transform of $x_i(k)$

$$X_i(n) = \sum_{k=0}^{N-1} x_i(k) e^{-j2\pi nk/N}$$

8. Compute the product

$$Y_i(n) = X_i(n)H(n)$$

9. Compute the inverse discrete transform of $Y_i(n)$

$$y_i(k) = \sum_{n=0}^{N-1} \left(\frac{1}{N} Y_i(n)\right)^{\#} e^{-j2\pi nk/N}$$

10. Repeat steps 6-9 until all sections are computed.
11. Combine the sectioned results by the relationships

$$\begin{aligned} y(k) &= y_1(k) & k &= 0, 1, \ldots, N-Q \\ y(k+N-Q+1) &= y_1(k+N-Q+1) + y_2(k) & k &= 0, 1, \ldots, N-Q \\ y(k+2(N-Q+1)) &= y_2(k+N-Q+1) + y_3(k) & k &= 0, 1, \ldots, N-Q \\ &\vdots \end{aligned}$$

Figure 13-16. Computation procedure for FFT convolution: overlap–add method.

compute the discrete Fourier transform of $h(k)$, setting the imaginary part of $h(k)$ to zero.

$$\begin{aligned} H(n) &= \sum_{k=0}^{N-1} h(k)\, e^{-j2\pi nk/N} \\ &= H_r(n) + jH_i(n) \end{aligned} \tag{13-27}$$

Next we form the complex function

$$p(k) = g(k) + js(k) \qquad k = 0, 1, \ldots, N-1 \tag{13-28}$$

and compute

$$P(n) = \sum_{k=0}^{N-1} p(k)\, e^{-j2\pi nk/N}$$
$$= R(n) + jI(n) \tag{13-29}$$

Using the discrete convolution theorem (7-8) we compute

$$y(k) = y_r(k) + jy_i(k) = p(k) * h(k) = \frac{1}{N} \sum_{k=0}^{N-1} P(n)\, H(n)\, e^{j2\pi nk/N} \tag{13-30}$$

From (10-25) and (10-26) the frequency function $P(n)$ can be expressed as

$$\begin{aligned} P(n) &= R(n) + jI(n) \\ &= [R_e(n) + R_0(n)] + j[I_e(n) + I_0(n)] \\ &= G(n) + jS(n) \end{aligned} \tag{13-31}$$

where

$$\begin{aligned} G(n) &= R_e(n) + jI_0(n) \\ S(n) &= I_e(n) - jR_0(n) \end{aligned} \tag{13-32}$$

Product $P(n)H(n)$ is then given by

$$P(n)H(n) = G(n)H(n) + jS(n)H(n) \tag{13-33}$$

and thus the inversion formula yields

$$y(k) = y_r(k) + jy_i(k) = \frac{1}{N} \sum_{n=0}^{N-1} P(n)H(n)\, e^{j2\pi nk/N} \tag{13-34}$$

where

$$\begin{aligned} y_r(k) &= \frac{1}{N} \sum_{k=0}^{N-1} G(n)H(n)\, e^{j2\pi nk/N} \\ jy_i(k) &= \frac{1}{N} \sum_{k=0}^{N-1} jS(n)H(n)\, e^{j2\pi nk/N} \end{aligned} \tag{13-35}$$

which is the desired result. That is, $y_r(k)$ is the convolution of $g(k)$ and $h(k)$, and $y_i(k)$ is the convolution of $s(k)$ and $h(k)$. If $g(k)$ and $s(k)$ represent successive sections as described in the previous section, then we have reduced the computing time by a factor of two by using this technique. One still must combine the results as appropriate for the method of sectioning being employed.

Now consider the case where it is desired to perform the discrete convolution of $x(k)$ and $h(k)$ in one-half the time by using the imaginary part of the complex time function as discussed in Sec. 10-10. Assume $x(k)$ is described by $2N$ points; define

$$\begin{aligned} g(k) &= x(2k) \qquad & k &= 0, 1, \ldots, N-1 \\ s(k) &= x(2k+1) \qquad & k &= 0, 1, \ldots, N-1 \end{aligned} \tag{13-36}$$

and let

$$p(k) = g(k) + js(k) \qquad k = 0, 1, \ldots, N-1 \tag{13-37}$$

But Eq. (13-37) is identical to (13-28); therefore

$$z(k) = z_r(k) + jz_i(k) = \frac{1}{N} \sum_{n=0}^{N-1} P(n)H(n)\, e^{j2\pi nk/N}$$

where the desired convolution $y(k)$ is given by

$$\begin{aligned} y(2k) &= z_r(k) \qquad k = 0, 1, \ldots, N-1 \\ y(2k+1) &= z_i(k) \qquad k = 0, 1, \ldots, N-1 \end{aligned} \tag{13-38}$$

As in the previous method we must still combine the results as appropriate for the method of sectioning being considered.

An ALGOL computer program for performing FFT convolution is described in [4].

13-5 APPLICATIONS SUMMARY

As discussed previously, the basic applications of the FFT are to compute the Fourier transform, convolution, and correlation integrals. It is futile to attempt to apply the FFT to problems such as digital filtering, systems analysis, and so on, unless the basics are thoroughly understood. Regardless of the desired FFT application, it is anticipated that the fundamental approach to problem solution is to be found within this book.

PROBLEMS

13-1. Given the functions $h(t)$ and $x(t)$ illustrated in Fig. 13-17, determine the optimum choice of N to eliminate overlap effects during convolution and correlation. Assume a sample period of $T = 0.1$ and a base 2 FFT algorithm. Graphically show how to restructure the data for efficient convolution computation.

13-2. Consider the functions $x(t)$ and $h(t)$ of Fig. 13-17. Graphically show how to apply the overlap-save and overlap-add sectioning techniques for computing the convolution of $x(t)$ and $h(t)$. Demonstrate your results analogously to Figs. 13-10, 13-11, 13-13, and 13-14.

13-3. Repeat Problem 13-2 for the correlation of $x(t)$ and $h(t)$.

13-4. Repeat Problem 13-3 for the functions $x(t)$ and $h(t)$ illustrated in Fig. 13-7.

PROJECTS

The following projects will require access to a digital computer.

13-5. Using the FFT program developed previously, duplicate the results shown in Figs. 13-8, 13-10, 13-11, 13-13, and 13-14. Apply the efficient FFT convolution techniques described in Sec. 13-4.

13-6. Repeat Problem 13-5 for the case of correlation of the two waveforms.

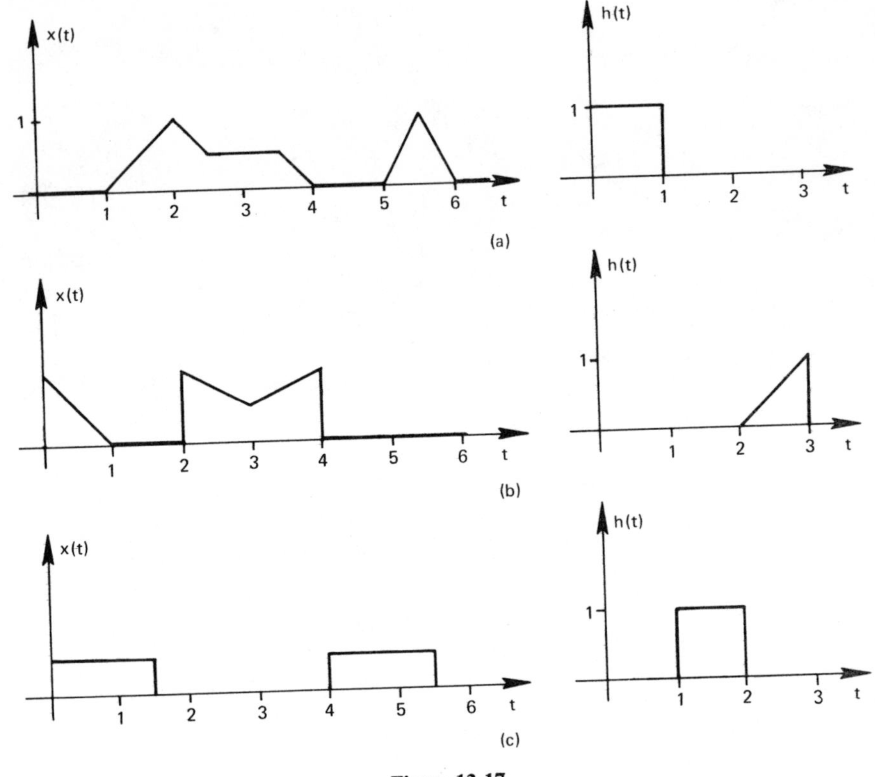

Figure 13-17.

REFERENCES

1. Cooley, J. W., "Applications of the fast Fourier transform method," *Proc. IBM Scientific Computing Symp. on Digital Simulation of Continuous Systems*. Yorktown Heights, N. Y.: Thomas J. Watson Research Center, 1966.
2. Cooley, J. W., P. A. W. Lewis, and P. D. Welch, "Application of the fast Fourier transform to computation of Fourier integrals, Fourier series, and convolution integrals," *IEEE Trans. Audio and Electroacoustics* (June 1967), Vol. AU–15, pp. 79–84.
3. Helms, H. D., "Fast Fourier transform method of computing difference equations and simulating filters," *IEEE Trans. Audio and Electroacoustics* (June 1967), Vol. AU–15, pp. 85–90.
4. Singleton, R. C., "Algorithm 345, an Algol convolution procedure based on the fast Fourier transform," *Commun. Assoc. Comput. Mach.* (March 1969), Vol. 12.

5. STOCKHAM, T. G., "High-speed convolution and correlation," *AFIPS Proc.*, 1966 Spring Joint Computer Conf., Vol. 28, pp. 229–233. Washington, D. C.: Spartan.

6. GENTLEMAN, W. M., and G. SANDE, "Fast Fourier transforms for fun and profit," *AFIPS Proc.*, 1966 Fall Joint Computer Conf., Vol. 29, pp. 563–578. Washington, D.C.: Spartan.

7. BINGHAM, C., M. D. GODFREY, and J. W. TUKEY, "Modern techniques of power spectrum estimation," *IEEE Trans. Audio and Electroacoustics* (June 1967), Vol. AU–15, No. 2.

8. COOLEY, J. W., P. A. W. LEWIS, and P. D. WELCH, "The finite Fourier transform," *IEEE Trans. Audio and Electroacoustics* (June 1969), Vol. AU–17, No. 2, pp. 77–85.

APPENDIX A

THE IMPULSE FUNCTION: A DISTRIBUTION

The impulse function $\delta(t)$ is a very important mathematical tool in continuous and discrete Fourier transform analysis. Its usage simplifies many derivations which would otherwise require lengthy complicated arguments. Even though the concept of the impulse function is correctly applied in the solution of many problems, the basis or definition of the impulse is normally mathematically meaningless. To insure that the impulse function is well defined we must interpret the impulse not as a normal function but as a concept in the theory of distributions.

Following the discussions by Papoulis [1, Appendix I] and Gupta [2, Chapter 2], we describe a simple but adequate theory of distributions. Based on this general theory, we develop those specific properties of the impulse function which are necessary to support the developments of Chapter 2.

A-1 IMPULSE FUNCTION DEFINITIONS

Normally the impulse function (δ-function) is defined as

$$\delta(t - t_0) = 0 \qquad t \neq t_0 \tag{A-1}$$

$$\int_{-\infty}^{\infty} \delta(t - t_0)\, dt = 1 \tag{A-2}$$

That is, we define the δ-function as having undefined magnitude at the time of occurrence and zero elsewhere with the additional property that area under the function is unity. Obviously it is very difficult to relate an impulse to a physical signal. However, we can think of an impulse as a pulse wave-

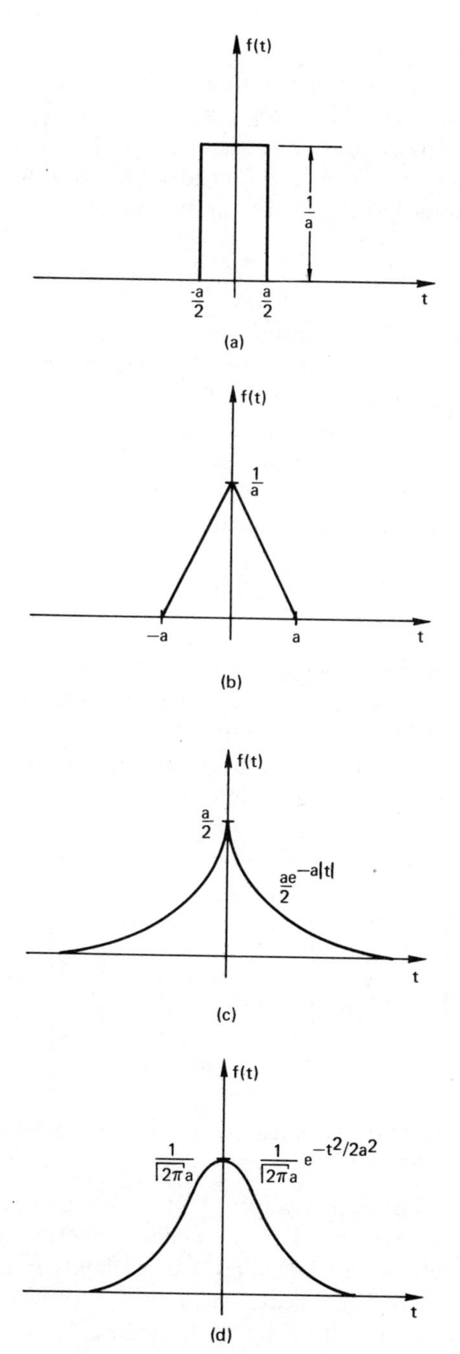

Figure A-1. Representations of the δ-function.

form of very large magnitude and infinitely small duration such that the area of the pulse is unity.

We note that with this interpretation we are, in fact, constructing a series of functions (i.e., pulses) which progressively increase in amplitude, decrease in duration, and have a constant area of unity. This is simply an alternate method for defining a δ-function. Consider the pulse waveform illustrated in Fig. A-1(a). Note that the area is unity and hence, we can write mathematically the δ-function as

$$\delta(t) = \lim_{a \to 0} f(t, a) \qquad \text{(A-3)}$$

In the same manner, the functions illustrated in Figs. A-1(b), (c), and (d) satisfy Eqs. (A-1) and (A-2) and can be used to represent an impulse function.

The various properties of impulse functions can be determined directly from these definitions. However, in a strict mathematical sense these definitions are meaningless if we view $\delta(t)$ as an ordinary function. If the impulse function is introduced as a generalized function or distribution, then these mathematical problems are eliminated.

A-2 DISTRIBUTION CONCEPTS

The theory of distributions is vague and, in general, meaningless to the applied scientist who is reluctant to accept the description of a physical quantity by a concept that is not an ordinary function. However, we can argue that the reliance on representation of physical quantities by ordinary functions is only a useful idealization and, in fact, is subject to question. To be specific, let us consider the example illustrated in Fig. A-2.

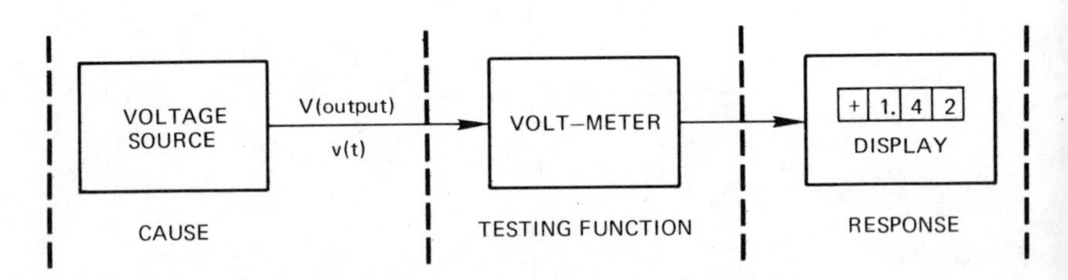

Figure A-2. Physical interpretation of a distribution.

As shown, the physical quantity V is a voltage source. We normally assume that the voltage $v(t)$ is a well defined function of time and that a measurement merely reveals its values. But we know in fact that there does not exist a voltmeter that can measure *exactly* $v(t)$. However, we still insist on defining the physical quantity V by a well defined function $v(t)$ even though we cannot measure $v(t)$ accurately. The point is that since we cannot measure the quantity V exactly, then on what basis do we require the voltage source to be represented by a well defined function $v(t)$?

A more meaningful interpretation of the physical quantity V is to define it in terms of the effects it produces. To illustrate this interpretation, note that in the previous example, the quantity V causes the voltmeter to display or assign a number as a response. For each change in V, another number is displayed or assigned as a response. We never measure $v(t)$ but only the response; therefore the source can be specified only by the totality of the responses that it causes. It is conceivable that there is not an ordinary function $v(t)$ which represents the voltage parameter V. But since the responses or numbers are still valid, then we must assume that there is a source V causing them and the only way to characterize the source is by the responses or numbers. We will now show that these numbers in fact describe V as a distribution.

A distribution, or generalized function, is a process of assigning to an arbitrary function $\phi(t)$ a response or number

$$R[\phi(t)] \tag{A-4}$$

Function $\phi(t)$ is termed a *testing function* and is continuous, is zero outside a finite interval, and has continuous derivatives of all orders. The number assigned to the testing function $\phi(t)$ by the distribution $g(t)$ is given by

$$\int_{-\infty}^{\infty} g(t)\,\phi(t)\,dt = R[\phi(t)] \tag{A-5}$$

The lefthand side of (A-5) has no meaning in the conventional sense of integration, but rather is defined by the number $R[\phi(t)]$ assigned by the distribution $g(t)$. Let us now cast these mathematical statements in light of the previous example.

With reference to Fig. A-2, we note that if the voltmeter is modeled as a linear system, then the output at time t_0 is given by the convolution integral

$$\int_{-\infty}^{\infty} v(t)\,h(t_0 - t)\,dt$$

where $h(t)$ is the time domain response of the measuring instrument. If we consider $h(t)$ as a testing function (that is, each particular voltmeter has different internal characteristics and as a result will yield a different response for the same input, we thus say that the meter *tests* or senses the distribution $v(t)$), then the convolution integral takes the form

$$\int_{-\infty}^{\infty} v(t)\,\phi(t, t_0)\,dt = R[\phi(t, t_0)] \tag{A-6}$$

Thus, for a fixed input V the response R is a number depending on the system function $\phi(t, t_0)$.

If we interpret (A-6) as a conventional integral and if this integral equation is well defined, then we say that the voltage source is defined by the ordinary function $v(t)$. But, as stated previously, it is possible that there does not exist an ordinary function satisfying (A-6). Since the response $R[\phi(t, t_0)]$ still exists, we must assume that there is a voltage source V which causes this response and that a method of characterizing the source is by means of the distribution (A-6).

The preceding discussion has cast the theory of distributions in the form of physical measurements for ease of interpretation. Based on the defining relationship (A-5), we now investigate the properties of a particular distribution; the δ-function.

A-3 PROPERTIES OF IMPULSE FUNCTIONS

The impulse function $\delta(t)$ is a distribution assigning to the testing function $\phi(t)$ the number $\phi(0)$;

$$\int_{-\infty}^{\infty} \delta(t)\,\phi(t)\,dt = \phi(0) \tag{A-7}$$

It should be repeated that the relationship (A-7) has no meaning as an integral, but the integral and the function $\delta(t)$ are defined by the number $\phi(0)$ assigned to the function $\phi(t)$.

We now describe the useful properties of impulse functions.

Sifting Property

The function $\delta(t - t_0)$ is defined by

$$\int_{-\infty}^{\infty} \delta(t - t_0)\,\phi(t)\,dt = \phi(t_0) \tag{A-8}$$

This property implies that the δ-function takes on the value of the function $\phi(t)$ at the time the δ-function is applied. The term *sifting property* arises in that if we let t_0 continuously vary we can *sift* out each value of the function $\phi(t)$. This is the most important property of the δ-function.

Scaling Property

The distribution $\delta(at)$ is defined by

$$\int_{-\infty}^{\infty} \delta(at)\,\phi(t)\,dt = \frac{1}{|a|}\int_{-\infty}^{\infty} \delta(t)\,\phi\left(\frac{t}{a}\right)dt \tag{A-9}$$

where the equality results from a change in the independent variable. Thus, $\delta(at)$ is given by

$$\delta(at) = \frac{1}{|a|}\delta(t) \tag{A-10}$$

Product of a δ-function by an Ordinary Function

The product of a δ-function by an ordinary function $h(t)$ is defined by

$$\int_{-\infty}^{\infty} [\delta(t)\, h(t)]\, \phi(t)\, dt = \int_{-\infty}^{\infty} \delta(t)\, [h(t)\, \phi(t)]\, dt \tag{A-11}$$

If $h(t)$ is continuous at $t = t_0$ then

$$\delta(t_0)\, h(t) = h(t_0)\, \delta(t_0) \tag{A-12}$$

In general, the product of two distributions is undefined.

Convolution Property

The convolution of two impulse functions is given by

$$\int_{-\infty}^{\infty}\left[\int_{-\infty}^{\infty} \delta_1(\tau)\, \delta_2(t-\tau)\, d\tau\right]\phi(t)\, dt = \int_{-\infty}^{\infty} \delta_1(\tau)\left[\int_{-\infty}^{\infty} \delta_2(t-\tau)\, \phi(t)\, dt\right] d\tau \tag{A-13}$$

Hence

$$\delta_1(t - t_1) * \delta_2(t - t_2) = \delta[t - (t_1 + t_2)] \tag{A-14}$$

δ-functions as Generalized Limits

Consider the sequence $g_n(t)$ of distributions. If there exists a distribution $g(t)$ such that for every test function $\phi(t)$ we have

$$\lim_{n\to\infty} \int_{-\infty}^{\infty} g_n(t)\, \phi(t)\, dt = \int_{-\infty}^{\infty} g(t)\, \phi(t)\, dt. \tag{A-15}$$

then we say that $g(t)$ is the limit of $g_n(t)$

$$g(t) = \lim_{n\to\infty} g_n(t) \tag{A-16}$$

We can also define a distribution as a generalized limit of a sequency $f_n(t)$ of ordinary functions. Assume that $f_n(t)$ is such that the limit

$$\lim_{n\to\infty} \int_{-\infty}^{\infty} f_n(t)\, \phi(t)\, dt$$

exists for every test function. This limit then is a number which depends on $\phi(t)$ and thus defines a distribution $g(t)$ where

$$g(t) = \lim_{n\to\infty} f_n(t) \tag{A-17}$$

and the limiting operation is to be interpreted in the sense of (A-15). If (A-17) exists as an ordinary limit, then it defines an equivalent function if we assume that we can interchange the order of limit and integration in (A-15). It is from these arguments that the conventional limiting arguments, although awkward, are mathematically correct.

The δ-function can then be defined as a generalized limit of a sequence of ordinary functions satisfying

$$\lim_{n \to \infty} \int_{-\infty}^{\infty} f_n(t)\, \phi(t)\, dt = \phi(0) \tag{A-18}$$

If (A-18) holds then

$$\delta(t) = \lim_{n \to \infty} f_n(t) \tag{A-19}$$

Each of these functions illustrated in Fig. A-1 satisfy (A-18) and define the δ-function in the sense of Eq. (A-19).

Another functional form of importance which defines the δ-function is

$$\delta(t) = \lim_{a \to \infty} \frac{\sin at}{\pi t} \tag{A-20}$$

Using (A-20) we can prove [Papoulis, p. 281] that

$$\int_{-\infty}^{\infty} \cos(2\pi f t)\, df = \int_{-\infty}^{\infty} e^{j2\pi f t}\, df = \delta(t) \tag{A-21}$$

which is of considerable importance in evaluating particular Fourier transforms.

REFERENCES

1. PAPOULIS, A., *The Fourier Integral and Its Applications.* New York: McGraw-Hill, 1962.
2. GUPTA, S. C., *Transform and State Variable Methods in Linear Systems.* New York: Wiley 1966.
3. BRACEWELL, R. M., *The Fourier Transform and Its Applications.* New York: McGraw-Hill, 1965.
4. LIGHTHILL, M. J., *An Introduction to Fourier Analysis and Generalized Function.* New York: Cambridge University Press, 1959.
5. ARSAC, J., *Fourier Transforms and the Theory of Distributions.* Englewood Cliffs, N. J.: Prentice-Hall, 1966.
6. FRIEDMAN, B., *Principles and Techniques of Applied Mathematics.* New York: Wiley, 1956.
7. ZEMANIAN, A. H., *Distribution Theory and Transform Analysis.* New York: McGraw-Hill, 1965.

BIBLIOGRAPHY

ALLEN, J. B., "Estimation of transfer functions using the Fourier transform ratio method," *American Institute of Astronautics and Aeronautics Journal.* Vol. 8, pp. 414–423, March 1970.

ALSOP, L. E., and A. A. NOWROOZI, "Fast Fourier analysis," *J. Geophys. Res.* Vol. 71, pp. 5482–5483, November 15, 1966.

ANDREWS, H., "A high-speed algorithm for the computer generation of Fourier transforms," *IEEE Trans. on Computers* (*Short Notes*). Vol. C–17, pp. 373–375, April 1968.

ANDREWS, H. C., "Multidimensional rotations in feature selection," *IEEE Symp. on Feature Extraction and Selection in Pattern Recognition.* pp. 10–18, Argonne, Ill., October 1970.

ANDREWS, H. C., and K. L. CASPARI, "Degrees of freedom and modular structure in matrix multiplication," *IEEE Trans. on Computers.* Vol. C–20, pp. 133–141, February 1971.

ANDREWS, H. C., and K. L. CASPARI, "A generalized technique for spectral analysis," *IEEE Trans. on Computers.* Vol. C–19, pp. 16–25, January 1970.

ANUTA, P. E., "Use of fast Fourier transform techniques for digital registration of multispectral imagery" and "Spatial registration of multispectral and multitemporal digital imagery using fast Fourier transform techniques," *IEEE Trans. Geoscience Electronics.* GE8, pp. 353–368, October 1970.

AOKI, Y., "Computer simulation of synthesizing images by digital phased arrays," *Proceedings of the IEEE.* Vol. 58, pp. 1856–1858, November 1970.

AOKI, Y., "Optical and numerical reconstructions of images from soundwave holograms," *IEEE Trans. Audio and Electroacoustics.* Vol. AU–18, pp. 258–267, September 1970.

Aoki, Y., and A. Boivin, "Computer reconstruction of images from a microwave hologram," *Proceedings of the IEEE.* Vol. 58, pp. 821–822, May 1970.

Aronson, E. A., "Fast Fourier integration of piecewise polynomial functions," *Proceedings of the IEEE.* Vol. 57, pp. 691–692, April 1969.

Bailey, J. S., "A fast Fourier transform without multiplications," *Proc. Symp. on Computer Processing in Communications.* Vol. 19, MRI Symposia Ser., New York: Polytechnic Press, 1969.

Bastida, E., and D. Dotte, "Applications and developments of the fast Fourier transform techniques," *Alta Frequenza.* Vol. 37, pp. 237–240, August 1968.

Bates, R. H. T., P. J. Napier, and Y. P. Chang, "Square wave Fourier transform," *Electron Letters* (*GB*). Vol. 6, pp. 741–742, November 12, 1970.

Benignus, V. A., "Estimation of the coherence spectrum and its confidence interval using the fast Fourier transforms," *IEEE Trans. Audio and Electroacoustics.* Vol. AU–17, pp. 145–150, June 1969; Correction, Vol. 18, p. 320, September 1970.

Benignus, V. A., "Estimation of coherence spectrum of non-Gaussian time series populations," *IEEE Trans. Audio and Electroacoustics.* Vol. AU–17, pp. 198–201, September 1969.

Bergland, G. D., "A fast Fourier transform algorithm for real-valued series," *Commun. ACM.* Vol. 11, pp. 703–710, October 1968.

Bergland, G. D., "A fast Fourier transform algorithm using base eight iterations," *Math. Computation.* Vol. 22, pp. 275–279, April 1968.

Bergland, G. D., "Fast Fourier transform hardware implementations. I. An overview. II. A survey," *IEEE Trans. Audio and Electroacoustics.* Vol. AU–17, pp. 104–108, 109–119, June 1969.

Bergland, G. D., "The fast Fourier transform recursive equations for arbitrary length records," *Math. Computation.* Vol. 21, pp. 236–238, April 1967.

Bergland, G. D., "Guided tour of the fast Fourier transform," *IEEE Spectrum.* Vol. 6, pp. 41–52, July 1969.

Bergland, G. D., "A radix-eight fast Fourier transform subroutine for real-valued series," *IEEE Trans. Audio and Electroacoustics.* Vol. 17, pp. 138–144, June 1969.

Bergland, G. D., and H. W. Hale, "Digital real-time spectral analysis," *IEEE Trans. Electronic Computers.* Vol. EC–16, pp. 180–185, April 1967.

Bergland, G. D., and D. E. Wilson, "A fast Fourier transform algorithm for a global, highly parallel processor," *IEEE Trans. Audio and Electroacoustics.* Vol. AU–17, pp. 125–127, June 1969.

Bertram, S., "On the derivation of the fast Fourier transform," *IEEE Trans. Audio and Electroacoustics.* Vol. AU–18, pp. 55–58, March 1970.

Bingham, C., M. D. Godfrey, and J. W. Tukey, "Modern techniques of power spectral estimation," *IEEE Trans. Audio and Electroacoustics.* Vol. AU–15, pp. 56–66, June 1967.

BLUESTEIN, L. I., *A linear filter approach to the computation of the discrete Fourier transform.* 1968 Nerem Record, pp. 218–219.

BOCKMAN, R., *Fast Fourier transform PMR studies of stereochemistry and amide hydrogen exchange rates in Polymyxin B1.* 62nd Annual Meeting of the American Society of Biological Chemists, San Francisco, Calif., 13–18 June 1971.

BOGART, B. P., and E. PARZEN, "Informal comments on the uses of power spectrum analysis," *IEEE Trans. Audio and Electroacoustics.* Vol. AU–15, pp. 74–76, June 1967.

BOGNER, R. E., "Frequency sampling filters—Hilbert transformations and resonators," *Bell Systems Tech. Jl.* Vol. 48, pp. 501–510, March 1969.

BOHMAN, H., "On the maximum deviation in random walks," *BIT* (Sweden). Vol. 11, No. 2, pp. 133–138, 1971.

BOND, W. H., and J. M. MYATT, "Investigation of distortion of diverse speech using power spectral estimates based on the fast Fourier transform," *CFSTI.* AD 707 729, p. 93, June 1969.

BOOTHROYD, J., "Complex Fourier series," *Computer J.* Vol. 10, pp. 414–416, February 1968; Correction, Vol. 11, p. 115, May 1968.

BORGIOLI, R. C., "Fast Fourier transform correlation versus direct discrete time correlation," *Proceedings of the IEEE.* Vol. 56, pp. 1602–1604, September 1968.

BRAYLEY, W. L., "A general signal processing program," *Marconi Review.* Vol. 33, pp. 232–238, 1970.

BRENNER, N. M., "Fast Fourier transform of externally stored data," *IEEE Trans. Audio and Electroacoustics.* Vol. AU–17, pp. 128–132, June 1969.

BRENNER, N. M., *Three Fortran programs that perform the Cooley-Tukey Fourier transform.* Tech. Note 1967–2, M. I. T. Lincoln Lab., Lexington, Mass., July 1967.

BRIGHAM, E. O., and R. E. MORROW, "The fast Fourier transform," *IEEE Spectrum.* Vol. 4, pp. 63–70, December 1967.

BRUCE, J. D., "Discrete Fourier transforms, linear filters and spectrum weighting," *IEEE Trans. Audio and Electroacoustics.* Vol. AU–16, pp. 495–499, December 1968.

BRUMBACH, R. P., *Digital computer routines for power spectral analysis.* Tech. Rept. 68–31, AC Electronics-Defense Research Labs., Santa Barbara, Calif., July 1968.

BUIJS, H. L., "Fast Fourier transformation of large arrays of data," *Applied Optics.* Vol. 8, pp. 211–212, January 1969.

BURCKHARDT, C. B., "Use of a random phase mask for the recording of Fourier transform holograms of data masks," *Applied Optics.* Vol. 9, pp. 695–700, March 1970.

BUTCHER, W. E., and G. E. COOK, *Application of fast Fourier transform to process identification.* 8th Annual IEEE region III convention, pp. 187–192, 1969.

BUTCHER W. E., and G. E. COOK, "Comparison of two impulse response identification techniques, *IEEE Trans. Automatic Control.* Vol. AC–15, pp. 129–130, February 1970.

BUTCHER, W. E., and G. E. COOK, "Identification of linear sampled data systems by transform techniques," *IEEE Trans. Automatic Control.* Vol. AC–14, pp. 582–584, October 1969.

CAIRNS, THOMAS W., "On the fast Fourier transform on finite Abelian groups," *IEEE Trans. on Computers.* Vol. C–20, pp. 569–571, May 1971.

CAPRINI, M. R., S. COHN-SFETCHI, and A. M. MANOF, "Application of digital filtering in improving the resolution and the signal to noise ratio of nuclear and magnetic resonance spectra," *IEEE Trans. Audio and Electroacoustics.* Vol. AU–18, pp. 389–393, December 1970.

CARSON, C. T., "The numerical solution of waveguide problems by fast Fourier transforms," *IEEE Trans. Microwave Theory.* Vol. MTT–16, pp. 955–958, November 1968.

CHANOUS, D., "Synthesis of recursive digital filters using the FFT," *IEEE Trans. Audio and Electroacoustics.* Vol. 18, pp. 211–212, June 1970.

CHIU, R. F., and C. F. CHEN, "Inverse Laplace transform of irrational and transcendental transfer functions via fast Fourier transform," *IEEE Symp. on Circuit Theory.* p. 6, December 8–10, 1969.

CHIU, R. F., C. F. CHEN, and C. J. HUANG, *A new method for the inverse Laplace transformation via the fast Fourier transform.* 1970 SWIEECO Record, pp. 201–203.

COCHRAN, W. T., et al., "What is the fast Fourier transform?" *IEEE Trans. Audio and Electroacoustics.* Vol. AU–15, pp. 45–55, June 1967; and *Proceedings of the IEEE,* Vol. 55, pp. 1644–1673, October 1967.

COFFY, J., *Manuel pratique d'analyse spectrale rapide.* Rept. E–12, 77 Fontainebleau, France: Centre d'Automatique de L'Ecole Nationale Superieure des Mines de Paris, 1968.

COOLEY, J. W., *Applications of the fast Fourier transform method.* Proc. IBM Scientific Computing Symp., Yorktown Heights, N.Y.: Thomas J. Watson Research Center, June 1966.

COOLEY, J. W., *Complex finite Fourier transform subroutine.* SHARE Doc. 3465, September 8, 1966.

COOLEY, J. W., *Harmonic analysis of complex Fourier series.* SHARE Program Library no. SDA 3425, February 7, 1966.

COOLEY, J. W., P. A. W. LEWIS, and P. D. WELCH, "The application of the fast Fourier transform algorithm to the estimation of spectra and cross-spectra," *Journal of Sound Vibration.* Vol. 12, pp. 339–352, July 1970; also *Proc. of Symp. on Computer Processing in Communications.* New York, pp. 5–20, 8–10 April 1969.

COOLEY, J. W., P. A. W. LEWIS, and P. D. WELCH, *The fast Fourier transform algorithm and its applications.* Research Paper RC-1743, IBM Corp., February 9, 1967.

COOLEY, J. W., P. A. W. LEWIS, and P. D. WELCH, "The fast Fourier transform algorithm: programming considerations in the calculation of Sine, Cosine, and Laplace transforms," *Journal of Sound Vibration.* Vol. 12, pp. 315–337, July 1970.

COOLEY, J. W., P. A. W. LEWIS, and P. D. WELCH, "The fast Fourier transform and its applications," *IEEE Trans. Education.* Vol. E–12, pp. 27–34, March 1969.

COOLEY, J. W., P. A. W. LEWIS, and P. D. WELCH, "The finite Fourier transform," *IEEE Trans. Audio and Electroacoustics.* Vol. 17, pp. 77–85, June 1969.

COOLEY, J. W., P. A. W. LEWIS, and P. D. WELCH, "Historical notes on the fast Fourier transform," *IEEE Trans. Audio and Electroacoustics.* Vol. AU–15, pp. 76–79, June 1967; also *Proceedings of the IEEE.* Vol. 55, pp. 1675–1677, October 1967.

COOLEY, J. W., P. A. W. LEWIS, and P. D. WELCH, "The use of the fast Fourier transform algorithm for the estimation of spectra and cross spectra," *Proc. Symp. on Computer Processing in Communications.* Vol. 19, MRI Symposia Ser., New York: Polytechnic Press, 1969.

COOLEY, J. W., and J. W. TUKEY, "An algorithm for machine calculation of complex Fourier series," *Math. Computation.* Vol. 19, pp. 297–301, April 1965.

CORINTHIOS, M. J., "Design of a class of fast Fourier transform computers," *IEEE Trans. on Computers.* Vol. C–20, pp. 617–622, June 1971.

CORINTHIOS, M. J., "A fast Fourier transform for high-speed signal processing," *IEEE Trans. on Computers.* Vol. C–20, pp. 843–846, August 1971.

CORINTHIOS, M. J., "A time-series analyzer," *Proc. Symp. on Computer Processing in Communications.* Vol. 19, MRI Symposia Ser., New York: Polythechnic Press, 1969.

CORYELL, D. A., *Address generator for fast Fourier transform.* IBM Technical Disclosure Bull., Vol. 12, pp. 1687–1689, March 1970.

CUSHING, R. J., *New techniques in fast Fourier transformation (FFT) processing of NMR data.* Pittsburgh Conference on Analytical Chemistry and Applied Spectroscopy, 28 February-5 March 1971.

CUSHLEY, R. J., *Sensitivity enhancement by fast Fourier transform Spectroscopy.* 160th National American Chemical Society Meeting, 13–18 September 1970.

DANIELSON, G. C., and C. LANCZOS, "Some improvements in practical Fourier analysis and their application to X-ray scattering from liquids," *J. Franklin Inst.* Vol. 233, pp. 365–380, 435–452, 1942.

DEPEYROT, M., *Fondements algebriques de la transformation de Fourier rapide.* Rept. E–11, 77 Fontainebleau, France: Centre d'Automatique de L'Ecole Nationale Superieure des Mines de Paris, November 1968.

DERE, WARREN Y., and D. J. SAKRISON, "Berkeley array processor," *IEEE Trans. on Computers.* Vol. C–19, pp. 444–446, May 1970.

DOLLAR, C. R., R. W. MIERILL, and C. L. SMITH, *Criteria for evaluation Fourier transform computational results.* 1970 Joint automatic control conference, pp. 66–71.

DRUBIN, M., "Computation of the fast Fourier transform data stored in external auxiliary memory for any general radix ($r = 2$ exp n, n greater than or equal to 1.)," *IEEE Trans. on Computers.* Vol. C–20, No. 12, pp. 1552–1558, December 1971.

DRUBIN, MEIR, "Kronecker product factorization of the FFT matrix," *IEEE Trans. on Computers.* Vol. C–20, pp. 590–593, May 1970.

DUMERMUTH, G., and H. FLUHLER, "Some modern aspects in numerical spectrum analysis of multichannel electoencephalographic data," *Med. and Biol. Engrg.* Vol. 5, pp. 319–331, 1967.

DUMERMUTH, G., P. J. HUBER, B. KLEINER, and T. GASSER, "Numerical analysis of electroencephalographic data," *IEEE Trans. Audio and Electroacoustics.* Vol. AU–18, pp. 404–411, December 1970.

ENOCHSON, L. D., and A. G. PIERSOL, "Application of fast Fourier transform procedures to shock and vibration data analysis," *Society of Automotive Engineers.* Rept. 670874.

EPSTEIN, G., "Recursive fast Fourier transforms," *AFIPS Proc.* 1968 Fall Joint Computer Conf., Vol. 33, p. 1, Washington, D.C.: Thompson, pp. 141–143, 1968.

FABER, A. S., and C. E. HO, "Wide-band network characterization by Fourier transformation of time domain measurements," *IEEE Journal of Solid-State Circuits.* Vol. SC–4, No. 4, pp. 231–235, August 1969.

FAVOUR, J. D., and J. M. LEBRUN, "Transient synthesis for mechanical testing," *Instruments and Control Systems.* Vol. 43, pp. 125–127, September 1970.

FERGUSON, M. J., "Communication at low data rates—spectral analysis receivers," *IEEE Trans. Commun. Tech.* Vol. 16, pp. 657–668, October 1968.

FISHER, J. R., *Fortran program for fast Fourier transform.* NRL–7041, CFSTI AD 706 003, 25 pp., April 1970.

FORMAN, M. L., "Fast Fourier transform technique and its application to Fourier spectroscopy," *J. Opt. Soc. Am.* Vol. 56, pp. 978–997, July 1966.

FRASER, D., "Incrementing a bit-reversed integer," *IEEE Trans. on Computers* (*Short Notes*). Vol. C–18, p. 74, January 1969.

FREUDBERG, R., J. DELELLIS, C. HOWARD, and H. SHAFFER, *An all digital pitch excited vocoder technique using the FFT algorithm.* Conf. on Speech Communication and Processing Reprints, pp. 297–310, November 1967.

G-AE Subcommittee on Measurement Concepts, "What is the fast Fourier transform?" *IEEE Trans. Audio and Electroacoustics.* Vol. AU–15, pp. 45–55, June 1967; also *Proceedings of the IEEE.* Vol. 55, pp. 1664–1674, October 1967.

GAMBARDELLA, G., "Time scaling and short-time spectral analysis," *Acoustical Society of America Journal.* Vol. 44, pp. 1745–1747, December 1968.

GENTLEMAN, W. M., "Matrix multiplication and fast Fourier transforms," *Bell Sys. Tech. J.* Vol. 47, pp. 1099–1103, July–August 1968.

GENTLEMAN, W. M., "An error analysis of Goertzel's (Watt's) method of computing Fourier coefficients," *Computer J.* Vol. 12, pp. 160–165, May 1969.

GENTLEMAN W. M., "Using the finite Fourier transform," *Proc. Symp. on Computer-Aided Engineering.* Waterloo, Ontario, Canada, pp. 189–205, 11–13 May 1971.

GENTLEMAN, W. M., and G. SANDE, "Fast Fourier transforms—for fun and profit," *AFIPS Proc.* 1966 Fall Joint Computer Conf., Vol. 29, Washington, D.C.: Spartan, pp. 563–578, 1966.

GILOI, W., "A hybrid special computer for high-speed FFT," *Proceedings of the IEEE.* 1970 International Computer Group Conference, Washington, D.C., pp. 165–170, June 16-18 1970.

GLASSMAN, J. A., "A generalization of the fast Fourier transform," *IEEE Trans. on Computers.* Vol. C–19, pp. 105–116, February 1970.

GLISSON, T. H., "Correlation using Fast Fourier transforms," *CFSTI.* AD676803, April 1968.

GLISSON, T. H., C. I. BLACK, and A. P. SAGE, "The digital computation of discrete spectra using the Fast Fourier Transform," *IEEE Trans. Audio and Electro-acoustics.* Vol. AU–18, pp. 271–287, September 1970.

GLISSON, T. H., and C. I. BLACK, "On digital replica correlation algorithm with applications to active sonar," *IEEE Trans. Audio and Electroacoustics.* Vol. AU–17, pp. 190–197, September 1969.

GOLD, B., I. LEBOW, P. MCHUGH, and C. RADER, "The FDP, a Fast programmable signal processor," *IEEE Trans. on Computers.* Vol. C–20, pp. 33–38, January 1971.

GOLD, B., and C. E. MUEHE, *Digital signal processing for range-gated pulse doppler radars.* XIX Advisory group for Aerospace Research and Development Avionics Panel Tech. Sym. AGARD66, Sec. 3, pp. 31–33, 1970.

GOOD, I. J., "The interaction algorithm and practical Fourier series," *J. Roy. Stat. Soc.* Ser. B, Vol. 20, pp. 361–372, 1958; Addendum, Vol. 22, pp. 372–375, 1960.

GOOD, I. J., "The relationship between two fast Fourier transforms," *IEEE Trans. on Computers.* Vol. C–20, pp. 310–317, March 1971.

GOODEN, D. S., *Use of fast Fourier transforms in reactor kinetics studies.* Atomic Industrial Forum 1970 Annual Conference/American Nuclear Society Winter Meeting, 15–19 November.

GRACE, O. D., "Two finite Fourier transforms for bandpass signals," *IEEE Trans. Audio and Electroacoustics.* Vol. AU–18, pp. 501–502, December 1970.

GROGINSKY, HERBERT L., "An FFT chart-letter, fast Fourier transform processors; chart summarizing relations among variables," *Proceedings of the IEEE.* Vol. 58, pp. 1782–1784, October 1970.

GROGINSKY, H. L., and G. A. WORKS, *A pipeline fast Fourier transform.* 1969 EASCON Convention Record, pp. 22–29; also *IEEE Trans. on Computers.* Vol. C–19, pp. 1015–1019, November 1970.

GUERRIERO, J. R., "Computerizing Fourier analysis," *Control Engineering.* Vol. 17, pp. 90–94, March 1970.

HARRIS, B., *Spectral Analysis of Time Series.* New York: Wiley, 1967.

HARTWELL, J. W., "A procedure for implementing the fast Fourier transform on small computers," *IBM Journal of Research and Development.* Vol. 15, pp. 355–363, September 1971.

HAUBRICH, R. A., and J. W. TUKEY, "Spectrum analysis of geophysical data," *Proc. IBM Scientific Computing Symp. on Environmental Sciences.* pp. 115–128, 1967.

HELMS, H. D., "Fast Fourier transform method of computing difference equations and simulating filters," *IEEE Trans. Audio and Electroacoustics.* Vol. AU–15, pp. 85–90, June 1967.

HOCKNEY, R. W., "FOUR 67, a fast Fourier transform package," *Computer Physics Communications* (Netherlands). Vol. 2, No. 3, pp. 127–138.

HONG, J. P., *Fast two-dimensional Fourier transform.* 3rd Hawaii International Conference on System Science, pp. 990–993, 1970.

HOPE, G. S., "Machine identification using fast Fourier transform," *IEEE Power Engineering Society 1971 Winter Meeting and Environmental Symposium.* 31 January-5 February 1971.

HUMPHREY, R. E., "A technique for computing the discrete Fourier transform of long time series," *CFSTI.* SCL-DR–69–73, p. 32, July 1969.

HUNT, B. R., *Spectral effects in use of higher ordered approximation for computing discrete Fourier transforms.* 3rd Hawaii International Conference on System Science, pp. 970–973, 1970.

"Industrial fast Fourier transform computer bows," *Electronics.* Vol. 42, pp. 171–172, October 27, 1969.

IZUMI, M., "Application of fast Fourier transform algorithm to on-line digital reactor noise analysis," *Journal of Nuclear Science and Technology* (Japan). Vol. 8, No. 4, pp. 236–239, April 1971.

JAGADEESAN, M., "n-dimensional fast Fourier transform," *Proceedings of the IEEE.* 13th Midwest Symposium on Circuit Theory, Minneapolis, Minn., 8 pp., 7–8 May 1970.

JENKINS, G. M., and D. G. WATTS, *Spectral Analysis and its Applications.* San Francisco: Holden-Day, 1968.

Johnson, N., and P. K. Bennett, *A study of spectral analysis of short-time records.* Final Rept., Signatron, Inc., Lexington, Mass., Contract N00014–67–C–0184, September 29, 1967.

Kahaner, D. K., "Matrix description of the fast Fourier transform," *IEEE Trans. Audio and Electroacoustics.* Vol. AU–18, pp. 442–450, December 1970.

Kaneko, T., and B. Liu, "Accumulation of round-off error in fast Fourier transforms," *Journal of the Association for Computing Machinery.* Vol. 17, pp. 637–654, October 1970.

Kaneko, T., and B. Liu, *Computation error in fast Fourier transform.* 3rd Asilomar Conference on Circuits and Systems, pp. 207–211, 1969.

Kido, K., "On the fast Fourier transform," *J. Inst. Electronics Commun. Engrs.* (Japan). Vol. 52, pp. 1534–1541, December 1969.

Klahn, R., and R. R. Shively, "FFT—shortcut to Fourier analysis," *Electronics.* pp. 124–129, April 15, 1968.

Kleiner, B., H. Fluhler, P. J. Huber, and G. Dummermuth, "Spectrum analysis of the electroencephalogram," *Computer Programs in Biomedicine* (Netherlands). Vol. 1, pp. 183–197, December 1970.

Knapp, C. H., "An algorithm for estimation of the inverse spectral matrix," *CFSTI.* Rept. U417–70–010, 34 pp., February 1970.

Kryter, R. C., "Application of the fast Fourier transform algorithm to on-line reactor diagnosis," *IEEE Trans. Nuclear Sci.* Vol. 16, pp. 210–217, February 1969.

Lanfenthal, I. M., and S. Gowrinathan, "Advanced digital processing techniques," *CFSTI.* 112 pp., May 1970.

Larson, A. G., and R. C. Singleton, "Real-time spectral analysis on a small general-purpose computer," *AFIPS Proc.* 1967 Fall Joint Computer Conf., Vol. 31, Washington, D.C.: Thompson, pp. 665–674, 1967.

Leblanc, L. R., "Narrow-band sampled data techniques for detection via the underwater acoustic communication channel," *IEEE Trans. Commun. Tech.* Vol. 17, pp. 481–488, August 1969.

Lee, W. H., "Sampled Fourier transform hologram generated by computer," *Applied Optics.* Vol. 9, pp. 639–643, March 1970.

Leedham, R. V., J. A. Barker, and I. F. D. Miller, *Spectrum equalization using the fast Fourier transform.* Colloquim on Engineering Applications of Spectral Analysis, London, England, 6 pp., January 7, 1969.

Leese, J. A., C. S. Novak, and V. R. Taylor, "The determination of cloud pattern motions from geosynchronous satellite image data," *Pattern Recognition* (GB). Vol. 2, pp. 279–292, December 1970.

Lesem, L. B., P. M. Hirsch, and J. A. Jordan, jr., "Computer synthesis of holograms for 3-D display," *Commun. ACM.* Vol. 11, pp. 661–674, October 1968.

LIANG, C. S., and R. CLAY, "Computation of short-pulse response from radar targets—an application of the fast Fourier transform technique," *Proceedings of the IEEE.* Vol. 58, No. 1, pp. 169–171, January 1970.

LIU, S. C., and L. W. FAGEL, "Earthquake interaction by fast Fourier transform," *Proceedings of the American Society of Civil Engineers.* Bibliog. diag., 97 (EM-4 No. 8324), pp. 1223–1237, August 1971.

MALING, G. C., JR., W. T. MORREY, and W. W. LANG, "Digital determination of third-octave and full-octave spectra of acoustical noise," *IEEE Trans. Audio and Electroacoustics.* Vol. AU–15, pp. 98–104, June 1967.

MARKEL, J. D., "FFT pruning," *IEEE Trans. Audio and Electroacoustics* (USA). Vol. AU–19, No. 4, pp. 305–311, December 1971.

MCCALL, J. R., and C. E. FRERKING, "Conversational Fourier analysis," *CFSTI.* N-68-26821, February 1968.

MCCOWAN, D. W., *Finite Fourier transform theory and its application to the computation of convolutions, correlations, and spectra.* Earth Sciences Div., Teledyne Industries, Inc., October 11, 1966.

MCDOUGAL, J. R., L. C. SURRATT, and L. F. STOOPS, *Computer aided design of small superdirective antennas using Fourier integral and fast Fourier transform techniques.* 1970 SWIEECO Record, pp. 421–425.

MCKINNEY, T. H., *A digital spectrum channel analyzer.* Conf. on Speech Communication and Processing Reprints, pp. 442–444, November 1967.

MCLEOD, I. D. G., "Comment on a high-speed algorithm for the computer generation of Fourier transforms," *IEEE Trans. on Computers*, Vol. C–18, p. 182, February 1969.

MERMELSTEIN, PAUL, "Computer-generated spectrogram displays for on-line speech research," *IEEE Trans. Audio and Electroacoustics.* Vol. AU–19, pp. 44–47, March 1971.

MILLER, S. A., *A PDP–9 assembly-language program for the fast Fourier transform.* ACL Memo 157, Analog/Hybrid Computer Lab., Dept. of Elec. Engrg., University of Arizona, Tucson, Ariz., April 1968.

MORGERA, S. D., and L. R. LEBLANC, "Digital data techniques applied to real-time sonar data," *Proc. Symp. on Computer Processing in Communications.* pp. 825–845.

NAIDU, P. S., Estimation of spectrum and cross-spectrum of astromagnetic field using fast digital Fourier transform (FDFT) techniques, *Geophysical prospecting* (Netherlands). Vol. 17, pp. 344–361, September 1969.

NEILL, T. B. M., *An improved method of analysing nonlinear electrical networks.* IEEE London International conference on computer aided design, pp. 456–462, 15–18 April 1969.

NEILL, T. B. M., "Nonlinear analysis of a balanced diode modulator," *Electronics Letters* (GB). Vol. 6, pp. 125–128, March 5, 1970.

NESTER, W. H., "The fast Fourier transform and the Butler matrix," *IEEE Trans. Antennas Propagations.* Vol. AP–16, p. 360 (correspondence), May 1968.

NICOLETTI, B., and S. CRESCITELLI, *Simulation of chemical reactors transients.* Proceedings of the 11th Automation and Instrumentation Conference, Milan, Italy, pp. 468–481, 23–25 November 1970.

NOAKS, D. R., and R. F. G. WATERS, *A digital signal processor for real-time spectral analysis.* Conference on computer science and technology, Manchester, pp. 202–209, June 30—July 3, 1969 (London IEE).

NUTTALL, ALBERT H., "Alternate forms for numerical evaluation of cummulative probability distributions directly from characteristic functions," *Proceedings of the IEEE.* Vol. 58, pp. 1872–1873, November 1970.

OBERFIELD, J. A., *Application of fast Fourier analysis to compression and retrieval of digitized magnetometers records.* 15th General Assembly of the International Union of Geodesy and Geophysics, Moscow, USSR, 30 July-14 August 1971.

O'LEARY, G. C., "Nonrecursive digital filtering using cascade fast Fourier transformers," *IEEE Trans. Audio and Electroacoustics.* Vol. AU–18, pp. 177–183, June 1970.

OPPENHEIN, A. V., "Speech spectrograms using the fast Fourier transforms," *IEEE Spectrum.* Vol. 7, pp. 57–62, August 1970.

OPPENHEIN, A. and K. STEIGLITZ, "Computation of spectra with unequal resolution using fast Fourier transform," *Proceedings of the IEEE.* Vol. 59, pp. 299–301, February 1971.

OPPENHEIM, A. V., and C. WEINSTEIN, "A bound on the output of a circular convolution with application to digital filtering," *IEEE Trans. Audio and Electroacoustics.* Vol. AU–17, pp. 120–124, June 1969.

OSTRANDER, L. E., "The Fourier transform of spine function approximations to continuous data," *IEEE Trans. Audio and Electroacoustics.* Vol. AU–19, pp. 103–104, March 1971.

PARZEN, E., *Statistical spectral analysis (single channel case) in 1968.* Tech. Rept. 11, ONR Contract Nonr-225 (80) (NR–042–234), Stanford University, Dept. of Statistics, Stanford, Calif., June 10, 1968.

PEASE, C. B., "Obtaining the spectrum and loudness of transients by computer," *CFSTI.* N–68–28799, December 1967.

PEASE, M. C., "An adaption of the fast Fourier transform for parallel processing," *J.ACM.* Vol. 15, pp. 252–264, April 1968.

PEASE, M. C., "Organization of large scale Fourier processors," *J. ACM.* Vol. 16, No. 3, pp. 474–482, July 1969.

PEASE, M. C., and J. GOLDBERG, *Investigation of a special-purpose digital computer for on-line Fourier analysis.* Special Tech. Rept. 1, Project 6557, Stanford Research Inst., Menlo Park, Calif., April 1967 (available from U.S. Army Missile Command, Redstone Arsenal, Ala., Attn: AMSMI-RNS).

PETERSEN, D., "Discrete and fast Fourier transformations on N-dimensional latices," *Proceedings of the IEEE.* Vol. 58, pp. 1286–1288, August 1970.

PIPES, L. A., and S. A. HOVANESSIAN, *Matrix-Computer Methods in Engineering.* New York: Wiley, 333 pp., 1969.

POLLARD, J. M., "Fast Fourier transform in a finite field," *Math. Computation* (Bibliog.). Vol. 25, pp. 365–374, April 1971.

PRIDHAM, R. G., and R. E. KOWALCZYK, "Use of FFT subroutine in digital filter design program," *Proceedings of the IEEE* (*Letters*). Vol. 57, p. 106, January 1969.

RABINER, L. R., and R. W. SCHAFER, *The use of an FFT algorithm for signal processing.* 1968 NEREM Record, pp. 224–225.

RABINER, L. R., R. W. SCHAFER, and C. M. RADER, "The chrip z-transform algorithm," *IEEE Trans. Audio and Electroacoustics.* Vol. AU–17 pp. 86–92, June 1969.

RADER, C. M., "Discrete Fourier transforms when the number of data samples is prime," *Proceedings of the IEEE* (*Letters*). Vol. 56, pp. 1107–1108, June 1968.

RADER, C. M., "An improved algorithm for high speed autocorrelation with applications to spectral estimation," *IEEE Trans. Audio and Electroacoustics.* Vol. 18, pp. 439–441, December 1970.

RAMOS, G., "Analog computation of the fast Fourier transform," *Proceedings of the IEEE.* Vol. 58, pp. 1861–1863, November 1970.

RAMOS, G. U., *Roundoff error analysis of the fast Fourier transform.* Stanford Comput. Sci Rep. STAN–CS–70–146, Stanford, Calif., February 1970.

READ, R., and J. MEEK, "Digital filters with poles via FFT," *IEEE Trans. Audio and Electroacoustics.* Vol. AU–19, pp. 322–323, December 1971.

REED, R. R., *A method of computing the fast Fourier transform.* M. A. thesis in electrical engineering, Rice University, Houston, Tex., May 1968.

REQUICHA, A. A. G., "Direct computation of distribution functions from characteristic functions using the fast Fourier transform," *Proceedings of the IEEE.* Vol. 58, No. 7 pp. 1154–1155, July 1970.

RIFE, D. C., and G. A. VINCENT, "Use of the discrete Fourier transform in the measurement of frequencies and levels of tones," *Bell System Tech. Journal.* Vol. 149, pp. 197–228, February 1970.

ROBINSON, E. A., *Multichannel Time Series Analysis with Digital Computer Programs.* San Francisco: Holden-Day, 1967.

ROBINSON, G. S., *Fast Fourier transform speech compression.* 1970 International Conference on Communications, pp. 26–33, 36–38, 1970.

ROTHAUSER, E., and D. MAIWALD, "Digitalized sound spectrograph using FFT and multiprint techniques," *Acoustical Society of America Journal.* Vol. 45, p. 308, 1969.

RUDNIK, P. "Note on the calculation of Fourier series," *Math. Computation.* Vol. 20, pp. 429–430, July 1966.

RUNGE, C., *Zeit fur Math. und Physik.* Vol. 48, p. 443, 1903.

RUNGE, C., *Ziet fur Math. und Physik.* Vol. 53, p. 117, 1905.

RUNGE, C., and KOENIG, "Die Grundlehren der mathematischen Wissenschaften," *Vorlesungen uber Numerisches Rechnen.* Vol. 11, Berlin: Springer, 1924.

SAIN, M. K., and S. A. LIBERTY, "Performance-measure densities for a class of LOG control systems," *IEEE Trans. Automatic Control.* Vol. AC–16, No. 5, pp. 431–439, October 1971.

SALTZ, J., and S. B. WEINSTEIN, "Fourier transform communication system," *Proc. ACM Symp. on Problems in the Optimization of Data Communications Systems.* pp. 99–128, 1969.

SHERING, G., and S. SUMMERHILL, *On-line high energy physics analysis with a PDP–9.* Decus Proceedings of the Spring Symposium 1969, Wakefield MA., pp. 61–68, 12–13 May 1969.

SHIRLEY, R. S., "Application of a modified fast Fourier transform to calculate human operator describing functions," *IEEE Trans. Man-Machine Systems.* Vol. MMS–10, pp. 140–144, December 1969.

SHIVELY, R. R., "A digital processor to generate spectra in real time," *Proc. 1st Ann. IEEE Computer Conf.* pp. 21–24, September 1967.

SILBERBERG, M., "Improving the efficiency of Laplace-transform inversion for network analysis," *Electronics Letters* (GB). Vol. 6, pp. 105–106, February 19, 1970.

SILVERMAN, H. F., *Identification of linear systems using fast Fourier transform techniques.* Report AFOSR–70–2263TR, AD711104, 135 pp., June 1970.

SILVERMAN, H. F., and A. E. PEARSON, "Impulse response identification from truncated input data using FFT techniques," *CFSTI.* Report AFOSR–70–2229TR, AD710650, 28 pp.

SINGLETON, R. C., "A method for computing the fast Fourier transform with auxiliary memory and limited high-speed storage," *IEEE Trans. Audio and Electroacoustics.* Vol. AU–15, pp. 91–98, June 1967.

SINGLETON, R. C., "On computing the fast Fourier transform," *Commun. ACM.* Vol. 10, pp. 647–654, October 1957.

SINGLETON, R. C., "Algol procedures for the fast Fourier transform," *Commun. ACM.* Vol. 11, pp. 773–776, Algorithm 338, November 1968.

SINGLETON, R. C., "An Algol procedure for the fast Fourier transform with arbitrary factors," *Commun. ACM.* Vol. 11, pp. 776–779, Algorithm 339, November 1968.

SINGLETON, R. C., "Remark on Algorithm 339," *Commun. ACM.* Vol. 12, p. 187, March 1969.

SINGLETON, R. C., *An algorithm for computing the mixed radix fast Fourier transform.* Research Memo., SRI Project 3857531–132, Stanford Research Inst., Menlo Park, Calif., November 1968.

SINGLETON, R. C., "An Algol convolution procedure based on the fast Fourier transform," *Commun. ACM.* Vol. 12, pp. 179–184, Algorithm 345, March 1969.

SINGLETON, R. C., "An algorithm for computing the mixed radix fast Fourier transform," *IEEE Trans. Audio and Electroacoustics.* Vol. AU–17, pp. 93–103, June 1969.

SINGLETON, R. C., and T. C. POULTER, "Spectral analysis of the call of the male killer whale," *IEEE Trans. Audio and Electroacoustics.* Vol. AU–15, pp. 104–113, June 1967; also Comments by W. A. Watkins and Authors' Reply, Vol. AU–16, p. 523, December 1968.

SLOANE, E. A., *An introduction to time-series analysis, Monograph 1: Concept of a data window and time windows and averages; Monograph 2: Fourier series and integrals; Monograph 3: Statistical windows and averages; Monograph 4: Applications.* Time/Data Corp., Palo Alto, Calif., 1966, 1967.

SLOANE, E. A., "Comparison of linearly and quadratically modified spectral estimates of Gaussian signals," *IEEE Trans. Audio and Electroacoustics.* Vol. AU–17, pp. 133–137, June 1969.

SPITZNOGLE, F. R., and A. H. QUAZI, "Representation and analysis of time-limited signal using a complex algorithm (discrete Fourier transform)," *Acoustical Society of America Journal.* Vol. 47, pp. 1150–1155, May 1970.

STARSHAK, A., *Fast Fourier transforms in industrial environment.* 5th Great Lakes Regional Meeting of American Chemical Society, Peoria, Ill., 10–11 June 1971.

STOCKHAM, T. G., "High-speed convolution and correlation," *AFIPS Proc.*, Vol. 28, pp. 229–233, 1966 Spring Joint Computer Conf., Washington, D.C.: Spartan, 1966.

STONE, H. C., "Parallel processing with the perfect shuffle," *IEEE Trans. on Computers.* Vol. C–20, pp. 153–161, February 1971.

STONER, R. R., "A flexible fast Fourier transform algorithm," Report ECOM 6046 (AD696431) *CFSTI.* 25 pp. August 1969.

STRACHEY, C., "Bitwise operations," *Commun. ACM.* Vol. 4, p. 146, March 1961.

STUMPFF, K., *Tafeln und aufgaben zur harmonischen Analyse und Periodogrammrechung.* Berlin: Springer, 1939.

SWICK, D. A., *Discrete finite Fourier transforms: a tutorial approach.* NRL Rept. 6557, Naval Research Labs., Washington, D.C., June 1967.

TAYLOR, A. D., *Fast Fourier transforms—real and complex, forward and inverse.* Computer Program, Environmental Sciences Services.

THEILHEIMER, F., "A matrix version of the fast Fourier transform," *IEEE Trans. Audio and Electroacoustics.* Vol. AU–17, pp. 158–161, June 1969.

THOMAS, J. G., "Phasor diagrams simplify Fourier transforms," *Electron Engineering* (Great Britain). Vol. 43, No. 524, pp. 54–57, October 1971.

THOMAS, L. H., "Using a computer to solve problems in physics," *Application of Digital Computers*. Boston, Mass.: Ginn, 1963.

THOMSON, D. J., "Generation of Gegenbauer pre-whitening filters by iterative fast Fourier transforming," *Proc. Symp. on Computer Processing in Communications*. New York, pp. 21–35, 8-10 April 1969.

THOMPSON, D. J., "Generation of Gegenbauer pre-whitening filters by fast Fourier transforming," *Proc. Symp. on Computer Processing in Communications*. Vol. 19, MRI Symposia Ser., New York: Polytechnic, 1969.

TILLOTSON, T. C., and E. O. BRIGHAM, "Simulation with the fast Fourier transform," *Instruments and Control Systems*. Vol. 42, pp. 169–171, September 1969.

VALLASENOR, A. J., *Digital spectral analysis*. Tech. Note D–1410, NASA, Washington, D.C., June 1968.

VERNET, J. L., "Real signals fast Fourier transform: storage capacity and step number reduction by means of an odd discrete Fourier transform," *Proceedings of the IEEE*. Vol. 59, No. 10, pp. 1531–1532, October 1971.

VOELCKER, H. B., "Digital filtering via block recursion," *IEEE Trans. Audio and Electroacoustics*. Vol. AU–18, pp. 169–176, June 1970.

VO-NGOC, B., and D. POUSSART, *An application of the FFT in automatic detection of sleep spindles*. Papers and presentation of Digital Equipment Computer Users' Society Spring Symposium, Atlantic City, N.J., pp. 297–300.

WEBB, C., *Practical use of the fast Fourier transform (FFT) algorithm in time-series analysis*. Report ARL—Tr–7022 (AD–713166), University of Texas, Austin, 205 pp., June 22, 1970.

WEINSTEIN, C. J., "Quantization effects in digital filters," *CFSTI*. TR–468–ESD–TR–69–348, AD706862, p. 96, November 1969.

WEINSTEIN, C. J., "Roundoff noise in floating point fast Fourier transform computation," *IEEE Trans. Audio and Electroacoustics*. Vol. AU–17, pp. 209–215, September 1969.

WELCH, L. R., *Computation of finite Fourier series*. Rept. SPS 37–37, Vol. 4, Jet Propulsion Labs, Pasadena, Calif., 1966; also, *A program for finite Fourier transforms*. Rept. SPS 37–40, Vol. 3, Jet Propulsion Labs, 1966.

WELCH, P. D., "The use of fast Fourier transform for the estimation of power spectra: a method based on time averaging over short, modified periodograms," *IEEE Trans. Audio and Electroacoustics*. Vol. AU–15, pp. 70–73, June 1967.

WELCH, P. D., "A fixed-point fast Fourier transform error analysis," *IEEE Trans. Audio and Electroacoustics*. Vol. AU–17, pp. 151–157, June 1969.

WESLEY, M., "Associative parallel processing for the fast Fourier transform," *IEEE Trans. Audio and Electroacoustics*. Vol. AU–17, pp. 162–164, June 1969.

WHELCHEL, J. E. W., and D. F. GUINN, *The fast Fourier Hadamard transform and its use in signal representation and classification.* Electronics and Aerospace Systems Record, pp. 561–573, 1968.

WHELCHEL, J. E., and D. F. GUINN, "FFT organizations for high speed digital filtering," *IEEE Trans. Audio and Electroacoustics.* Vol. AU–18, pp. 159–168, June 1970.

WHITE, P. H., "Application of the fast Fourier transform to linear distributed system response calculations," *Acoustical Society of America Journal.* Vol. 46, Pt. 2, pp. 273–274, July 1969.

WILSON, J. C., *Computer calculation of discrete Fourier transform using the fast Fourier transform.* Center for Naval Analyses, Arlington, Va., OEG Research Contrib. 81, June 1967.

YAVNE, R., "An economical method for calculating the discrete Fourier transform," *AFIPS Proc.* Vol. 33, Pt. 1, pp. 115–125, 1968 Fall Joint Computer Conf., Washington, D.C.: Thompson, 1968.

YOUNG, R. C., *The fast Fourier transform and its application in noise signal analysis.* Report THEMIS–UM–69–6 (AD689847). Available from *CFSTI.*

ZORN, J., "The evaluation of Fourier transforms by the sampling method," *Automatica* (GB). Vol. 4, pp. 323–335, November 1968.

ZUKIN, A.S., and S. Y. WONG, "Architecture of a real-time fast Fourier radar signal processor," *AFIPS Proc.* Spring Joint Computer Conference, Vol. 36, pp. 417–435, 1970.

INDEX

A

B

C

D

E

F

L

M

N

O

P

R

S

T

U

W